# Intelligent Transportation Systems Architectures

For a complete listing of the *Artech House ITS Library*,
turn to the back of this book.

# Intelligent Transportation Systems Architectures

Bob McQueen
Judy McQueen

Artech House
Boston • London

**Library of Congress Cataloging-in-Publication Data**
McQueen, Bob.
Intelligent transportation systems architectures / Bob McQueen, Judy McQueen.
p. cm. — (Artech House intelligent transportation systems library)
Includes bibliographical references and index.
ISBN 0-89006-525-X (alk. paper)
1. Transportation—United States—Planning. 2. Intelligent Vehicle Highway Systems—United States. I. McQueen, Judy. II. Title III. Series.
HE206.2.M38 1999
388'.0973—ddc21 99-10826
CIP

**British Library Cataloguing in Publication Data**
McQueen, Bob
Intelligent transportation systems architectures. — (Artech House intelligent transportation systems library)
1. Intelligent Vehicle Highway Systems — Designs and plans 2. Electronic traffic controls 3. Transportation engineering I. Title II. McQueen, Judy
388.3'12

ISBN 0-89006-525-X

**Cover design by Lynda Fishbourne**


**685 Canton Street**
**Norwood, MA 02062**



International Standard Book Number: 0-89006-525-X
Library of Congress Catalog Card Number: 99-10826

10 9 8 7 6 5 4 3 2 1

# Contents

# Preface

If we were to sum up the reasons for writing this book in as simple terms as possible, we would say that we are trying to be helpful. The intelligent transportation system (ITS) field has some unique and challenging aspects. We wanted to write a book that identified these challenges and offered a range of helpful advice on how to approach and manage them. This resulted in a book that from its title, seems to be all about technology, but is in fact all about people helping people to get what they want from the application of information and communications technologies to transportation.

There was also an unexpected side effect to writing this book: We helped ourselves in the process. The act of putting thoughts and experiences down on paper in a structured way, with an audience in mind, transformed our perspectives on many of the concepts and ideas.

The concepts and ideas originate from the diverse range of people we have encountered in the course of our work in the transportation profession and business world. In our minds, this diversity sums up the world of ITS. It is a confluence point for many disciplines and multiple backgrounds. If we can learn from each other and apply the superb range of knowledge and expertise to our transportation needs, objectives, issues, and problems, then surely we can make things better.

The development program for the U.S. National Architecture for ITS was a significant point in our lives. It represented an awakening to the possibilities of combining formal structured system engineering techniques with traditional approaches to the planning and development of transportation initiatives and infrastructures. On a personal level it was also a turning point. We met each other as a direct consequence of the development program. We suppose you could view it as an unexpected deliverable from the program.

Consequences and the inter-related impacts of the actions of one on the other are strong themes in this book. It seems to us that the development of a regional ITS architecture (known to us as an *ITS Future Big Picture*) is all about the identification, understanding, and management of such consequences. It may not be possible to foresee every likely cause and effect, but we have found the development of an ITS Future Big Picture to be an enormous help in this regard. The ability to lay out the big picture in a way that can be shared with others and studied cooperatively is extremely valuable.

We hope that this book proves to be of use to you and that you enjoy reading it. After all, reading the book requires an investment of your resources, and our primary aim is to make sure you get what you want and expected from that investment.

# Acknowledgments

There have been so many influences on the development of the material in this book that it would be impossible to list all the people involved. Here are a few of the many:

David Abraham
Olu Adeyinka
Gini Ainley
Steve Albert
Roemer Alfelor
Gad Bachrach
Philip Banks
Richard Barber
Bob Barrett
Rena Barta
Chris Bausher
Susan Beaty
Bob Behnke
Brien Benson
George Beronio
Adam Brand
Wei Bin Zhang
Dick Bishop
Tom Black
Jim Blain
Arnie Bloch
Jacov Blum
Phil Blythe
Peter Bonsall
Richard Bossom
Marc Bounds
Serge Bueno
Ian Catling
Chester Chandler
Sompol Chatusripitak
Kan Chen
Mel Cheslow
Jon Cheyne
Allan Clelland
Belle Cole
John Collura
Christine Cox
Brad Dennard
Khaled Dessouky
John Earp
Bob Easter
Bruce Eisenhart

Don Erwin
Gary Euler
Jorge Figueredo
Gordon Fink
Elizabeth Fischer
Stig Franzen
Jonathan Gifford
Rick Glasco
Bob Glass
Greg Hatcher
Glenn Havinoviski
Ron Heft
Cliff Heiss
Bengt Henoch
Martin Herbert
Carl Herbert Rokitansky
Tom Horan
Tom Humphrey
David Humphreys
Bill Hyman
Carol Jacoby
Rob Jaffe
David Jeffrey
Allan Jepps
Christine Johnson
Bill Jones
Job Klinjhout
Ken Kneipp
Ron Knockheart
Jurgen Kossack
Walter Kraft
Mike Krueger
Tom Lambert
Steven Landau
Barney Legge
Eva Lerner-Lam
Mac Lister
Steve Lockwood
Shelley Lynch
Vito Mauro
John Mcgowan
Mike Mcgurrin
John Miles
Khairun Zainal Mokhtar
Martha Morecock
Steve Mortensen
John Olson
Patricia Palumbo
Bob Parsons
Raman Patel
Chris Patton
Rudolf Peelen
Bob Phen
Andrew Pickford
Laurie Pickup
Mike Pietrzyk
Jerry Pittenger
Vic Poteat
Dan Powell
David Prais
Pierre Pretorious
Jeff Randall
Arlee Reno
Jim Reynold
Frik Rheeder
Tony Ridley
Marlin Ristenbatt
Helen Roach
Douglas Robertson
Siegfried Rupprecht
Eric Sampson
Don Savitt
Lyle Saxton
Neil Scales
George Schoene
Rick Schuman
Ed Seymour
Mike Shagrin

Benny Shalita
Sameer Sharma
Steven Shladover
Lee Simmons
Loyd Smith
Ted Smith
Jerry Sobetski
Heinz Sodeikat
Bill Spreitzer
Kenk Staats
Ray Starsman
Henk Stoelhorst
Gloria Stoppenhagen
Roger Stough
Roy Sumner
Larry Swartzlander
Phil Tarnoff
Paul Taylor
Frank Tobin
Cary Vick
Evelyn Wagner
Charlie Wallace
Chip White
Ladon White
Paul White
Malcolm Williams
John Wootton
Hal Worrall
John Wright
Karl Wunderlich
Oliver Yandle

We acknowledge all the other people we have learned from and been influenced by over the course of the development of the material in this book, with a special mention to all the participants in the National Architecture for ITS Development Program, ISO TC204 Working Group 1 and CEN TC 278 Working Group 8.

Contact details:

Bob McQueen – PBS&J (407) 647-7275 x328
e-mail: bobmcqueen@pbsj.com

Judy McQueen – Coherence Incorporated (407) 654-7861
e mail: coherence@msn.com

# 1

# Introduction

## 1.1 Introduction

*Intelligent Transportation Systems Architectures* has its roots in transportation. Although it deals with subjects involving the application of advanced technologies such as information and communications technologies, it addresses these topics from a transportation point of view. The contents represent a very personal view of ITS development, planning, and implementation, based on the experiences of both authors over the past few years.

We hope that this book accomplishes the following:

- Explains the cooperative approach to ITS development, planning, and implementation;
- Provides some assistance in dealing with the multidisciplinary nature of ITS;
- Assists the transportation community in bridging the gap between transportation and system engineering;
- Promotes the development and deployment of smarter ITS;
- Helps stakeholders to consider, prepare, and deliver effective input to ITS development projects;
- Identifies and addresses the "people" issues associated with ITS.

Both authors worked as part of the program management team during the three years it took to create the National Architecture for ITS. This framework defines the major subsystems and the data flows that work harmoniously toward ITS implementation at a national level. Although the architecture was intended for development and deployment only in the United States, it holds great significance for transportation professionals worldwide. We believe that the many valuable lessons and important techniques identified and developed during the course of the program offer considerable value to the transportation community in effective deployment of ITS. These lessons and techniques revolve around a combination of two components: transportation experience that supports a detailed understanding of the ITS user and the context within which the ITS will operate and sound system engineering techniques that have evolved over many years of practice in nontransportation fields such as aerospace, defense, and commercial systems development.

During the program activities and in subsequent projects, we have applied these principles in several locales around the world, evolving in an approach to ITS planning development and implementation that embraces the principles of the National Architecture for ITS and supports a cooperative approach to the definition of ITS solutions. Our experiences suggest that a book describing this approach, which we call the "cooperative development approach to ITS," and how to implement it may help people to get the most from ITS.

The idea of writing this book arose from the realization that, outside of a small core of highly involved people, the great majority of the transportation profession—certainly the majority of the public or stakeholders required to provide input to the process—are really not familiar with ITS development processes or the issues involved in deploying a successful ITS.

Professional people can be technically knowledgeable, highly specialized, and experienced in transportation (or their own particular fields) but at the same time be what we call "nonarchitecturally aware." In other words, the approach, tools, and techniques applied within an ITS Future Big Picture development process can be quite alien to even the most seasoned transportation professionals. Accordingly, this book aims to spread knowledge of the ITS Cooperative Development Methodology.

We want to provide readers with a general introduction to ITS technology application and to provide transportation professionals, the general public, and user groups involved in stakeholder meetings or consensus formation teams associated with ITS implementations with a good awareness and understanding of the ITS Future Big Picture development process. This includes introducing the terms and steps involved in the process and pointing to other references for further information.

We would also like to state clearly the things the book will not do for you: It is not a "how to" guide that will enable its readers to go off and develop an

ITS implementation. Similarly, it is not intended to be a detailed guide on designing or procuring ITS. Those subjects are briefly discussed in Chapters 15 and 16. For additional information, readers can turn to more substantial texts on information technology in the development of large-scale systems that offer more details.

This book is written as a nontechnical introduction to the cooperative development approach to ITS. It will be particularly useful to transportation professionals trying to harness the power of ITS. It will also help others, including members of the public involved as stakeholders in consensus or user input processes associated with ITS plans and implementations.

To begin, we will answer a few questions readers might have, listed as follows and answered in Sections 1.2–1.5.

- Who should read this book?
- Why should you read this book?
- What will reading the book do for you?
- How should you read this book?

## 1.2 Who Should Read This Book?

We had a number of reader groups in mind when we started to write this book. To help us to keep you in mind, we gave each group a label. These labels—stakeholders, students, transportation professionals, and consultants—are discussed in Sections 1.2.1–1.2.4.

### 1.2.1 Stakeholders

This is an abbreviation for the reader group containing members of interest groups or stakeholders participating in the planning, design, and development of an ITS implementation.

This group could consist of members of the transportation community and the general public. Accordingly, it might contain people representing special interest groups such as environmentalists, residents' associations, trucking company associations, and public transportation advocacy groups. We remembered that since this group of readers would be drawn from the general public and might not necessarily have a technical background, its members might be interested in learning what ITS technologies are, what they can do, and why we should bother to take the time to understand their application. We also considered that this group of readers would be involved in some kind of ITS development process and would like to identify and understand the steps involved in the development process.

### 1.2.2 Students

While students differ, we use this as a short descriptor for the reader group comprising transportation planning and traffic engineering students.

For this reader group, we hope to provide an introduction to a cooperative development approach to ITS planning and design. This is intended to supplement such readers' knowledge of transportation and traffic, while enhancing their view of ITS planning and development. We assume that this group of readers is interested in understanding what ITS is, what it can do, and how to apply it effectively and successfully within the overall transportation context.

### 1.2.3 Transportation

Although this is a widely used term in the United States and the rest of the world, in Britain it previously was used to mean the process of dispatching criminals to the penal colonies in Australia. We certainly do not mean to imply that this is suitable treatment for national and local government officials engaged in planning, development, procurement, and design of integrated ITS, but we have adopted transportation as a short label for this reader group.

We view this group of readers as traffic and transportation professionals drawn from national and local government agencies. We expect that the backgrounds and experience of these readers will vary quite considerably, including training, education, and experience and such disciplines as engineering, planning, and finance. This experience might also vary according to the role readers fulfill in the transportation community and the mode of transport most closely related to their everyday activities.

While we expect this group of readers to be knowledgeable in technical and institutional issues related to their particular fields, they might not have a detailed appreciation of, or exposure to, the cooperative development approach to ITS or ITS architectures.

### 1.2.4 Consultants

This label is used as a short descriptor for transportation consultants, system engineers, and system integrators engaged in the development and implementation of ITS. This group of readers is considered to be well-versed in transportation issues and, in some cases, experienced in system design and implementation. These are the trusted advisors to which implementing agencies turn for specialist advice and additional resources for transportation projects and initiatives. For this group of readers, we try to provide an overview of the latest philosophy being applied to ITS requirements definition, system planning, and engineering.

For each of the above reader groups, we wanted to provide a practical, no-nonsense guide to the application of ITS within a structured framework, or ITS Future Big Picture.

If you do not see yourself in any of the above reader groups but still find at least some of the book's contents interesting or useful, please read on and drop us a line, telling us who you are and what you found interesting or useful.

## 1.3 Why Should You Read It?

There are many reasons, but one of the most important relates to getting the most from ITS: We are convinced that better ITS will result if ITS specifiers, procurers, designers, and developers are able to combine transportation experience with proven system engineering techniques.

Some early experiences in ITS development and deployment have proved to be very traumatic for the people involved. Based on our own experience, we believe that the cooperative development approach described in this book has a great deal to offer by way of potential solutions to some of the most concerning and prevalent problems.

These include a range of issues, many of which are not unique to ITS but may be new to the transportation field. Based on our experiences, Sections 1.3.1–1.3.5 describe some of the most significant problems, issues, and concerns that we have encountered in ITS planning, development, and implementation.

### 1.3.1 Project Management

This includes direction and control of ITS projects ensuring that they are delivered on time and within budget. The original requirements stated by the customer have to be met. Consequently, the following issues are typically encountered.

- *Concern for timely implementation:* ITS has become a high-profile subject area. We have successfully convinced political decision makers to invest resources to achieve benefits. We must now implement ITS as swiftly and efficiently as possible and show a return on this investment.

- *Cost and schedule control:* ITS projects tend to be very complex, with many separate components to be developed and integrated. While the transportation community has successfully implemented very large and complicated projects in the past (the U.S. Interstate Highway Network is a prominent example), some aspects of ITS are new and unfamiliar, requiring original management approaches and additional expertise to stay on budget and on time.
- *Scope creep:* In many cases, ITS projects are subject to *scope creep* where the endpoint of the project seems to keep moving back due to changing and expanding implementing agency requirements. This makes it difficult to finalize the project.

### 1.3.2 Requirements

ITS is a new subject area for many people. As a result, the definition and analysis of requirements must be handled very carefully. In the course of defining and agreeing to requirements, the following issues will often emerge.

- *Wide variation in requirements:* We have also experienced a wide and sometimes unnecessary variation in requirements identified and expressed by similar implementing agencies. Instead of taking advantage of shared development costs and lower production expenses due to volume of sales, many agencies have insisted on developing a fully customized ITS solution.
- *Identifying and understanding wants and needs—both now and in the future:* In many cases, the transportation community has struggled to clearly identify and understand the complete set of requirements to be addressed by a proposed ITS implementation. Sometimes the requirements are not defined at the beginning and instead are allowed to unfold over the course of the development. On other occasions, requirements have been defined rapidly with no buy in from the proposed users and no account taken of the possibility that requirements may evolve as the users gain a fuller understanding of ITS capabilities. It can be very costly to accommodate changes in requirements once software development has started and hardware has been procured.
- *Wide variations in objectives of different stakeholders:* ITS implementations typically involve a broad cross-section of stakeholders or key participants whose objectives must be addressed and satisfied. In some cases, these objectives can vary so widely that it seems impossible to satisfy all parties.

- *Clearly defining requirements:* In many cases, the implementing agencies and the end users of the system simply do not know what they want. It is easy to get into a circular or "chicken and egg" situation in which the system developer asks "what do you want?" and the users reply "what can we have?" We have also encountered on many occasions a situation in which the users are not only struggling to find answers to their questions but are also oblivious to the questions they should ask to uncover the answers.
- *Getting started:* It can sometimes be very difficult to decide on the best first step toward applying ITS. There is a wide range of potential ITS solutions, and it can be very difficult to determine the best options for both now and the future. This occasionally leads to "analysis paralysis" in which lengthy studies are undertaken, yet no implementations result.

### 1.3.3 Technology

Many of the technologies utilized and applied in ITS are new to transportation professionals. This leads to a number of issues that have to be managed, such as the following.

- *Varying perceptions of ITS technology capabilities:* Many ITS technologies are new and different to the transportation community. In some cases, this can lead to almost theological debates about the relative advantages and disadvantages of one technology over another. Perceptions of technology capabilities and belief systems regarding ITS are often driven by sales literature, rather than practical experience.
- *Future-proofing:* As ITS is relatively new, there are few standards and only a small range of "off-the-shelf" products. This makes it difficult to make a wise long-term investment. It is also hard to avoid getting "locked in" to proprietary systems, thereby closing down future flexibility and growth paths. Combine these issues with the immaturity of the ITS market and lack of formal Measures of Effectiveness and concern for future-proofing becomes a critical issue.
- *Expandability:* This is related to future-proofing as it addresses future flexibility. It is difficult to see ahead and determine whether the approach you take to short-term ITS deployment will provide you with good options for future expansion of both geographical coverage and system capabilities.
- *Successful integration of ITS technologies:* How do you take a number of ITS technologies developed by different manufacturers to different

specifications and make them work together in a seamless way that provides the solution you require? Another related issue revolves around ensuring that the ITS deployed by one agency in a region or area is compatible with that being developed or implemented by another agency in the same geographic area.

- *Incorporating existing systems and protecting sunk investment:* We have often compared the planning and development of ITS to trying to put a saddle on a running horse. It is very rare to be able to start from scratch with a "green field" situation in which there are no existing systems or transportation initiatives that will affect and influence the proposed ITS. This is a difficult job, but in many cases, a failure to clearly demonstrate the way in which previous implementations and committed investments will be embraced and utilized by the new system can be a "showstopper" for the proposed ITS implementation.

### 1.3.4 Institutional/Organizational

These are the issues that involve individuals and organizations. For an ITS to operate effectively, the "people" issues have to be effectively addressed. These include the following.

- *Achieving consensus across the transportation community:* It can be very difficult to attain the level of consensus required across the whole transportation community in a particular region. Unfortunately, ITS requires a high level of consensus and cooperation if the full benefits and synergy are to be obtained. The way transportation is organized and operated historically lends itself to polarization of interests and narrow mode-specific objective setting. This often leads to agencies adopting a defensive uncooperative stance forming the basis for turf battles and interagency conflicts.
- *Identifying roles and responsibilities for all key participants:* It can be difficult to comprehend the potential impact of ITS on the local transportation scene. In particular, the task of identifying and assigning roles and responsibilities to all key participants forming an institutional and organizational framework that will effectively support the chosen technical approach can be daunting. It is also necessary to effectively answer key questions. For example, who will perform the following tasks?

    1. Operate;
    2. Assume responsibility and liability;

3. Maintain;
4. Fund;
5. Benefit.

Such questions need to be answered for each individual component of the system and for the system as a whole.

- *Developing a long-term transportation vision:* Many transportation agencies have short-term objectives that are difficult to achieve due to funding and other resource constraints. Consequently, it is challenging to develop a longer term vision. The emphasis is placed on short-term expediency, and this is reflected in the type of procurements and implementations attempted.

### 1.3.5 Cost and Funding

ITS does not typically fall within the normal range of activities for transportation professionals. Consequently, there may be no precedent or prior model that can be followed when identifying funding opportunities. There may also be difficulty in establishing accurate costs for new hardware and software. The following issues typically arise.

- *The cost of pioneering:* In many cases, the first entity to try something new pays more to cover the cost of breaking new ground or pioneering. There is a major concern about how to manage this aspect of ITS effectively, ensuring that the pioneering costs are constrained to a relatively small level compared to the benefits achieved. From a central and local government point of view, it is also important to ensure that the mechanisms exist to enable "pioneering" agencies to share their experiences and lessons learned with others following behind. This ensures that public expenditure on pioneering elements is fully amortized across multiple agencies, preventing a situation in which multiple agencies make duplicate investments in similar pioneering elements of ITS.
- *Making accurate cost estimates:* This is again related to the newness and unfamiliarity of ITS. It can sometimes be difficult to accurately assess the real cost of implementation. This can cause all sorts of problems in the procurement process, especially when bid prices turn out to be substantially higher than the estimates used as the basis for set asides and budgets.

- *Value:* How do you ensure that you get a value commensurate with the cost of procuring new technologies in an emerging market? There is often very little to use for comparison purposes when making value judgments on ITS technologies and applications.
- *Project funding:* When the project you are proposing is substantially different from the average transportation project, how do you justify the expenditure and seek financial support for the proposal? It can be difficult to obtain funding when the proposed project does not fit within existing guidelines or procedures.
- *Eliciting private-sector participation and investment and leveraging public-sector funds:* There is a great hope that many ITS applications will be interesting to the private sector as investment and business development opportunities. Many applications, such as traveler information, are amenable to direct subscription payment mechanisms, enabling the private sector to make a business while satisfying public-sector transportation policy objectives. We have described this concept of public/private cooperation to the public sector in terms of "getting your policy objectives satisfied by leveraging other people's money." It is an important and potential source of funding for ITS but requires a commercially oriented approach to project definition and plan development that can be alien to the public sector.

## 1.4 What Will Reading This Book Do for You?

As well as trying to provide good practical answers and possible approaches to address the issues and problems described in Sections 1.3.1–1.3.5, we want to address something we call the *synergy dividend.* ITS exists in an interesting "interface zone" at the boundaries between transportation, system engineering, computing, electronics, electrical engineering, information technology, the general public, nontechnical political decision makers, and others. If we can find a way to manage this interface zone and engage all the participants in an effective manner then we can unlock the synergy dividend, producing results that could not be achieved by any discipline alone.

This book describes some techniques and approaches that may be useful in managing the interface zone. For transportation professionals in particular, ITS may well present a type of cultural challenge in that many of the concepts and ideas are different from those traditionally adopted in the transportation field. Yet there are many other techniques and approaches that have a great deal of similarity to those already used in transportation.

The transportation community also has a detailed understanding of the needs and issues associated with the end users of ITS implementations—the traveling public. As a consequence, we believe that it is very important that the transportation professional be aware of the special needs of ITS planning, development, design, and implementation, enabling the combination of this awareness and transportation experience to support an end-user driven approach to ITS. The transportation community also deals with developing and designing the conventional infrastructure—things like new roads, bridges, light rapid transit systems, and other transportation infrastructure that will have to be integrated alongside ITS solutions. Accordingly, it is clear that the transportation community needs to become a lot smarter about these specialized disciplines. We want to help transportation people discover and understand some of the techniques and approaches used in the development of large information technology systems.

Successful development of an ITS, or, in fact, any information technology-based system, requires that user needs be captured very carefully. It is essential to bring the user to the development loop in the beginning so that the user's requirements are taken fully into account and explored so that the user gains a sense of ownership in the system at an early stage. Related to this is our belief that a system can be defined as being not just hardware, software, and data but hardware, software, data, and people. We firmly believe that if you leave the people out of the loop, you might as well not bother dealing with the other components.

Bringing users into the loop places a substantial burden on the developer and makes it necessary to explain in fairly straightforward terms what will probably be quite complicated development processes and to explain the meaning of substantial pieces of terminology and possible jargon. It is also worthwhile noting at this point not just what this book intends to be but also what it definitely is not going to be. The book is a straightforward user or transportation community guide to the application of architecture-based approaches to successful ITS implementation. It is not intended to be a system engineering handbook; there are many detailed texts available on that subject.

## 1.5 How Should I Read This Book?

This seems like a really stupid question at first glance, doesn't it? However, when you think about it a little and watch how people actually read books, you start to understand that we are all different. Ideally, we would have liked to have put all the material in this book on a CD-ROM or on the Internet and then provide readers with a browser tool to cut their own paths through the material, from one connection to another as they see fit. Unfortunately, we have not done that (yet!), so you will have to read it the way we wrote it.

In a perfect world, you would read the whole book from start to finish in chapter order—and we encourage you to do this if you want to get the most from the experience. However, a theme running through this book is that the world of ITS is not ideal and that you have to come up with ways to manage it.

To help manage your approach to this book, we thought it would be helpful to provide some advice on the order in which you may wish to tackle the chapters. First, we briefly describe the contents of each chapter.

- *Chapter 1:* This chapter aims to provide a basic orientation to the contents of the book and a characterization of the intended audience. It provides the background to the authors' decision to write the book and gives you some idea of what you should expect to gain from the time investment in reading the book. This is recommended reading for all readers as it will help you to find your way around the rest of the book.
- *Chapter 2:* This chapter, which is intended for the reader who is unfamiliar with ITS, presents a basic introduction to ITS technologies and applications. It provides the grounding required to understand the techniques and methods explained in subsequent chapters. The final part of the chapter provides a structured view of ITS technologies and applications that clearly explains the sometimes confusing array of technologies and how they all fit together. Even those already familiar with ITS technologies will find this chapter useful as a summary and reminder of the ITS technologies that are described and referred to in later chapters.
- *Chapter 3:* This chapter introduces the concepts and philosophy on which the ITS Cooperative Development Methodology is based. It describes the basics of structured engineering approaches to requirements definition and system development. It also explains what an ITS Future Big Picture or regional ITS architecture is by providing a formal definition and some examples and analogies. Reasons for investing time and resources in developing it are explored along with an explanation of the relationship between ITS Future Big Picture development and design. The steps involved in developing a regional framework are also introduced and defined. Readers who are developing an ITS Future Big Picture or participating in a consensus or stakeholder group associated with the development of an ITS Future Big Picture will find Chapter 3 useful.

- *Chapter 4:* This chapter builds on Chapter 3 by providing a detailed description of the ITS Cooperative Development Methodology. The ITS Cooperative Development Methodology, which is the central theme of the book, is explained in terms of a process diagram illustrating the separate steps required to support a cooperative, user-driven approach to ITS development and design. The key concepts of ITS Objectives Statement, ITS Vision Statement, ITS User Services, and ITS Market Packages are introduced, explained, and justified in this chapter. Like Chapter 3, you will find Chapter 4 useful if you are trying to develop a regional ITS framework, or taking part in consensus formation or stakeholder involvement activities. The benefits of adopting this approach and some practical lessons learned in applying it are also described.

Chapters 5 to 14 explain specific parts of the ITS Cooperative Development Methodology in some detail.

- *Chapter 5:* This chapter provides some guidance and advice on the development of ITS User Services for capturing requirements. These, along with the ITS Objectives Statement, ITS Vision Statement, and Strawman ITS Future Big Picture form the needs model for the development. Chapter 5 provides some information on ITS User Service development techniques and addresses the development of the Strawman ITS Future Big Picture.
- *Chapter 6:* Explains the ITS Objectives Statement and ITS Vision Statement and how to go about developing them. It also explains how these components of the needs model are used in practice.
- *Chapter 7:* Explores the important aspects of ITS technology selection and review. Building on the description of the structured view of ITS technologies and applications in Chapter 2, Chapter 7 explains the techniques used to review and select the most appropriate ITS technologies for a given deployment or application. It describes the use of ITS Market Packages to characterize available technologies and relate them to actual needs.
- *Chapter 8:* Addresses the difficult, yet vital area of integrating a new ITS deployment with existing systems and initiatives. Techniques for assessing the value and feasibility of including existing systems in the deployment are reviewed and explained.

- *Chapter 9:* Explores institutional and organizational issues associated with the development and deployment of ITS. Issues are defined, explained, and illustrated by example.
- *Chapter 10:* Explains the steps involved in developing a logical view of the ITS Future Big Picture for an ITS development program. It also explains the need for the logical view and the application of the logical view in subsequent steps of the process.
- *Chapter 11:* Explains the development of the physical architecture, or ITS Future Big Picture. This includes the definition of subsystems and interfaces required to provide a complete solution.
- *Chapter 12:* Builds on Chapter 11 by discussing the topics to be considered when developing an ITS Standards Application Plan for an ITS Future Big Picture. The plan describes how to approach the incorporation of current standards and those currently under development to help support the required data flows and interfaces.
- *Chapter 13:* Discusses the essential elements of a successful outreach program for an ITS project. This includes the definition of the message to be projected and characterization of the various audiences and techniques to support the various levels of outreach required.
- *Chapter 14:* Looks at the phased deployment of a comprehensive ITS solution, over both time and space. There is a complicated range of factors, technical and nontechnical that must be taken into account when defining the best migration path from today's situation to tomorrow's comprehensive solution. Chapter 14 examines the key aspects of ITS Implementation Strategy development and provides some practical advice on how to develop one.
- *Chapter 15:* Explores some of the issues associated with the development of successful detailed designs for ITS deployments, following from the development of an ITS Future Big Picture. This includes the identification and description of some of the key elements of design and techniques that can be adopted to maximize the effectiveness of the design process.
- *Chapter 16:* Explores some of the procurement options open to public agencies in deploying ITS technologies. This includes a discussion on popular contractual mechanisms and things to consider when selecting the most appropriate one for a specific agency.
- *Chapter 17:* Describes some of the issues associated with commercial aspects of ITS developments and deployments. This includes a

discussion on the difference between a public-sector view of ITS and a private-sector commercial perspective. Some of the difficulties encountered by the private sector in dealing with the public sector in ITS are described.

- *Chapter 18:* Provides a description of one potential approach to the economic evaluation of ITS technologies and applications.
- *Chapter 19:* Includes a brief overview of ITS simulation models. This is followed by a discussion of the key issues involved in selecting and using mathematical simulation models for ITS.
- *Chapter 20:* Illustrates the practical application of the Cooperative Development Methodology through the use of stories from three previous studies. These are actual implementations with which the authors have had personal involvement over the past few years. The experiences have been selected to highlight specific aspects of the Cooperative Development Methodology, rather than to provide comprehensive examples of applications of the entire process. The following tools and concepts are illustrated in Chapter 20:
  - ITS Objectives Statement;
  - ITS Vision Statement;
  - ITS User Services;
  - Measures of Effectiveness;
  - Strawman ITS Future Big Picture;
  - ITS Market Packages;
  - ITS conceptual design.

These have been illustrated using three stories from previous projects with which we have been involved. We hope that this will clarify our earlier explanations of the various components of the ITS Cooperative Development Methodology.

- *Chapter 21:* Briefly summarizes the essential concepts and information contained in this book.

Now that we have given you a preview of the book, in what order should you read the chapters? This varies according to who you are, of course, so we have provided advice for each of our four reader groups. We have called these "grasshoppers," as they define a path from chapter to chapter, hopping over some chapters and including others. Readers can use this advice in one of two ways: Either start the book by following the appropriate grasshopper, get oriented, and then go back and read the omitted chapters, or just stop at the end of the grasshopper trail if you have read enough. Figures 1.1–1.4 illustrate the "grasshoppers."

The stakeholder grasshopper introduces ITS and the cooperative approach to ITS development and then jumps to a description of effective ITS outreach programs so that readers can see how their input fits into the overall picture.

The student grasshopper assumes that readers are transportation planning or traffic engineering students, moving through the majority of chapters, hopping over those with greater depth in system engineering techniques.

The transportation grasshopper provides the same introduction but then covers some important issues for local transportation officials including legacy

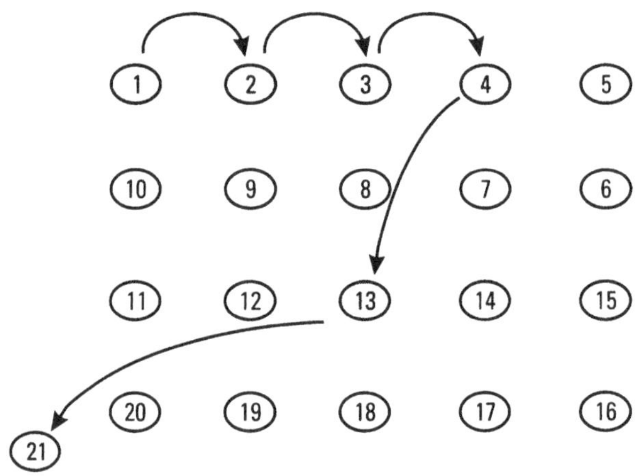


**Figure 1.1** The stakeholder grasshopper.

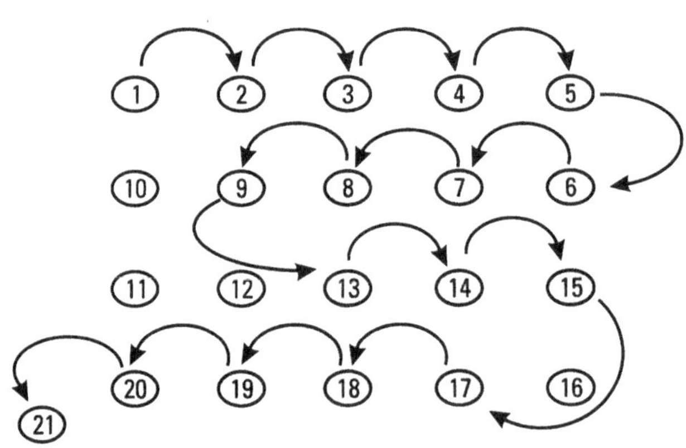


**Figure 1.2** The student grasshopper.

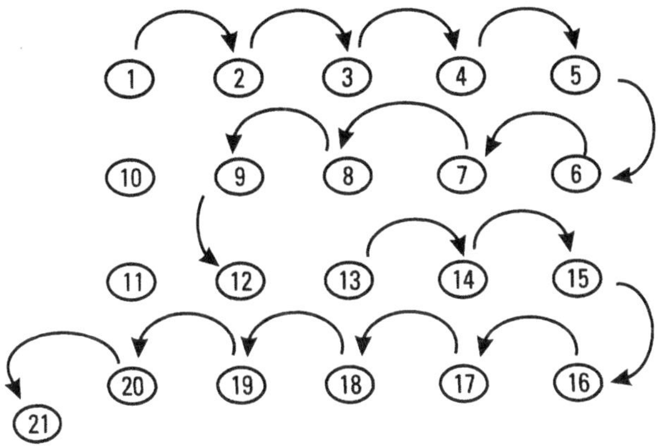


**Figure 1.3** The transportation grasshopper.

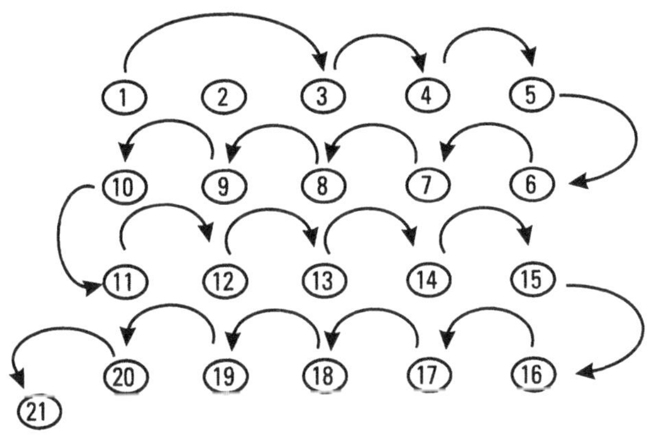


**Figure 1.4** The consultant grasshopper.

systems incorporation, protection of sunk investment, and institutional/organizational issues. It then leaps to outreach programs and implementation strategies before finishing on a sequence of chapters covering ITS procurement, commercial aspects, economics, simulation, and case studies.

The consultant grasshopper visits almost every chapter, although readers mainly interested in transportation may wish to hop over Chapters 10, 11, 12, and 13 on the first pass through the book.

As a final word of introduction, it is also worth pointing out what this book will not do: It is not designed as a "how to" guide to develop a regional ITS Future Big Picture from a system engineering perspective. There are many other texts available on detailed system engineering. The book is in fact designed for transportation practitioners and members of stakeholder groups, user groups, or consensus groups who are expected to give input into the system architecture development process. It is written from the point of view of these nonarchitecturally aware people, avoiding a great deal of system engineering detail.

Furthermore, we have no pretensions about providing magic answers on how to do detailed design, actual deployment, or smart procurement of ITS. Although there are chapters on these subjects, they aim to explain practical steps to a nontechnical audience from an observer's perspective rather than a doer's perspective.

# 2

# What Are Intelligent Transportation Systems?

## 2.1 Introduction

ITS is a global phenomenon, attracting worldwide interest from transportation professionals, the automotive industry, and political decision makers. The Japanese seem to have initiated the whole modern day notion of ITS with the work carried out in the 1980s. It must be said that the United States was also addressing the application of ITS at an early stage in the course of the Electronic Route Guidance Project (ERGS) in the 1970s. At that time, the Japanese had not coined a specific name for ITS as it was considered part of traffic control. The European Union picked up the theme, partly based on a reaction to the Japanese and partly spurred by Siemens' pioneering work in the Ali-Scout route guidance project in Berlin. The Europeans referred to the subject as road transport informatics. The United States then picked up the theme in the late 1980s, referring to the subject as Intelligent Vehicle Highway Systems (IVHSs), a term that we believe was originated by our good friend, Professor Kan Chen at the University of Michigan. We recently asked Kan to tell us how he came up with the term. He said that the acronym both described the intelligent aspects of the vehicle and the highway and included the letters VHS, reminding people of the VHS videotape, which was a winning technology (beating rival Betamax). The moniker also reflects a bit of one-upmanship over the Japanese Intelligent Vehicle System (IVS) program operated in the mid

1980s under the aegis of the Japanese Ministry of International Trade and Industry (MITI).

Not to be outdone, the Europeans came back with another new term during their second round of funding. This time the expression was Advanced Transport Telematics (ATT), reflecting their evolving view that the application of information and communications technology to transport was part of a wider application to other fields such as health and education, known collectively as telematics.

The United States then came up with yet another term, ITS, giving recognition to the wider application of technology to transit systems as well as private cars and highways. Fortunately, this name seems to have stuck, with many international organizations adopting it. In fact, the series of annual World Congresses, cosponsored by Europe, the United States, and Japan has been making use of the ITS name since the first congress, which was held in Paris in 1994. On the other hand, the International Standards Organization (ISO), the body responsible for international standards development, has decided to use yet another term in its standards development activities—Road Traffic and Transport Telematics. Did somebody mention standards?

Wherever you live and whatever you decide to call it, you are probably referring to "the application of information and communications technologies to the planning and operation of transportation systems." For the purposes of this book, we will adopt this definition and refer to it as ITS.

An ITS consists of smart roads and routes, smart cars, smart buses, smart trains, smart cargo, smart baggage, and smart travelers all working together in a cohesive system. There are many technologies that can be applied to ITS. You may be familiar with some of them already.

## 2.2 Why Bother With ITS?

As you can see from the amount of activity that has been going on for a number of years, there has been and continues to be a very high level of interest in this thing called ITS. Why is ITS so important?

Transportation planners and engineers have always faced a critical dilemma with respect to the transportation process. On one hand, transportation is considered to be a vital part of society, providing benefits such as mobility and accessibility. On the other hand, transportation is perceived as a major cause of environmental damage.

There is a fundamental relationship between the emergence of reliable transportation and our evolution from self-sufficient generalists who possessed enough skills and the ability to do everything needed to survive at some proficiency to specialist groups, trading with other groups and hence able to specialize in particular skills and obtain the kind of synergy that only comes through teamwork.

This concept of specialization and trading—the ability to use your skills and specialize, then trade with other specialists to provide the services you require to complete the picture—is also one of the main themes of this book. The definition and development of a regional ITS architecture supports the ITS equivalent of specialization and trading. Each agency and organization involved in transportation within a region defines the areas in which it will specialize and the data it will "trade" with other specializing agencies, in order to complete the entire transportation picture for the region. This lays the foundation for a team approach to regional ITS deployment.

Getting back to the "why bother?" question, transportation has the ability to provide some powerful benefits to society. In addition to supporting specialization, transportation provides us with the sort of mobility and accessibility we need to live our lives in the way we want to live them. Mobility enables us to separate home from work and visit friends and family, as well as to allow us to do business across a wider region. Accessibility opens up job opportunities.

Generally, there is a widely accepted link between economic well-being and good transportation.

However, the picture is not all rosy. There is a price to pay for good transportation. This comes in the form of undesirable side effects such as environmental impacts, energy consumption, land take, congestion, casualties, and, of course, money required to build infrastructure.

Growing concern about the impact of these undesirable side effects has influenced most developed countries to move away from the "build it and they will come," infrastructure-intensive, capital-intensive transportation strategies, toward more balanced and sustainable transportation solutions.

This is where ITS comes in. More than any other development in the transportation field in recent years, ITS holds the promise of sustainability. ITS appears to be a balanced solution that gives us the mobility and accessibility we desire, while avoiding or mitigating the undesirable side effects (the "price"). Please note that we do not see ITS as a "magic bullet" that will provide the answer to all the world's transportation problems. Instead, we see ITS as a major new element in the transportation professional's portfolio of approaches and solutions to transportation problems.

ITS presents an opportunity for better management of existing resources and infrastructure, through the provision of information to travelers and transportation professionals and through new control possibilities.

## 2.3 A Few Examples of ITS Applications

ITS is not brand new; it has been slowly creeping into our lives over the past couple of decades. In fact, if you consider electric traffic signals to be ITS (and we think they could be considered the "original" ITS), then ITS has been around since the 1930s. The history of the traffic signal is quite interesting and illustrates another point we would like to make. Apparently, the first traffic signal anywhere in the world was installed outside the houses of parliament in the late 1800s. The mechanism consisted of a semaphore arm with a gas-powered lamp to illuminate the signal. Unfortunately, the system exploded shortly after implementation, injuring a policeman and putting the British off the whole idea of traffic signals for decades. The first electric powered traffic signals were installed in Cleveland, Ohio in the 1930s.

It is interesting to note that the failed experiment in London was carried out in such a high-profile location and had such disastrous results that it effectively created a vacuum for similar technologies for a long time afterward. Maybe the lesson for ITS is that early attempts to implement untested technology should be carried out in low-profile circumstances. It might also suggest the existence of a rather unfortunate paradox centered around the early stages of any ITS implementation. The beginning of your deployment, when you are having the most trouble making it all work and getting people accustomed to using the system, is also the time when the users are at their most critical. At this point, impressions are being created, opinions are formed, and a close evaluation of your system is carried out. Get it wrong at this point and you may "scorch the Earth" for future implementations, no matter how good they are.

We will return to this point in Section 3.3, as it is yet another reason for including the development of an ITS architecture in your approach planning and system development activities.

Apart from the traffic signal example, there are many other applications of ITS around us. Sections 2.3.1–2.3.5 describe a few generic applications. Then, some real world examples of ITS are explored in Section 2.4.

### 2.3.1 Smart Buses

This might sound strange, but buses can be smart. Imagine a bus that knows exactly where it is and can communicate with a control center for instructions and announce the name of the next bus stop as it approaches. As it nears traffic signals, the lights change to green to let it through, and it tells passengers at bus stops further up the route when it will arrive. Does this sound like science fiction? Well, actually all of these features are already fitted on buses across the United States. Automatic vehicle location can be used to pinpoint buses to within a few feet utilizing satellite navigation, or radio beacons fitted along the routes, and cellular telephones or other advanced radio techniques enable buses to send and receive data from a central control center and allow the driver to talk to the dispatcher. Onboard, voice synthesizers, triggered by the position of the bus can announce the name of the next bus stop, while communication with the computer controlling the traffic signals can be achieved through the control center or via dedicated short-range communications directly with the local traffic signals. The control center can also communicate with smart bus stops with information displays indicating the time to the next bus (or the time the last bus left) and other schedule information to the traveler.

### 2.3.2 Smart Cars

Those smart people in Detroit, Toyota City, and Stuttgart have been working on this one for a long time, apparently attracted by the idea of selling you more electronic content when you buy a car. Many automotive industry observers have predicted that the proportion of automobiles' value attributed to electronics will increase substantially over the next decade. Maybe one of these days they will be selling us electronics packages and throwing in the wheels, engine, and transmission for free!

Smart cars can have many different capabilities, including in-car information and navigation systems, electronic payment and toll collection capabilities, intelligent engine management systems, and even driver support and control systems.

In-car information systems are an interesting starting point. We remember reading an article in proceedings from a conference a few years ago in which some Japanese researchers reported the results of a study of how drivers communicate with each other. They had figured out that when drivers are sealed into their nice well-insulated cars, there are only five ways in which they can communicate with the outside world, listed as follows.

- Hand signals;
- Indicators;
- Horn;
- Hazard warning lights;
- Headlights.

They then looked for a comparison and concluded that under these circumstances the driver has the same communications capabilities as an insect. This could, of course, explain a lot about driver behavior.

Seriously though, this constrained communications environment in which the driver has little information about the overall travel context and has poor abilities to communicate with fellow travelers is where ITS can be very powerful. Consider in-car information systems that can advise drivers on the best route to take, accounting for current traffic conditions and any nonrecurring congestion or incidents that have occurred. These systems provide the type of contextual information that enables drivers to make better decisions. They do, however, require some kind of communications link between the vehicle and the roadside and from the roadside to some sort of control center. Consequently, they are still a bit expensive at the moment and depend on the availability of roadside infrastructure such as transceivers and wireline communications. It does not take a leap of imagination to see how these systems may evolve in the future into general purpose information and reservation systems, enabling the driver to reserve parking spaces in advance, obtain "yellow pages" information, and even make payments for goods and services.

In the meantime, autonomous in-car information systems that do not need the communications link are already on the market. These are independent systems that are able to locate your car on a map display on the dashboard and provide you with turn-by-turn directions to your selected destination. The routes suggested are, of course, based on historical rather than current or dynamic information.

Looking further ahead, some technologies being tested today have the potential to assist the driver and perhaps even take over some of the control from the driver in certain circumstances. It is probably difficult to imagine a car that can keep itself in the appropriate lane—so that drivers do not have to steer—and one that will automatically keep a car the appropriate distance from the vehicle in front, while continuously sensing and detecting potential obstacles. Even under manual control, the car looks after its driver. For example, when the driver looks at the side mirror to check whether it is safe to change lanes, an indicator on the mirror turns red if there is another car within the danger zone that is obscured from view. These seemingly fantastic capabilities are becoming available today, and we will see them incrementally applied to cars in the near future.

### 2.3.3 Smart Highways

Smart highways have features that enable drivers to communicate with a control center, with roadside devices, and with other drivers. They also have features that enable road operating agencies to inform drivers about traffic conditions, hazardous road conditions, and diversion routes.

Interestingly, there are smart highways and even smarter highways. Smart highways exist today. They are no longer just asphalt, concrete, and steel but encompass communications sensing, display, and processing capabilities.

Traffic sensors of various types are used to determine current traffic flow data; communication systems are used to get this data back to a central control; and variable message signs are used to provide advice and guidance to drivers. There are many examples of this in practice in the United States, Europe, and Japan right now, and many people are obtaining substantial benefits in terms of congestion avoidance and incident management. Traffic sensors are limited, however, in that they tell all travelers the same information and cannot tailor the information to the needs of the individual. That is where even smarter highways come into play. These have the same capabilities as smart highways but offer additional infrastructure to support one- or two-way communication between the car and the roadside. Information to drivers can then be displayed inside the car and tailored to a set of predefined requirements. These highways will also be using the communications link to identify vehicles at key points along the highway in order to assess average speeds and journey times.

This same communications link can also be utilized for electronic toll collection, obviating the need to slow down and stop at toll plazas, reducing congestion, and making money handling more efficient for the toll operators. Drivers can either prepay the tolls into a special account or be billed in the same manner as credit card billing.

### 2.3.4 Smart Trucks

These are like smart cars, but they have additional features that are of particular use to truck drivers and trucking companies. Smart trucks use the same type of communications links and vehicle location technologies as their car counterparts, while also providing support for administrative processes such as electronic preclearance for border crossings and roadside safety inspections.

### 2.3.5 Smart Travelers

Smart travelers depend on good information to make intelligent decisions about their travel plans. Just imagine if you could find out before you left home how long it would take you to get to work at exactly that time of day and be able to see comparative statistics on different routes and different modes of travel. Furthermore, imagine being able to determine whether a 20-minute delay to the start of your journey might still get you to work on time since you have missed the peak traffic congestion. This is all possible through the use of

ITS. Advanced traveler information systems on home or office computer terminals provide travelers with these options.

## 2.4 Project Examples From Around the United States

So far, we have described ITS applications in fairly general terms. Fortunately, there are already some specific examples in existence that are worth using as additional examples. These examples, presented in Sections 2.4.1 and 2.4.2, illustrate the application of ITS technologies in combinations that are designed to solve particular problems and address specific needs.

### 2.4.1 TranStar, Houston, Texas

Our first real-life example is TranStar, the Greater Houston Transportation and Emergency Management Center. TranStar, a $13.5 million, 52,000-square-foot intermodal transportation management facility, is intermodal as it addresses the management of private car traffic, public transport, and transfer between the two modes. An important aspect of TranStar is that it involves the cooperation of multiple agencies such as the Texas Department of Transportation, the Metropolitan Transit Authority of Harris County, the City of Houston, and Harris County. In other words, the agencies responsible for private cars, public transport, and traffic operation and enforcement have all agreed to cooperate in this deployment.

TranStar is a multipurpose control center that utilizes a sensor, control, and communications network to manage transportation. Closed-circuit television (CCTV) surveillance and ramp metering are utilized to manage over 300 miles of freeway. TranStar also controls and coordinates the Houston Regional Traffic Signal System, which consists of over 2,800 traffic signals. This includes the day-to-day management of traffic flows to optimize the use of the entire network and allow traffic to flow as smoothly as possible.

TranStar uses inductive loops and video detection technologies to monitor traffic conditions on a continuous basis. Variable message signs (VMS) are then utilized to inform and advise drivers about road and traffic conditions. This is particularly useful for situations in which an unexpected event causes nonrecurring congestion on a part of the highway network. The VMS can be used to inform drivers of suitable alternative routes to help reduce or avoid congestion.

TranStar also manages a network of 105 miles of high-occupancy vehicle lanes, enabling public transport vehicles and cars with at least two occupants to benefit from dedicated infrastructure and to attain faster journey times during peak periods.

In addition, TranStar is equipped to function as an emergency management operations center for evacuations and disasters.

An interesting feature of the TranStar application is the use of a technique known as *probe vehicle data.* This takes advantage of the fact that around 80,000 vehicles within Houston are fitted with transponders or tags for electronic toll collection purposes. Additional tag readers have been installed at strategic locations around the freeway network, enabling equipped vehicles to act as probes for the TranStar data collection system. Each equipped vehicle can be recognized at these reader locations, and the time taken for them to reach the next reader can be deduced. This information is used to determine average speeds for all vehicles currently on the network. Hence, the vehicle acts as a sensor or probe, indicating how quickly traffic is moving through the freeway network. Temporary identity labels are used for each probe vehicle. These are randomly generated and thrown away after each use, preventing anyone from tracking the vehicle.

TranStar also uses a technique known as Highway Advisory Radio (HAR) to provide information to drivers. A special radio frequency is used to provide a continuous broadcast of travel information such as road conditions, hazards, and delays. Roadside signs advise drivers the HAR frequency to which they should tune their car radios to receive the broadcast information.

#### 2.4.1.1 Ramp Meters

We mentioned in Section 2.4.1 that TranStar uses ramp meters to manage the flow of traffic on the freeways. At freeway on-ramps, special traffic signals are used to control the flow of traffic from the ramp onto the freeway. Drivers are allowed to enter the freeway at the rate of one car per green. Traffic sensors on the main freeway in the vicinity of the ramp provide traffic flow data to a roadside computer that determines the optimum flow of traffic from the on-ramp, then adjusts the ramp meter signal timings accordingly. Thus, the traffic entering the freeway from the ramp is *metered* onto the freeway at the most appropriate rate.

### 2.4.2 Atlanta Showcase Project

As part of the 1996 Olympic Games in Atlanta, the U.S. Department of Transportation sponsored a major application of ITS technologies, known as the Atlanta Showcase. This was an important initiative designed to demonstrate the capabilities of ITS to an international audience during this major sporting event. It is also interesting to note that one of the reasons for the choice of Atlanta as the showcase site was that the city already had an advanced traffic management system under development. This was at an advanced stage of implementation and incorporated an extensive traffic surveillance and data collection capability. The availability of extensive, high-quality traffic and transportation data is considered to be a prerequisite for workable traveler information systems. This is a good example of the synergy between different ITS applications, under which the very existence of one application makes another different application easier to support—or feasible. We will return to the subject of synergy in Chapter 3, as it is an important theme running through the book.

The Showcase project actually involves five separate applications of ITS that collectively provide traveler information to a range of information users, using a number of communication and information delivery techniques.

#### 2.4.2.1 On-line Computer Information Services

The recent explosive growth in the use of the Internet and the World Wide Web means that many readers may have already used one of these services. Personal computers (PCs) are no longer used solely for business applications in the office; now, they are also used to run home-based information and entertainment products. Such products are sometimes placed in the *infotainment* category. Through the use of telephone communications combined with a home PC, users can access a wide range of information and a number of services such as home banking and shopping. The Showcase project—recognizing that growing numbers of households have home PCs and make use of on-line information services—provided home PC users with access to pretrip information on current traffic conditions and transportation schedules and even allowed them to make reservations online through the Internet and other channels.

#### 2.4.2.2 Personal Digital Assistants

In the same way that the home PC has become popular, there has been a growth in the use of small portable personal computers known as personal digital assistants (PDAs). These are small enough to slip into your pocket or purse and can be linked to an Internet-based information service using either telephone lines or a wireless communications technology like cellular telephones. This enables travelers to get current travel information on the go, so they can confirm that their current plans are the best ones or get new information to cope with a change in itinerary or schedule.

#### 2.4.2.3 Interactive Television

Another variation on pretrip information delivery is the provision of current travel information to hotel guests in their rooms, utilizing interactive television techniques. You may have stayed in hotels that use this technique to enable you to check out from your room or review the current status of your account. The same technique can be used to allow guests to work through an on-screen menu and gain access to current travel information.

#### 2.4.2.4 Cable Television

Like the interactive TV application described in Section 2.4.2.3, cable television broadcasts current traffic conditions to viewers. However, there is no interaction, and the viewer is not able to make requests for specific information. The information is available to all regular TV viewers and covers the entire Atlanta area, using announcers and graphics to indicate current traffic and transit operating conditions.

#### 2.4.2.5 In-Vehicle Navigation

In this application, in-vehicle navigation devices are being enhanced to provide current travel information and best route choices, rather than ones based on historic data. Computers and information displays inside the vehicle provide the driver with this information.

### 2.4.3 Trafficmaster, United Kingdom

This is a rather unique in-vehicle information system, developed and operated by a private sector company in the United Kingdom. Under this program, sensors have been installed on the entire national motorway network in the United Kingdom, enabling current traffic speeds to be measured and data sent back to a control center in the southeast of England. This is collated and then sent out to subscribers who receive the information on one of two types of specially developed in-vehicle units. The first of these, Trafficmate, is a low-cost in-car device that provides current information on traffic conditions utilizing voice synthesis techniques. It can provide audible warnings of traffic congestion either within a ten-mile radius of the vehicle, or up to two intersections ahead of the vehicle. This enables the driver to make rerouting decisions before encountering the appropriate intersection.

The second device is called Trafficmaster YQ. This is a more sophisticated unit that incorporates a liquid crystal display (LCD) enabling the driver to view graphical images of the road network and receive up-to-date information on traffic delays. These in-vehicle units are portable, enabling their use in pretrip planning.

While this development has been pioneered in the United Kingdom, Trafficmaster is growing outside of the United Kingdom in international markets.

## 2.5 Structured View of ITS

You probably think you have a reasonable picture of ITS applications by now, but we have only just scratched the surface. There are many, many more applications and technologies. It would be very difficult to explain ITS if we continued in this random fashion, as it might seem confusing or be too much for readers to handle. Instead, we can take advantage of the fact that much work has been done to understand how all these technologies work together and how to go about selecting the ones you will need. This enables us to present a more structured view, grouping ITS applications into areas and technologies into groups. There are many classification systems in use both in the United States and overseas. We have decided to adopt the ITS market areas approach presented in Section 2.5.1, since it is straightforward and takes a user-oriented approach that aligns nicely with some other points we want to make in Chapter 3. It is based on a sort of "atomic structure" for ITS as shown in Figure 2.1.

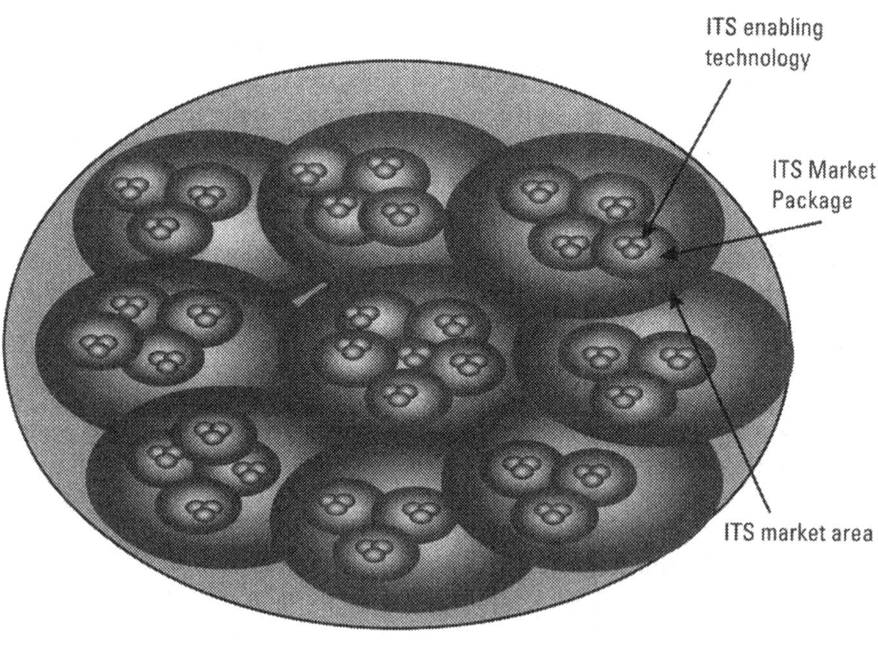


**Figure 2.1** The world of ITS.

### 2.5.1 ITS Market Areas

There are several different ways to classify and segment the ITS field. In the beginning, it was popular to use technology groupings, splitting ITS into different technology applications such as communications, display technologies, navigation technologies, and vehicle location technologies. These are certainly very helpful from a designer point of view, but they have not provided the necessary focus on the requirements of ITS users.

Another way of looking at ITS that solves this problem is to take a market view that identifies buyers and users of the ITS and tries to group them into those having similar needs or objectives. ITS market areas can be defined that represent groups of ITS users, operators and procurers with similar needs and buying characteristics. Companies supplying ITS equipment and services might typically divide the entire field into such market areas when defining customers, developing products, and carrying out marketing and sales strategies. Taking this view also supports an end-user driven approach to the planning and development of ITS by recognizing the way in which transportation is currently operated and organized and observing practical institutional and organizational boundaries rather than theoretical technological ones.

We make use of the ITS market areas illustrated in Figure 2.2.

#### 2.5.1.1 Area 1: Traffic Management

In this market area, the customers are primarily local and central government agencies. These are the people responsible for using tax money to improve the transportation process, making it safer and more efficient. Their main motivation is to manage the entire road network as well as possible on behalf of the general public. This is achieved through collecting traffic data and influencing and managing traffic flowing on the network

#### 2.5.1.2 Area 2: Emergency Management

The people responsible for getting us out of trouble are in this market area. This group responds to incidents and emergencies and has primary responsibility for identifying problems, deciding on appropriate responses and resources, and then managing the situation. This encompasses the following:

- Fire;
- Police;
- Ambulance;
- Courtesy patrol.

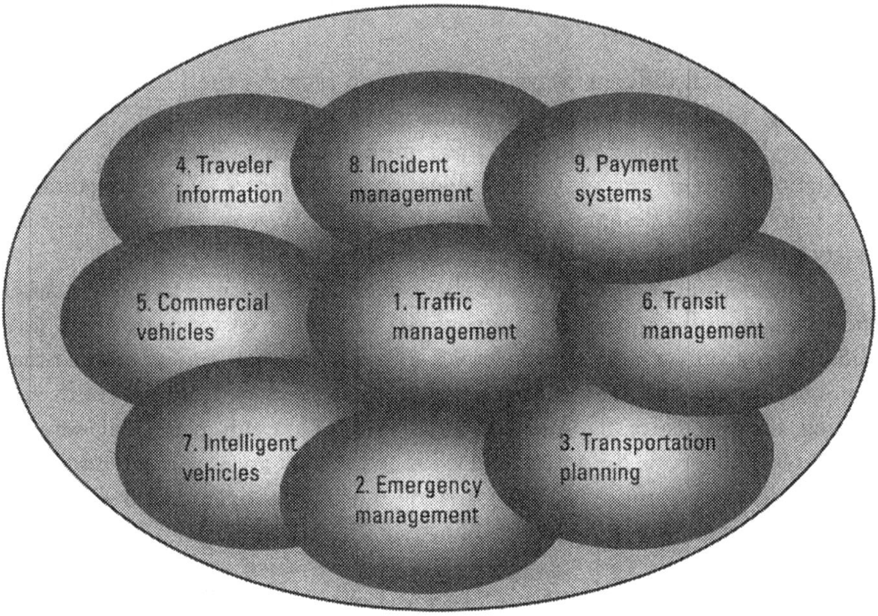


**Figure 2.2** ITS market areas.

#### 2.5.1.3 Area 3: Transportation Planning

Customers in this market area try to match transportation supply with demand both now and in the future. These are the people who try to determine current transportation use patterns and make predictions about future use patterns. They include local, state, and federal transportation planners, traffic planners, and transit planners.

#### 2.5.1.4 Area 4: Traveler Information

In this market area, the customers for ITS are the providers and users of travel information. Although central, state, and local transportation agencies can play a role in this area, it is likely that the significant users will be private-sector information service providers and subscribers.

#### 2.5.1.5 Area 5: Commercial Vehicles

There are two primary groups of customers in this area—the trucking industry and the local agencies that regulate them. The trucking industry is comprised of potential customers for in-vehicle devices providing travel information and fleet management services. The regulatory authorities procure infrastructure to support electronic data interchange and automated licensing and inspection procedures.

#### 2.5.1.6 Area 6: Transit Management

This area is populated by those who plan and operate our transit systems in both urban and rural areas. They are concerned with increasing the operational efficiency of all transit modes as well as achieving a tangible improvement in the attractiveness of these service offerings.

#### 2.5.1.7 Area 7: Intelligent Vehicles

The automobile manufacturers, automotive electronics manufacturers, and truck and transit vehicle manufacturers all belong in this category. They are all concerned with enhancing the capability of road vehicles through the use of electronics, sensors, communications, and control actuator technologies. Vehicle drivers also belong in this group, as it is they who will be the eventual consumer or end user of the various ITS applications.

#### 2.5.1.8 Area 8: Incident Management

This area, again, consists of the central, state, and local government officials charged with improving the efficiency of the roadway network. The people responsible for incident and emergency management tend to have different needs and objectives and belong to different organizations than the traffic managers; accordingly, they have their own market area.

We should make it clear at this point that the market areas are not mutually exclusive. There is substantial overlap between many of the market areas because of the integrated nature of transportation demand. The market areas should be considered as a way of focusing on the needs of a particular group of people, rather than a means of rigid distinction.

#### 2.5.1.9 Area 9: Payment Systems

This area encompasses all the people that try to take money from you in return for providing a service. It includes toll road operators, transit agencies, car parking operators, and all the customers and users of ITS involved in the process of fee payment for transportation services. The payment systems area is also expanding to include people responsible for general purpose payment services beyond the transportation realm. This includes people involved in the development of applications such as Pay-per-view TV, telephone cards, and general purpose debit cards. There is a growing overlap between transportation and nontransportation needs because of the potential for synergy in the payment systems area. For example, the same smart card could be used to pay for bus fares and make telephone calls within a region.

## 2.6 ITS Market Packages

Now that we have defined the market for ITS products and services in terms of ITS market areas, we can consider these to be targets for the application of ITS. First, we need to define groups of ITS technologies to satisfy the needs of each market area. We call these groupings ITS Market Packages. We use this expression because they represent bundles or groups of ITS technologies that have been brought together into a product or service in response to an identified and defined market need.

The traffic signals with which most of us are familiar are an example of an ITS Market Package. A group of technologies has been bundled together to meet a defined need within a market area. In this case, the need relates to managing traffic flows at intersections by separating conflicting traffic flows using *time-sharing*. Vehicles in the conflicting flows share a common part of the highway at different times. The time-share is managed through the use of a package consisting of display, communications, data storage and processing, and sensor technologies.

ITS Market Packages can be viewed as the fundamental building block—at a planning and development level—for ITS implementations. ITS Market Packages are technology-dependent but not technology-specific and indicate the types of technologies, products, and services that need to be grouped to address the needs, issues, problems, and objectives, but they do not have the definition required to support detailed design or procurement.

## 2.7 ITS Enabling Technologies

There is a range of information and communications technologies that are applied in combination to transportation problems. We refer to these as ITS enabling technologies because their existence enables the development of ITS. This section explains what they are and what they do.

### 2.7.1 What Are They?

These are advanced technologies such as fiber optics, CD-ROM, electromagnetic compasses, global positioning systems (GPSs), laser sensors, digital map databases, and display technologies like cathode ray tubes and LCDs that enable ITS applications.

### 2.7.2 What Do They Do?

ITS-enabling technologies have two primary roles. First, as the name suggests, they enable ITS to function by providing functionality and capability. Second, they act as elements in the ITS Market Packages, allowing us to address defined market needs and service requirements.

ITS-enabling technologies are numerous and varied, but we can group them into broad categories according to what they do to make it easier to comprehend the whole picture.

Note that many of the enabling technology examples described here could in fact be considered as bundles or groupings of more basic or fundamental technologies. To keep things simple, we consider the enabling technology to be the smallest "atomic unit" in the world of ITS, since our objective is to show you how to apply the technologies rather than focus on how they work.

Table 2.1 presents some examples of prevalent enabling technologies that you may encounter in ITS implementations.

**Table 2.1**
Enabling Technologies Summary

| | |
|---|---|
| Communications | Wireless<br>Wireline |
| Data storage and processing | Compact Disc<br>Magnetic storage media—magnetic stripe cards, hard disks, floppy disks, and data cartridges<br>Smart Cards |
| Database management systems | Data warehousing<br>Expert systems |
| Information displays | Cathode ray tubes (CRTs)<br>LCDs<br>Variable message signs |
| Location | Dead reckoning<br>Map matching<br>GPS<br>Beacon-based vehicle location |
| Sensors | Inductive loops<br>Infrared beams<br>Microwave (RADAR)<br>LIDAR<br>Vision-based sensors<br>Acoustic<br>Scanning laser |
| Actuators | Gates and displays |

### 2.7.3 Communications: "Pass it On"

Figures 2.3 to 2.5 illustrate the range of communications technologies available in terms of a "family tree" of communications technologies. Within each branch there are also variations in technology. For example, under wireless digital you could adopt either cellular or wide-area packet data radio technology, while under wireline you could use either coaxial or fiber optic media.

### 2.7.4 Data Storage and Processing: "Keep it for Later and Work Out the Details"

These technologies lie at the heart of the information technology revolution that has been developing for the last 20 years and promises to impact almost every aspect of our lives. The ability to store massive amounts of data and

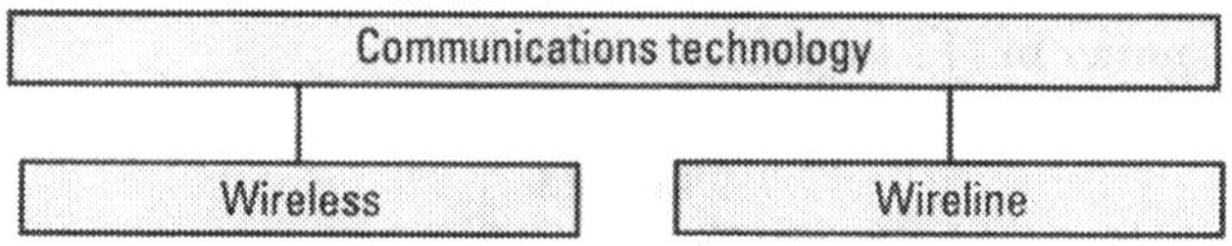


**Figure 2.3** The communications technology family tree level 1.

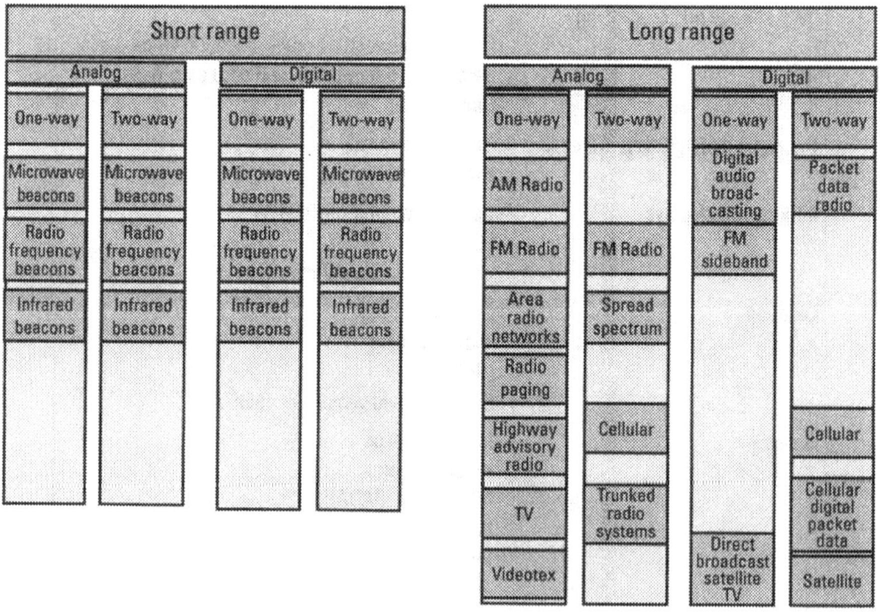


**Figure 2.4** The communications technology family tree level 2.1, wireless.

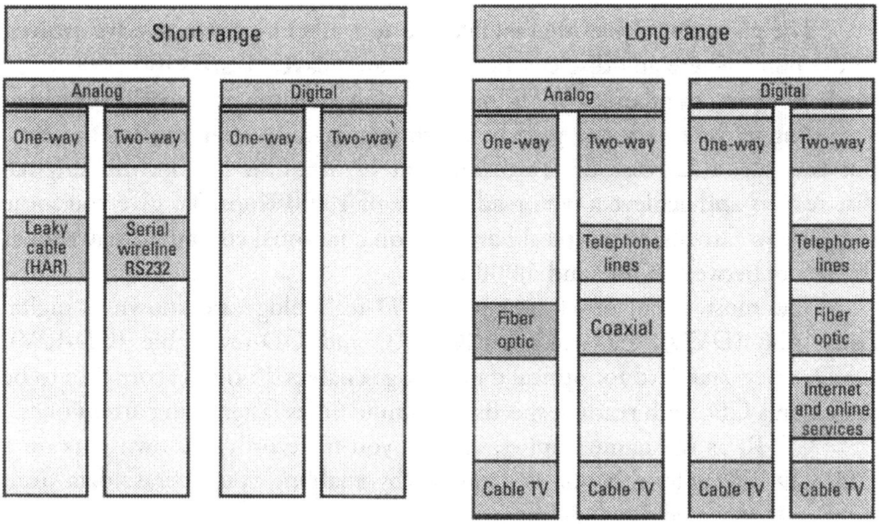


**Figure 2.5** The communications technology family tree level 2.2, wireline.

process it quickly and efficiently to produce the desired information is central to the application of ITS.

Sections 2.7.4.1–2.7.4.3 present some examples of data storage technologies that are being utilized today for ITS applications.

### 2.7.4.1 Compact Disc

This is really a family of optical data storage technologies that evolved from an original technology developed by Philips in the 1970s. The technology makes use of laser technology to read the surface of a spinning disk. The disk surface is specially treated to have some highly reflective areas and some less reflective areas. This causes the laser light reflected back from the disk surface to vary in a way that can be sensed, measured, and processed into digital data. This data can represent audio, video, photographic images, computer data, or a combination of all of these.

The family of compact disc (CD) technologies includes compact disc-interactive (CDI) photo CD and compact disc-read only memory (CD-ROM). These use the same laser-based technology but have different data recording formats governing the structure of the data on the disk. Any one of these disks can store more than 650 megabytes of data or 74 minutes of audio. In comparison, a typical floppy disk can store only 1.14 megabytes. They are all *read-only* storage devices that require extensive production facilities, making them suitable for mass distribution of data.

The players or drives that enable you to read compact discs have evolved to the point where any disc can be read by any player, thanks to the development of open standards by Philips and Sony. Drives are usually classified according to the rotational speed at which they operate when reading the data. For example, a *12x* disk drive will rotate at 12 times the speed of the original disc readers and achieve a data reading rate of 1,800 Kbps. To give you some idea of how fast this is, a typical hard disk on a personal computer may read at the rate of between 3,000 and 4,000 Kbps.

The most recent developments in CD technology are known as digital video disc (DVD), CD-recorder (CD-R), and CD-rewritable (CD-RW). DVD, a new standard for optical data storage enables 25 times more data to be stored on a CD, with reading speeds up to nine times faster than current ones.

CD-R, as the name implies, enables you to record your own data on a CD. CD-RW takes this one step further by enabling you to erase data from previously recorded discs and record new data.

These last two members of the CD family are now emerging into the market.

#### 2.7.4.2 Magnetic Storage Media: Magnetic Stripe Cards, Hard Disks, Floppy Disks, and Data Cartridges

There is a large family of data storage technologies based on the use of ferromagnetic materials to store and retrieve data. They all make use of the same basic principle involving coating a flat surface such as a metal or plastic disc, or a length of tape, with a material containing ferromagnetic particles. These particles act like small bar magnets and can be realigned on the surface with the application of a magnetic field. The particles then stay in the new orientation until another magnetic field is applied.

This family includes the familiar music cassettes, videotapes, and specialized data storage tapes. These products all come in a variety of formats and sizes. Two of the more popular data storage tape formats are quarter-inch cartridge (QIC) and Travan. QIC's name refers to the width of the magnetic tape used in the cartridge. Travan tape cartridges are QIC variants with larger data storage capacities. Magnetic strip cards are very popular for credit card, automated teller machine (ATM), and common stored value ticketing systems. The strip can be supported on either a plastic or paper substrate depending on the tradeoff between durability and cost. Figure 2.6 shows the anatomy of a typical magnetic stripe card application.

#### 2.7.4.3 Smart Cards

Smart cards promise to be the most ubiquitous of the new information and communication technologies to be applied to transportation. Although we have listed them here under data storage devices, they can process data as well.

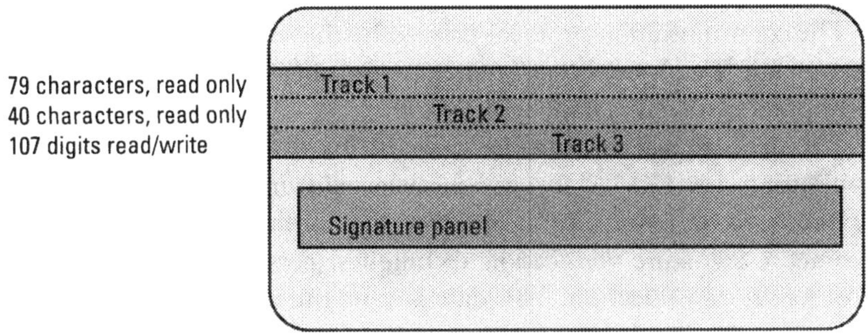


**Figure 2.6** Anatomy of a magnetic stripe card.

A smart card is essentially a miniature computer, the same size as the ISO standard magnetic stripe bank card used by American Express, VISA, and major banks. Unlike the magnetic stripe card, smart cards have an onboard processor and memory capability that can be used to store and manipulate user data, provide security keys, and provide authentication algorithms that are inaccessible to the user. The smart card can perform transactions with host systems using a card acceptor device. The card can be credited with value, or loaded with data, that can then be transferred or decremented as appropriate. A typical smart card can store approximately 2–3 Kbytes of data.

The technology utilized for the smart card is basically a microchip embedded into the standard bank card. Data storage is effected by means of electronically erasable programmable read only memory (EEPROM) technology that enables all memory locations to be read and written under the control of the onboard operating system. Figure 2.7 illustrates a typical smart card operating system. For some reason, the vendors often refer to this as the "mask."

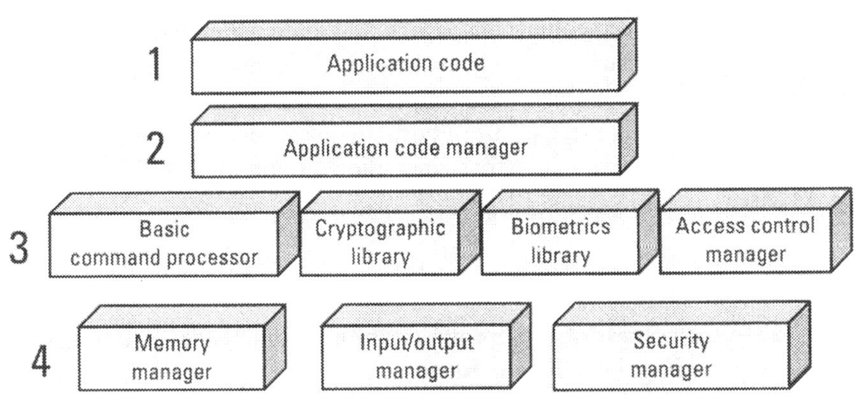


**Figure 2.7** Smart card operating system.

The communication protocol utilized by the smart card can be translated to be compatible with any host system, by means of a card acceptor device, such as a computer system or a basic terminal.

The data storage area on the card can be divided into a maximum of 15 application data files (ADFs) or functions, allowing the potential for multiple applications for a single card. Access to each of the ADFs can be controlled by biometric signature verification techniques, personal identification numbers, or password protection. This data is stored in a separate area of memory that is only accessible to the operating system, rendering it invisible and out of bounds to the user.

The card acceptor devices have until recently been based on the use of a contact plate to activate the smart card, requiring that the acceptor be designed in the shape of a card holder. However, technology has now been developed to enable contactless smart card operation, utilizing a communications link between the card and the acceptor, thus obviating the need for physical contact between card and acceptor. The communications link is achieved by using a serial data link when the card is brought within a few centimeters of the acceptor device.

The capacity for contactless communication increases the manufacturing cost of the smart card but offers some advantages. Removing the need for contact gives greater flexibility in the design of the card acceptor, enabling it to be molded into a card holder as would be required for the contact version or implanted into a flat surface that is then designated and marked as the reading area. This is useful for public transit applications as there is good potential to make the unit vandal-proof, and the contactless communication method should also speed vehicle boarding. More recent advances in technology have enabled the development of what is known as the *combi card.* The combi card has both contact and contactless capabilities on the same card, supporting interoperability between contact and contactless approaches in the same region. For example, if the transit company has chosen the contactless approach, while the local bank has opted for a contact approach, the combi card would enable the cardholder to obtain services from both organizations in a seamless manner. Some versions even have magnetic stripes allowing interoperability with existing ATMs.

### 2.7.5 What Can Smart Cards Do?

Used within the context of an appropriate information system, smart cards are capable of providing the following functions:

- *Stored credit:* Users can purchase the smart card complete with prestored credit and return the smart card when all credit has been used. This enables smart cards to be interrogated centrally to extract stored management information and allows close control of the circulation of smart cards. Alternatively, smart card charging stations can be provided at selected locations such as bus stations, ticket offices, or large trip generators such as shopping malls. Users can then be allowed to restore the credit on the smart card by inserting cash into the charging station, which adjusts the balance on the card and transfers any management information back to the main system.
- *Automatic capture of passenger information and journey profiles:* Smart cards can store copies of all the transaction records related to the use of the card. This data can be retrieved and utilized as management information. The type of data that can be stored includes service usage and trip patterns. This data can also be used to provide card users with a summary of transport usage for either business or personal use.
- *Multiple usage:* Because of the ability to partition the smart card memory into 15 discrete areas protected by a secure operating system, it is possible to use a single card for multiple applications. These applications could be public transit related—pertaining to different operators or different services—or nontransport related—pertaining to the purchase of goods and services from other participating organizations. This offers the prospect of the transport operator taking over the role that is traditionally held by banks and credit card companies, where multiple organizations participate in the smart card application, with transactions cleared through the transport operators' system.

There is also a variation of the smart card that uses optical data storage techniques to store large volumes of data. These utilize storage techniques similar to the CD technologies described in Section 2.7.4.1.

Data storage capabilities have increased in leaps and bounds over the past 10 years, partly because of the development of the physical techniques described above but also due to some major advances in data processing or compression techniques.

These fall on the boundary between data storage and processing and include a wide range of signal processing techniques that are used to compress data before recording and to re-expand the data on playback. The adoption of these techniques has also had a significant impact on the capacity of communications systems as well as storage devices. This has made the delivery of digital high-definition TV to homes technically feasible and allowed entire movies to be recorded on a single CD.

Before we leave the subject of smart cards, we'd like to tell you a story about the societal effects of the ubiquitous card technology. Bob was born and spent a good part of his early career in Glasgow, Scotland. Apart from the wonderful scenery, Glasgow has a culinary reputation that is second to none. Probably the favorite food of the Glaswegian is deep-fried fish and chips. This is preferably eaten after an extended visit to a local pub or hostelry and is most authentic when eaten from a newspaper wrapping while walking along the street.

After working late one evening (as he has been known to do, on occasion) Bob was leaving his office in Glasgow when he bumped into a fellow Glaswegian with a card technology problem. The other Glaswegian—let's call him Jimmy as most people in Glasgow seem to be called this anyway—had followed the time-honored ritual of the extended visit to the pub, followed by the outdoor consumption of the fish and chips. Jimmy was standing next to an ATM and on seeing Bob, asked an interesting question: "Hey pal, do you have card for this ATM?" Bob, being a cautious Scot (especially when it comes to financial matters) answered with another question: "Why do you want to know?" Jimmy then explained that he had been walking along, eating his fish and chips when he noticed that his shoe lace had come undone. As he approached the ATM, another Glaswegian had just finished using it, and Jimmy noticed that the ledge in front of the ATM would make an ideal resting place for his fish and chips, while he tied his shoelaces. Placing his fish and chips on the ledge, Jimmy tied his shoe laces then got up to retrieve his fish and chips. To his horror, he found that a glass security screen had descended over the ledge, trapping his fish and chips inside. That was why he was asking if Bob had a card for that ATM. He wanted to get his fish and chips back!

Bob obliged and noted that card technology can sometime have unpredictable side effects.

### 2.7.6 Database Management Systems

Moving on to data processing techniques, it obviously would not help us very much if we could store all this data but did not have the ability to process and retrieve it quickly and efficiently. This is where data management and processing technologies come into play. These technologies allow us to structure and organize data so that we can bring it back from storage on demand, navigate our way through the data until we find what we want, and extract information from the data. They enable us to identify patterns and trends in the data and allow us to produce reports, tables, graphs, and summaries that provide the information on which to base good decisions regarding transportation planning and management.

We are sure that just describing and explaining what is currently being developed and applied in this technology field could be the subject of a whole series of books, so we will not even start to scratch the surface of this one. The use of relational database management techniques supported by structured query language (SQL)—sometimes pronounced sequel—has provided us with a flexible approach to the management of large volumes of data. This allows the rapid establishment and relatively easy operation and maintenance of comprehensive traffic and transportation databases, such as digital maps. Reports can be defined and generated quickly, either by a professional programmer, or by a knowledgeable end user.

### 2.7.6.1 Data Warehousing

This more recent innovation exploits the latest abilities of information technology to collect a variety of data into a single, useful information source. In recent years, there has been an explosion in the use of data warehousing techniques with many proprietary software systems now available.

In essence, the approach involves the collection of all data for an organization into a central repository or warehouse. The data usually originates from many sources, including existing information management systems and is subject to "cleaning" processes to ensure that the data is consistent and reliable.

Another significant feature of the data warehousing approach is the use of user-friendly tools to enable data and information users to search and navigate the data warehouse unassisted by information technology specialists. Using graphical interfaces and browser-type technologies, these tools enable the user to carry out what is known as *data mining*. This describes the process in which the user navigates around the data warehouse looking for patterns or trends in the data and identifying relationships between the data elements. Using these tools, an end user can quickly navigate through large volumes of data and establish the desired data summaries and information outputs.

### 2.7.6.2 Expert Systems

Sometimes referred to as artificial intelligence (AI), this technology supports sophisticated approaches to programming or software development for computers. With this technology, it is possible for a computer to be programmed to "learn" as it carries out preprogrammed tasks and becomes more functional. It is also possible to define a set of rules that describe expertise in an area of transportation enabling the computer to become an "expert" in this subject.

### 2.7.7 Information Displays: "Show Me"

One of the most important ways in which systems relate to users is through the use of a visual interface or display. There are a number of information display technologies that can be utilized in ITS.

#### 2.7.7.1 Cathode Ray Tube (CRT)

The cathode ray tube is probably the most popular of information display technologies as it is the technology upon which television relies for displaying the pictures in your home. It is also the technology utilized by most home and desktop personal computers. This is quite a complicated technology, but here is the "nutshell" version of how it works. A cathode generates a stream of electrons that are focused into a thin beam of electrons, using a strong magnetic field. The cathode is usually heated to make it give up the electrons, producing the "warming up" period when you switch on your TV or computer monitor.

This beam of electrons carries the picture information, which is represented by varying the intensity of the electron stream over time. The beam is focused and scanned across a phosphor-coated vacuum tube, making the phosphor dots glow to varying degrees, in proportion to the intensity of the electron beam, thus making a still picture image on the screen.

Different phosphor formulations that glow either red, green, or blue when bombarded by the electrons are utilized in combination to make color pictures. The still picture images are shown in rapid sequence (25 or 30 frames per second) providing the illusion of movement.

In the transportation context, CRT or TV technology is used to display information to operators and travelers on single-screen and multiscreen display configurations. The multiscreen displays are popular in traffic control centers and are sometimes referred to as *video walls*.

#### 2.7.7.2 Liquid Crystal Displays (LCDs)

You may very well be wearing one of these right now, as they are used in those ubiquitous low-cost digital watches. This technology is part of a larger family commonly referred to as *flat panel displays,* since they tend to be thin and flat compared with other display technologies. LCDs exploit the properties of certain crystalline materials that can remain in one of two states, either blocking or passing light, and can be switched between states through the application of a small electric current. LCDs are also referred to as *nonemissive display devices,* as they do not give off any light of their own but just block or let light through.

Each element of the LCD is switched independently to either the block light or black state or pass light or white state. Each picture element or pixel is combined to make up the image or character to be displayed.

#### 2.7.7.3 Variable Message Signs

There are a number of technologies that enable traffic sign messages to be altered. These include mechanical roller blinds, rotating planks, rotating prisms, magnetic flip discs, and matrices of light bulbs or light-emitting diodes (LEDs). This last type is typically deployed on freeway VMS, as they provide the greatest flexibility for message display. The bulbs or LEDs can be illuminated in many combinations making it possible to display just about any text character or message. Some are even capable of displaying colored graphical images.

### 2.7.8 Location: "Where Am I?"

This is one of the most important areas of ITS technology. Determining where a vehicle is located is fundamental to the success of ITS applications. Fortunately, there are many technological solutions.

#### 2.7.8.1 Dead Reckoning

The simplest of the location technologies is known as dead reckoning. This technology is very old, with documented examples of the Chinese using this technique calling it the *south-pointing chariot.* It involves the measurement of wheel rotation to determine distance traveled from a known point. Measuring relative wheel rotation as well as total distance traveled enables one to determine the direction traveled as well as the distance. These days, dead reckoning is carried out with the help of some sort of compass. Typically, an electromagnetic compass is used to determine the direction of travel or bearing, and wheel rotation is used for distance only. Electromagnetic compasses take advantage of the fact that there are lines of magnetic flux encircling the earth from the North Pole to the South Pole. As the vehicle travels along the highway, it cuts across these flux lines, inducing an electric current that can be measured and used to determine the direction in which the vehicle is traveling.

#### 2.7.8.2 Map Matching

This technology greatly improves the accuracy and reliability of dead reckoning by correlating the computed vehicle path with a digital map of the known road network. Map matching compensates for cumulative error incurred in the dead reckoning, by comparing the path taken with a digital map (commonly stored on CD-ROM). So long as this map is up to date, map matching cannot only increase accuracy but may also perform the location calculation if the dead-reckoning system becomes ineffectual.

#### 2.7.8.3 Global Positioning System (GPS)

The U.S. military has established a constellation of 24 satellites spanning the Earth. Each satellite continuously transmits very precise time signals, that can be received anywhere on the globe, with a special GPS receiver. The GPS receiver picks up signals from three or four satellites simultaneously, then compares time differences to determine its position on the globe, thus offering high-precision location capabilities on a global basis.

A full system is now in operation with two levels of accuracy available, Precise Positioning System (PPS) and Standard Positioning System (SPS). Quoted accuracies are 16 meters and 60 meters SEP, respectively. However, the U.S. Department of Defense (which owns the satellites) retains a block on quality by barring civilian access to the PPS.

At the present time, only SPS is available for civil applications, with the military reserving the right to introduce a degradation signal that increases the error to 100 meters. However, actual errors have turned out to be lower than those quoted (30 meters), with the use of differential GPS techniques. This gets around the degradation by introducing a timing signal from a known location on the Earth and using this in the signal comparison.

GPS receivers, which are about the size of a car radio, can be purchased for approximately $1,000 or less and can be easily fitted to a car or truck. No additional infrastructure is required.

#### 2.7.8.4 Beacon-Based Vehicle Location

This is a technology employed by the commercial vehicle and transit industries to monitor vehicles in transit, most commonly by the secure and hazardous goods transport industries. There are two primary approaches—long-range and short-range. Under the long-range approach, tracking systems compute current vehicle location using a network of terrestrial beacons, usually low-frequency radio wave transceivers. Each vehicle is fitted with a radio transceiver, and position is calculated using triangulation, whereby the distance from any three of the receiving beacons is calculated and compared. Utilizing the short-range approach, each vehicle is identified as it passes dedicated short-range communications transceivers at the roadside. The exact geographic location of the roadside infrastructure is known, enabling the current vehicle location to be established. The vehicle location precision can be refined by the use of algorithms that determine the point at which the vehicle leaves the communications zone, pinpointing the vehicle at that time.

### 2.7.9 Sensors: "Seeing, Hearing, Smelling, Feeling"

These are the technologies that enable us to sense and measure the current state of traffic on the highway and the status of the transportation system as a whole. Using these technologies, we can detect the presence of vehicles, weigh them, measure them, determine how fast they are going, check out the levels of emissions from them, and even count the number of passengers onboard.

Here are a few examples of technologies that belong in this group.

#### 2.7.9.1 Inductive Loops

Inductive loops consist of loops of copper wire embedded in slots in the road surface which carry an electric current. This sets up an electromagnetic field that is cut when a vehicle passes over the top of the loops. The action of cutting the field induces (hence the name inductive loop) a change in the electrical current passing through the loops. This can be measured very accurately and used to determine the presence of, and to count, vehicles. Sequences of loops can also be configured to measure the speed of vehicles passing across them. These are sometimes referred to as *speed discrimination loops*.

#### 2.7.9.2 Infrared Beam

These utilize a thin beam of infrared light transmitted across the highway to a receiver on the other side. When a vehicle, or a pedestrian, crosses the beam and breaks it, the receiver registers this and signals a roadside control box containing electronics that processes this signal and counts vehicles.

#### 2.7.9.3 Microwave (Radar)

*Radio detection and ranging* (radar) technology has been around since World War II and has become very sophisticated with signal processing and other techniques used to enhance the basic capability of the radar.

Radar essentially operates by sending out a beam of microwave energy from a transmitter, then listening with a receiver for the energy as it is reflected back from metal objects. In the transportation context, radar is used to detect the presence of vehicles, measure the speed of vehicles, and, more recently, determine the distance between vehicles.

#### 2.7.9.4 LIDAR

Light detection and ranging (LIDAR), the light equivalent of radar, works on the same transmission, reflection, and reception principle as radar but uses a concentrated beam of light instead of microwave energy. This beam of light is typically generated by a laser light source. In the transportation context, this technology can be used for vehicle presence detection, speed measurement, and classification. It can also be used to measure the emissions from vehicles as they pass the sensor.

#### 2.7.9.5 Vision-Based Sensors

These are sensors that take advantage of recent progress in video image processing technology to harness CCTV cameras as vehicle detectors and classifiers. Images from the cameras are processed utilizing specially developed hardware and software that can detect changes in light levels on the millions of tiny dots or elements that make up each picture. Changes from one frame to the next can be compared and evaluated to detect vehicles. This technology is typically utilized to count and classify vehicles, but it can also be very effective in detecting stalled or parked vehicles in a road scene or recording the presence of pedestrians at a crossing.

#### 2.7.9.6 Acoustic

Some sensors on the market listen to passing traffic, compare the noise profile to a database of noise profiles, and then determine the type of vehicle that just passed. Ultrasonic sensors have also been in use in Japan for a number of years.

#### 2.7.9.7 Scanning Lasers

These are diode laser-based vehicle detectors and classifiers. A pulsed, scanning laser mounted above the road is used to determine accurate three-dimensional vehicle profiles that are then compared with profiles stored in a database.

### 2.7.10 Actuators: "Make It Happen"

These are the corollary to sensors since they make things happen, rather than sense things that happen. They take instructions from another part of the system, or from a human operator, then carry out a specified action or sequence of actions. For example, when you ring a doorbell, your instructions are carried along a copper wire to an actuator inside the house. In this case, the actuator is a chime or bell unit that goes through a sequence of actions designed to make a noise.

Actuators come in many guises, including servo motors for opening and closing gates and barriers at toll booths and car parks, mechanisms that control brakes and accelerator functions on automobiles under cruise control, and others that open and close doors on buses.

There are also mechanisms that change the message displayed on VMS at the side of the road.

We believe that actuators will be an exciting area for new development in the future. There is considerable research and development effort going on in an area known as *micromachines*. This involves the integration of microscopic actuators, such as levers, motors, and valves, with processors on a single chip. As the results of this research filter through to product development, we should see some interesting new actuators for a range of ITS applications including VMS and traffic management systems.

## 2.8 Summary

We hope that this chapter provides a good overview of ITS technologies. We started with a general introduction including some examples of current applications, then moved on to defining and describing a structured view of ITS that we have found to be useful in practice. Readers that would like more information on ITS technologies should refer to the technical paper, *ITS Functions and Concepts* by Kan Chen and Jens E. Pedersen, published by ITS World Congress (Berlin, 1998).

# 3

# ITS Cooperative Development Concepts

## 3.1 Introduction

Now that you have learned about ITS and have some idea of the wide range of technologies and how these can be readily understood through a structured view, you probably have a few questions:

- How do we take all these powerful technologies and harness them so that they work in harmony and provide the results we expect and desire?
- How do we determine our future direction and then set about getting there in small manageable steps?
- How do we make sure that what we do today fits with what we want to do tomorrow?
- How do we preserve the existing and planned systems and infrastructure that make up our transportation heritage?
- How do we identify and define the organization and administration required to support the technical solutions we are planning to implement?

Perplexing questions like these have been the subject of much experiment and study over the past 10 years or so. We have learned a great deal, not just about the power of individual applications, but about the supremacy of integrated systems. Interestingly, these lessons have not come mainly from the transportation field—although we are pretty good at planning—but from the fields of computing, business systems, and aerospace. People in these fields have had a tremendous amount of early experience in the development of very big systems and have learned, sometimes the hard way, the best ways to go about system development and what to avoid.

There are many, many techniques and methods that have been developed to support large-scale integrated system development. There are also many great reference books on the subject, so we make no pretense at comprehensive coverage of these topics. However, there are several concepts that have found their way into ITS and provide some amazing value for the ITS developer and implementer. They have rather unfortunate names that are fairly meaningless to transportation people but have significance to systems people. Therefore, we will introduce them using their "proper" names and then introduce our own transportation-specific terminology.

We intend to answer the questions by introducing an approach to ITS planning and development that we call the ITS Cooperative Development Methodology. In this chapter, we explain a number of key concepts that represent the philosophy behind the methodology. The concepts are not original; they have been developed and applied over a number of years in the course of major system development projects and have been described and documented in other texts [1]. What is new in this book is that we have taken the concepts and, based on our experiences in ITS planning, development, and implementation, configured them to the transportation context. Each of the concepts is introduced and explained in terms of the nature of the concept and why it has evolved to be of importance to ITS planning and development. The concepts covered in this chapter are listed as follows:

- The "what?/how?" cycle;
- Requirements analysis and exploration;
- ITS User Services;
- ITS Market Packages;
- ITS Equipment Packages;
- ITS Future Big Picture.

The whole basis for the ITS Cooperative Development Methodology is to ensure that the user gets what he or she wants from ITS. This sounds simple but is probably one of the hardest things to do in ITS. Part of the difficulty revolves around the fact that ITS technologies are relatively unknown and strange to transportation professionals and the general public. Consequently, users not only do not know the answers but often do not know the questions to ask to uncover the required information. In our experience, the art of getting what you want from ITS requires a lot of early patience and the adoption of a methodical approach incorporating some key concepts. The remainder of this chapter explores and explains these concepts.

## 3.2 "What?/How?" Cycle

One of the most powerful concepts we have come across in our attempts to improve the process of ITS planning and development is the concept we call the *what?/how? cycle.* This involves the deliberate and clear separation of "what?" and "how?" while requirements are being explored at the start of a project. In other words, the users are asked to initially focus on what they want from ITS by agreeing on the problems, needs, issues, and transportation policy objectives to be addressed by a successful ITS implementation. Only when this has been fully discussed and agreed on by all major stakeholders is the second question "how?" addressed. This question relates to the identification and selection of appropriate combinations of ITS technologies to satisfy the previously defined "whats," or requirements.

You have probably already guessed that it is not possible to deal with "what?" and "how?" separately for very long. This gets us into a subject we call the *what?/how? cycle.* Figure 3.1 illustrates this mechanism.

We have found that in 99 cases out of 100 the requirements set initially identified by the user group are never the final requirements set. This is not because users are devious or untrustworthy. They do not change their minds on purpose; it just seems to be a natural part of making a major purchasing decision. We have thought a great deal about this and believe that there are a couple of factors at work. The first of these has a lot to do with a fairly fundamental circular relationship between what you want and what you can have. When you decide to buy something you really need—a new car, for example—there may be many ways to go about the purchase. You may have a particular model already in mind based on advertisements you have seen, word of mouth from friends and relatives, or a previous purchase. Perhaps you have not gotten that far and just want to go to the dealer and see what is available.

No matter how you approach it, you probably move at least a little from the original concept of what you want sometime during the buying process, as you learn more about the choices available within your budget constraints, and refine the information you already have. Sometimes you may completely redefine what it is you want based on some new information about what you can have. This shifting in requirements is a key aspect of ITS too, and as there is a

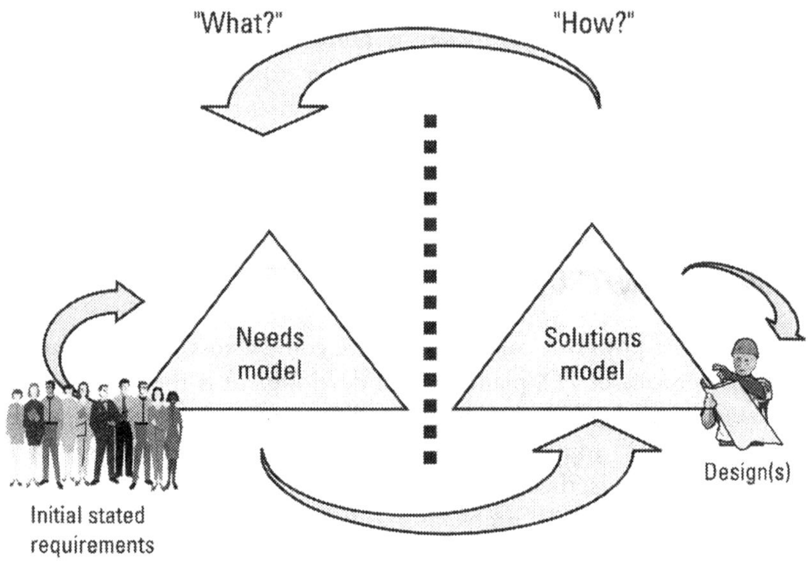


**Figure 3.1** The *what?/how? cycle.*

complex range of choices, you probably start off with a lot less information than you will have once you begin talking to a system developer or consultant.

Here is something else to think about: Have you ever had a bad experience when you bought something based on a salesperson convincing you that you really needed it and when you got it home, you did not really understand why you bought it? This can happen in ITS too, especially if you have no clear understanding of what you want. In ITS, this is called *technology push,* describing a situation in which the user purchases an ITS system based on what the technology can do and what the salesman says you need. Ideally, we want the opposite to happen. We want *user pull* to drive the purchase and acquisition of ITS.

One way to get around these problems with the many purchases you make in life is to buy things on a sale or return basis, with the option of taking them back if it turns out that they do not do the job or your requirements change. Unfortunately, this is very rarely an option in ITS or transportation as items usually have to be specially installed or even customized for you and often require a significant investment in infrastructure and accommodation works that are impossible to reverse. Consequently, you usually get only one shot at buying an ITS. That is why requirements analysis and user-driven system development are so important. We need to get the requirements right before we unleash the power of ITS. There is a sort of equivalent to "sale or return" for

ITS; it's called architecture or ITS Future Big Picture development, in which you explore your needs and lay out the whole system on paper before committing any investment in hardware, software, or infrastructure. This will be discussed in more detail in Chapters 5 and 11.

The what?/how? cycle appears to be a feature of many transportation and system development projects. We believe that it is particularly prevalent in ITS projects due to the learning curve that users climb as potential solutions are described to them. Consequently, the initial "what" defined by the users is altered and enhanced when they are exposed to the initial "how." The converse is also true in that the potential ITS solutions or "hows" will also change to fit the changing requirements. This could, of course, be viewed as one of those endless circles with no beginning and no end. However, we believe that the most sensible way to enter the circle is from the user point of view. Then, the exit from the circle becomes clear as both requirements and solutions stabilize after a number of iterations.

Another insight we have had with regard to system development and people stating initial requirements is that it would appear that most people are better at telling you how they would tackle the problem, than they are at telling you what they want. Maybe it is a feature of engineering, where people tend to be very action-oriented and practical, but we have found that people stating requirements seem to be more comfortable describing what they want in terms of how they envision the solution. This leaves the system developer, if he or she is a good one, with the job of "cracking the code," or translating the "hows" back into "whats." Consider the following example.

A telephone subscriber calls her local telephone company and says that she wants to subscribe to a Caller ID service. What she is really stating is a need for a means of identifying the callers from whom she receives calls. There could be more than one technical solution to this—some cheaper than others.

By stating the need in terms of a proposed solution, the customer immediately reduces the field of view of possible solutions and can sometimes obscure the real need. Section 3.3 addresses this issue.

## 3.3 Requirements Analysis and Exploration

The concept of requirements analysis and exploration is central to the successful adoption of a user-driven approach to ITS development and planning. The key to successfully managing user requirements in an ITS development project is to understand the natural iterative cycle that will take place and to make the best use of it. Many developers make the mistake of adopting a linear, sequential approach to requirements definition. They take no account of the what?/how? cycle described in Section 3.2 and proceed to develop software and procure hardware on the basis of initially stated user requirements. Inevitably, the what?/how? cycle will still take its course, but in this case the loop between "what" and "how" will be loaded with expensive resources and a high investment in preliminary solutions. This investment forces the system developer to be rigid and start to influence users toward requirements that can be satisfied with the already defined solutions. A way to avoid this is to keep the what?/how? cycle "fast and light" by developing both a needs model and a high-level conceptual solution framework or ITS Future Big Picture, then use the iteration between "what" and "how" to remove ambiguity from the view of requirements.

Exploration of requirements as the what?/how? cycle progresses enables the real needs, objectives, problems, and issues to be identified and understood. One way to look at requirements is shown in Figure 3.2.

Figure 3.2 shows requirements as consisting of two components—what you want and what you do not want. Unfortunately, in real life, the picture looks more like Figure 3.3.

The line between the two is usually shrouded in mist. We need to explore the boundary in order to dispel this mist. Incidentally, this way of looking at requirements was inspired by a wonderful book entitled *Exploring Requirements: Quality Before Design* [2].

The need for early removal of ambiguity in requirements is a very important lesson that system engineers and developers have learned over the past 20 years or so in executing numerous large business and commercial system implementations. They noticed that the cost of fixing an error caused by misunderstood requirements (ambiguity) rises at a frightening pace throughout the project development period. As more and more resources (person hours) are

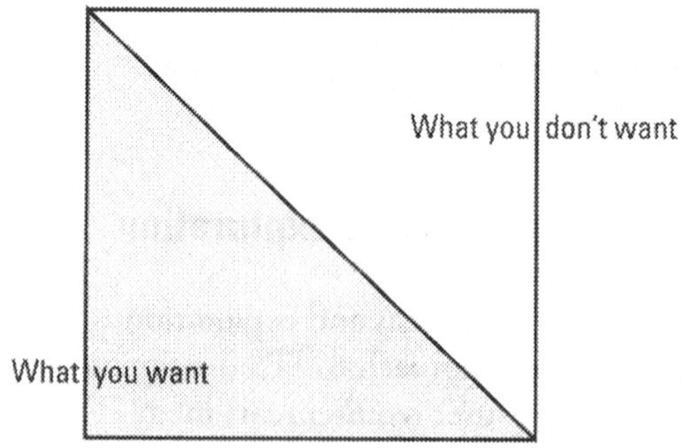


**Figure 3.2** What you want and what you do not want.

**Figure 3.3** Mist over the boundary between what you want and what you do not want.

committed to a particular design direction, and as the system grows and becomes more complex and interconnected, the time and effort required to make a change increases. One reason for this is that large systems are typically composed of many inter-related components. If you change something in one part of the system, it may cause a consequential change in another part of the system. These days, system developers are getting much better at coping with this. Systems are now designed in a very modular fashion, perhaps utilizing *object-oriented* approaches to development. Nevertheless, it still burns disproportionate amounts of resources if needed changes are discovered late in the day. Another factor is that as we get deeper and deeper into system development, so many technology tradeoffs and design choices have already been made that it becomes more and more difficult to be flexible. The system developer gets "painted into a corner."

Figure 3.4 illustrates this point rather well; it shows the relative cost of fixing an error caused by ambiguity in requirements at various stages in project development.

It follows that systems can be developed more economically and more effectively if time and resources are invested wisely at the beginning of a project through the application of requirements analysis techniques that support the user and the analyst in identifying, understanding, and confirming the user requirements for the system.

## 3.4 ITS User Services and ITS Market Packages

To support the iterative requirements process we have called the what?/how? cycle, we make use of two of the tools defined in the National Architecture for

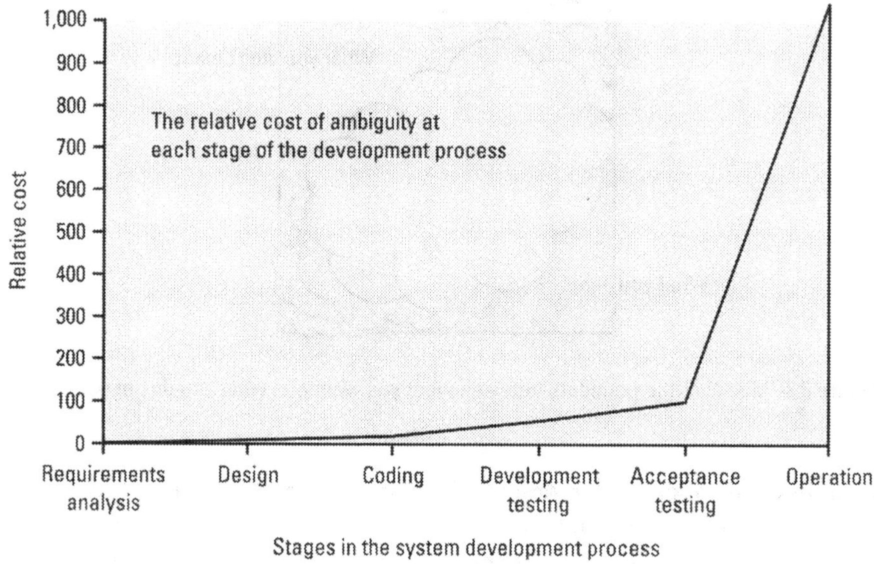


**Figure 3.4** The relative cost of fixing ambiguity at different stages in the system development life cycle [3].

ITS, ITS User Services and ITS Market Packages. We have discovered that this terminology does not sit well with most transportation professionals and sometimes causes confusion. Despite the awkward names, the concepts are fairly straightforward. As illustrated in Figure 3.5, ITS User Services characterize the "what" and ITS Market Packages characterize the "how" as we go through the cycle.

Chapter 5 provides more detail on ITS User Services. We have already discussed ITS Market Packages in Chapter 2, and we will revisit them in Chapter 7.

## 3.5 ITS Equipment Packages

ITS Equipment Packages is another tool or term which we need to introduce you to in order to provide the whole picture. An equipment package is an ITS Market Package that has grown up. ITS Market Packages do not have enough detail for design and implementation and do not obey the yet-to-be-defined subsystem boundaries. It is difficult to buy an ITS Market Package as these products typically straddle subsystem interfaces, putting buyers in the

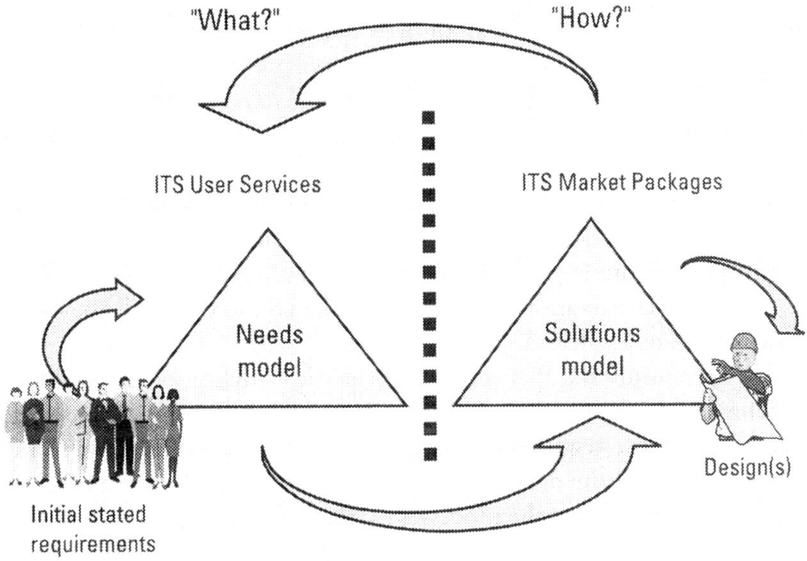


**Figure 3.5** The what?/how? cycle (with ITS User Services and ITS Market Packages).

impractical situation in which they would purchase one ITS Market Package that would represent several incomplete subsystems. The answer is to convert ITS Market Packages to ITS Equipment Packages. This happens during the detailed design stage. You then have units that can be specified, procured, and implemented. Chapter 7 describes ITS Equipment Packages in more detail.

## 3.6 ITS Architecture

We have encountered many difficulties when discussing the concept of a *regional ITS architecture* with highly technical transportation professionals and nontechnical user groups alike. Perhaps this is because we all have a preconception that the word *architecture* relates only to buildings, or maybe it is such an abstract concept that most people do not see enough potential value in it to justify spending the time understanding it. However you consider it, the system architecture concept has been a difficult one to get people to embrace and like. It certainly seems to trouble the majority of people. Even the system engineering community that adopted the term in system development methodology can get into quite lengthy debates about what it actually means. We refer to our ITS version as the ITS Future Big Picture, since the ITS architecture is a framework defining the technical, institutional/organizational, and commercial features of the future system in an outline and graphical format, showing how all the individual subsystems and components will work together.

It is worth taking a bit of time at this point to consider exactly what the ITS Future Big Picture is and why it is so useful. Starting with the U.S. National Architecture for ITS as a reference model, it is possible to develop and configure a regional architecture or ITSFBP that meets regional needs and priorities. This is the approach being proposed in the United States, since the National Architecture for ITS was designed to support this. For international applications, the U.S. National Architecture for ITS can optionally be used as a reference model or as a learning tool.

The ITS Future Big Picture provides us with a definition of all the major subsystems within the overall ITS. The subsystems are developed and selected to exactly match the requirements identified in the requirements exploration and analysis phase of the project. The definition of each subsystem is provided in terms of a description of the processes that have to be supported by each subsystem and the type of technologies to be deployed. It is also a vital component of an iterative design/requirements cycle that takes account of shifting user needs and perceptions of needs and enables a design to be developed through consensus between the user and the developer.

Developing and utilizing the ITS Future Big Picture is the design engineer's way of saying, "We're still listening to you and you have more to tell us, but here is a rough picture of what the system will look like based on our current understanding of your needs." The engineers have not invested such a high level of resources that they have become entrenched and intransigent to the users' changing needs. As a result, the users believe that they play an active role in the development process.

We have to tell you that many engineers find this counterintuitive, since they have been trained to hide the designs they are working on until they are polished and perfect and can be unveiled as masterpieces to the user. Developing an ITS Future Big Picture is a way of externalizing, from the head of the engineer or developer, a partial solution that has all the necessary attributes required to satisfy the initially stated needs. It is not yet a deployable design, as further design decisions and choices and technical tradeoffs between options need to be made. The detailed design is where engineers really earn their keep as they have the experience to efficiently plow through all the what ifs and all the detailed analysis work required using their experiences from other projects to guide them. They are the indisputable experts at finding and selecting the best technologies and the best configurations. However, users are the indisputable experts in what they need and want. The ITS Future Big Picture is a balancing act. On one hand, it provides enough details to ensure that user needs are fully satisfied. On the other hand, it is flexible enough to be capable of accommodating multiple design approaches and evolving technologies. This seems like an impossibility, but it is exactly what a good ITS Future Big Picture will do for you.

A good ITS Future Big Picture will do much more than describe the technical framework for the proposed solution: It will also define the institutional and organizational arrangements required to support the chosen technical subsystems and the data flows required between them. This will typically be in the form of memoranda of understanding and operational procedures required to support the operation of each subsystem ensuring that they work together as a harmonious single entity.

In addition, the ITS Future Big Picture will provide a commercial solution that underpins the proposed technical implementation. The commercial aspects of ITS are often overlooked, but they are crucial to the success of most ITS implementations. This would include a definition of how the proposed ITS implementation is to be funded both in terms of capital investment and recurring operational and maintenance costs. This might indicate where private-sector investments will be applied and where public-sector subsidies will be required. It will also show the flow of revenue around the whole ITS, indicating where revenues will be generated and where investments will be made and illustrating subscription mechanisms.

The ITS Future Big Picture will also provide a clear and unambiguous definition of the key interfaces to which standards are to be applied. These might be international, national, regional, local, or product standards.

The logical part of the ITS Future Big Picture that identifies where the processing should be carried out and what data needs to be shared between each processing center, also provides a first-class platform for software development. Appropriate modules for software development and operation can be identified in the ITS Future Big Picture, and the logic flow of the entire software development can be defined at this point.

A major advantage of developing the ITS Future Big Picture is to see the whole picture laid out in some level of detail, to understand the consequences of actions taken by you and your organization on others, and to be equipped to work in harmony. This also enables risks to be identified and either avoided or mitigated through planned action.

At this point you may well ask, "If it is so difficult, then why bother with it?" In light of our experiences in the application of ITS in the United States and overseas, we can give you an emphatic answer to the question: The development of an ITS architecture as part of the planning phase of an ITS deployment provides exceptional value and benefits.

The ITS Future Big Picture concept is so useful and flexible that it becomes difficult to explain it in a general way that all audience groups will understand. This could be one of the reasons why it has been such a difficult subject area to explore.

Since the ITS Future Big Picture means slightly different things to different people, then the best course of action is to describe and highlight the nature of the ITS Future Big Picture from as many meaningful perspectives as possible. One way of thinking about this is to consider the ITS Future Big Picture as a comprehensive framework with many interconnected elements and multiple facets. Until it is explained to you, it might as well be in a darkened room. Accordingly, our job is to shine a flashlight on the ITS Future Big Picture to illuminate the aspects of it that are meaningful to you. You will, of course, also see features on the periphery and in the shadows. These are the other parts of the ITS Future Big Picture that do not seem so important and do not really interest you, but you will know that they are there and perhaps understand the need for these parts to interact with the parts you decide to use.

One set of perspectives to consider is from the viewpoints of different people with varying backgrounds and experiences. Depending on your perspective, your experience, or your background, you will probably have a different view of the ITS Future Big Picture than the other people on your development team or within your consensus group. Sections 3.6.1–3.6.7 present a few varying perspectives on the ITS Future Big Picture.

### 3.6.1 Communication Engineer

Communication engineers see the ITS Future Big Picture in terms of connections between subsystems. They want to know what systems will be connected and what form of communications will be required to support data flows. They want to get to the detail required to select the communications media and protocols required. They really do not want to see the whole picture—only the connections.

### 3.6.2 System Engineer

These are probably the only people who want to see the whole picture all the time; their interests center on getting to the detail required to ensure that user needs are fully satisfied and understanding the issues associated with integration of the various system components into the final configuration.

### 3.6.3 Administrator

Looking at it from the organizational and institutional perspective, these users are trying to figure out what agreements would have to be struck between agencies. They are looking at the required organizational structures and agencies that should be involved if the planned organization is to be effective in supporting the proposed technical solution. This group might also be focused on the legal aspects of the proposed system and potential liability issues.

### 3.6.4 Traffic Engineer

From mode-specific traffic management perspectives, you need to know about the management and control possibilities that the new system will offer and what new data can be provided to enable better traffic management reflecting current traffic conditions. Other typical questions from this perspective include the following:

- What will the new ITS offer to the drivers using the network in terms of improvements in safety and operational efficiency?
- What new infrastructure will be required along the highway?
- How much will it take away from my existing traffic signal budget?
- With what other agencies should I be exchanging data and with what data will they provide me?
- Will I lose control of my existing organization under the proposed institutional and organizational arrangements?

### 3.6.5 Transportation Planner

This group will be very concerned with how the ITS will fit alongside conventional transportation initiatives such as new transportation facilities and infrastructure. They will ask questions like the following:

- How will the operation of the ITS affect the land use patterns and travel patterns on which current forecasts and plans are based?
- What data and information will the ITS provide that will improve our understanding of demand for travel and the dynamics of transportation use?

### 3.6.6 Transit Manager

This user group is primarily focused on encouraging the use of public transit and improving the efficiency and attractiveness of public transportation facilities and services. They will have the following questions:

- How can I attract more customers to my transit service?
- How can I improve the efficiency of my transit service and close the gap between the farebox revenue and the operational cost?
- How can I leverage private-sector investment into my transit system?
- What data should I share with traffic managers and what data can they give me?
- How can we optimize the operation of the whole regional transportation system?

### 3.6.7 Private Sector

This group is naturally commercially oriented. Private-sector entities want to know where the investment opportunities are and something about the nature of them. They want to rank and prioritize potential opportunities according to profit, revenue generation potential, and associated risks. They need to know with whom they should negotiate and what the public sector long-term objectives and policies are and will be in the future. They want to see how the revenue will flow around the connected parts of the system, understand the size of the investments required, and identify possibilities for direct subscription payment mechanisms to recover initial outlays and generate profit.

They are also interested in understanding their role relative to the public sector and have a good understanding of how the eventual system will be organized and operated and with what constraints they may have to carry out business.

Another way to explain an ITS Future Big Picture is to make comparisons with more familiar items that have similar features to an architecture. One of these is the ubiquitous automobile.

A car has the same architecture or overall framework no matter who made it or what model it is. In order to fulfill its purpose as an efficient mode of transportation, a car needs the following features:

- Wheels;
- Steering wheel;
- Brakes;
- Accelerator;
- Headlights;
- Tail and brake lights;
- Indicators;
- Horn.

Although these are all standard and in fact governed by regulations in most countries, there are narrow ranges of variability in these elements. Automobile manufacturers take these features and build on this basic architecture to create unique designs or models. This raises the question of what the difference is between an ITS architecture (or ITS Future Big Picture) and a design.

## 3.7 What Makes a Design a Design and Why Is it Different From an Architecture?

You can see from our representation of an ITS architecture for an automobile that while it specified all the major parts and how they are connected, it does not give us enough detail to build one. We have not chosen the slant of the windshield, the location and number of doors, or the size and location of the engine. (Will it be at the front or at the back, mounted longitudinally, or transversally?)

Suspension systems, transmission systems, and other major components can all have design variations, making each model unique. These are all design decisions that add more detail to turn an architecture into a deployable design.

## 3.8 So What Is the Point of Having an Architecture if it Is not Deployable?

This is a good question! A regional system architecture is really just an interim step on the way to design. It is a place to pause and take stock of what you have done so far and decide whether it is still what the user wants. In fact, many system engineers believe that they can skip the architecture step altogether—since engineers are so good at collecting and understanding user needs, they believe that they can go straight to design and get it right the first time. We suspect that those who take this approach often find themselves in the situation in which they are trying to bend the requirements to fit a preconceived solution.

Another good analogy that most of us probably think of when someone talks about *architecture* is the house. We are all familiar with the assembly of wood, bricks, cement, and plaster that we call home. Have you ever thought of it as a system? A house has multiple subsystems to provide the comforts and conveniences we all take for granted and require in our homes. A house's architecture has to describe the main elements and their relationship to each other if the subsystems are to work together in harmony to provide the desired outcome—a comfortable habitat. Multiple subsystems must work together to achieve this objective.

These subsystems include the following:

- *Plumbing:* For drinking water, bathing, and waste disposal;
- *Electricity:* For heat, light, and appliances;
- *Walls:* For protection from the elements and to hold the roof up;
- *Roof:* To stop the rain from coming in.

Some homes have additional subsystems such as cable TV and air conditioning. Each subsystem has to perform two roles. First, it has to do its own job and carry out the functions it was designed to do in the most efficient manner. Second, and just as important, it has to work in harmony with the other subsystems to provide the kind of habitat we expect. It is also highly likely that the various subsystems were designed by different vendors and manufacturers and installed by a number of independent specialists. You would not ask the plumber to install the electrical subsystem.

## 3.9 The Value of the ITS Future Big Picture

Why should we go to the bother of developing a ITS Future Big Picture? Our primary motivation should be the fact that the development and use of the ITS Future Big Picture adds considerable value to the overall development process in several areas.

### 3.9.1 Technology Maturity and User Drive

The ITS Future Big Picture supports user pull rather than technology push as described in Section 3.2. It could be argued that technology push is most appropriate when technologies are relatively primitive, the market is immature, and technology capability is the major constraint. However, as technology matures and choice, flexibility, and capability all expand, it becomes more important for the user to drive development in order to prevent the development and adoption of a suboptimal solution.

Developing an ITS Future Big Picture also supports close integration of ITS with conventional transportation initiatives by providing the information required to support influencing conventional deployments, as well as providing the guidance for future ITS deployment. This can also provide valuable input to planning and deploying future conventional solutions and illustrate how they will work together.

### 3.9.2 Performance Balancing

An ITS is composed of subsystems and components that must work together effectively. There is no point in having high-performance capabilities in one component if the performance of other components constrain the overall system performance. A balanced approach to component selection must be taken, accounting for overall performance of the whole system. The ITS Future Big Picture enables you to take this holistic view. There would be no point in having an exceedingly expensive, high-quality monitor on the end of your fiber optic cable, for example, if the cable itself were not capable of producing a signal that would give you the full quality from the video display. Similarly, if you had an inferior camera producing the video signal along the fiber optic to the display at the other end, there would not be any point in having a high-quality monitor. Therefore, matching components to produce sensible procurement strategies and sensible levels of performance is another aspect of putting together an ITS Future Big Picture.

### 3.9.3 Risk Management

Externalizing and sharing your view of the future is a great way to approach the subject of risk. Risk plays a major role in any new implementation. Some risks are unavoidable, while others can be avoided if they are identified and understood. Whether they fall into the former or the latter category, we believe that it is very important to identify as many of the potential risks as possible early in the planning and development of an ITS. The unavoidable risks can be defined and strategies developed to minimize their impact. It can also be recognized that these risks may be associated with particular approaches or choices. The risk factors associated with each approach can then be taken into account when making choices between approaches. Avoidable risks can be identified and strategies developed to navigate around them.

There are numerous risk management strategies that become viable options when a ITS Future Big Picture is developed. These include the following.

- *Expansion of market sizes for ITS products:* Knowing that the choice of a particular product or service will increase the overall size of the market for that product or service can be an important factor in managing the use and application of new technologies. In this case, the technology vendor or supplier that has invested in the development of the product or service who is hoping for a return, also shares the risk.

- *Promotion of ITS standards development:* The ITS Future Big Picture also identifies the standards that will be required at local, regional, national, and product levels. In the U.S. context, the National Architecture for ITS identifies the needs for standards development and promotion, forming the platform for standards development activities. Defining the different subsystems and the data that has to flow between them provides a good idea about the interfaces that have to be standardized and identifies the crucial elements that should be the subject of standards on at least one of the aforementioned levels. Some people believe that implementing agencies for ITS should not consider a technology or product or service until a standard is available. In line with our user-driven focus, we do not subscribe to this view. Instead, we believe that it is important to the creation and agreement of worthwhile and effective standards that the implementers get involved in the standards development process. Our reasoning is simple: You should figure out what you want to the best of your ability, then get your sleeves rolled up and play an active part in standards development. You might not get what you need and want if you leave standards development to the vendors. We do not want to paint a black picture of vendors, as good ones will always be listening to what their market and the implementers really want anyway. However, we are sure that better standards result when implementers or buyers are actually involved in the standards development process. Another reason for thinking this way is that in our experience, implementers do not wait for standards to be developed and accepted, especially when there are large potential benefits to the community. It then becomes vital for the implementers to be involved in the standards development process, as interim arrangements will have to be devised until the standard is ready. Implementers under these circumstances can also bring a wealth of practical experience to the standards development table since they have tried to apply the technologies involved in a real-life situation. This input can also result in the definition of interim standards and migration paths from current practice to future standards.
- *Aligning for synergy:* Once you have developed your ITS Future Big Picture, you are equipped to engage in what we call *aligning for synergy.* That is, you can open discussions with other organizations and agencies and seek to develop the level of cooperation and understanding required to realize your common transportation objectives. Agencies can share plans and agendas more easily when the full picture is defined and all the various repercussions have been identified.

- *Identifying and confirming needs, objectives, problems, and issues:* We talk an awful lot about this subject in lots of places in the book, because we believe that it is crucial. We just cannot see how you can do a good job of identifying and defining requirements unless you lay out the whole picture for the developer and the customer to review and revise.
- *Linking ITS to the transportation planning process:* Linking or *mainstreaming* ITS into the conventional transportation planning process has become a subject of much interest in recent times. ITS started off life as a standalone subject with dedicated funding and a devoted band of followers. As ITS becomes more widely understood and generally accepted as another tool in the transportation professional's portfolio, it becomes vital that ITS be more integrated into the process used to plan conventional transportation initiatives. This typically means that the proposed ITS has to be compared with some sort of long-range plan for transportation within a region or area. An ITS Future Big Picture helps to support this type of integration by enabling the full effects of the proposed system to be identified and relationships with conventional transportation solutions to be identified, engineered, and managed. This should provide an advantage for both ITS and conventional transportation plans. Having an ITS Future Big Picture in which all objectives are identified and laid out for all to see can also be very useful in terms of laying out conventional transportation plans and seeing how the conventional transportation plans merge with the ITS deployments.

## References

[1] Hatley, D.J., and I. Pirbhai, *Strategies for Real-Time System Specification,* NY: Dorset House Publishing, 1987.

[2] Gause, D.C., and G. M. Weinberg, *Exploring Requirements: Quality Before Design,* NY: Dorsett House Publishing, 1989.

[3] Boehm, B.W., *Software Engineering Economics,* Englewood Cliffs, New Jersey: Prentice Hall, 1988.

# 4

# ITS Cooperative Development Methodology

## 4.1 Introduction

This chapter introduces the ITS Cooperative Development Methodology, illustrating how the concepts explained in Chapter 3 would be applied by a good ITS developer or consultant. We also hope that it will explain how adoption of the methodology, or a similar methodology, will help readers highlight their role and determine why it is important to the overall success of the project.

This chapter should provide readers with an overall understanding of the ITS Cooperative Development Methodology process and how it all fits together to form a complete approach—as well as some insight into the usefulness of each step. It should also give readers an appreciation of how the philosophy described in Chapter 3 can be applied in the real world.

The concepts described in Chapter 3 will work well when used independently. They each add something toward the efficiency and effectiveness of ITS planning and development. However, when the various concepts are utilized within a framework, or methodology, the sum of the parts becomes greater due to synergy between them. Here then is a proposed methodology for utilizing each of the concepts in harmony and unison to help in planning, developing, and designing an ITS.

We have called it the ITS Cooperative Development Methodology, because it is primarily designed to support a high degree of cooperation between the client and user of the system and the system developer. It also supports cooperation between users in the form of consensus building between various transportation agencies and user groups.

As stated in Chapter 3, we have tried, as far as possible, to avoid jargon when discussing concepts and techniques in this book. We have tried to carry this through to the labeling of the steps in the process. Consequently, system engineers reading this chapter will find some interesting new names used for steps and techniques that are more commonly known by well-established system engineering terms.

We acknowledge the enormous debt this book owes to system engineers and system engineering, but we believe that the use of terms meaningful to stakeholders and transportation people is crucial to the adoption of such techniques in ITS. As a result, we have placed effectiveness in communicating ideas and concepts before formality and rigor in labeling.

Figures 4.1 and 4.2 illustrate the ITS Cooperative Development Methodology.

Figure 4.1 shows a high-level view of the process indicating where the ITS needs and ITS solutions models fit in. Figure 4.2 shows a detailed description of the process with all the steps shown. The aim of adopting the ITS Cooperative Development Methodology is to address a series of objectives, listed as follows:

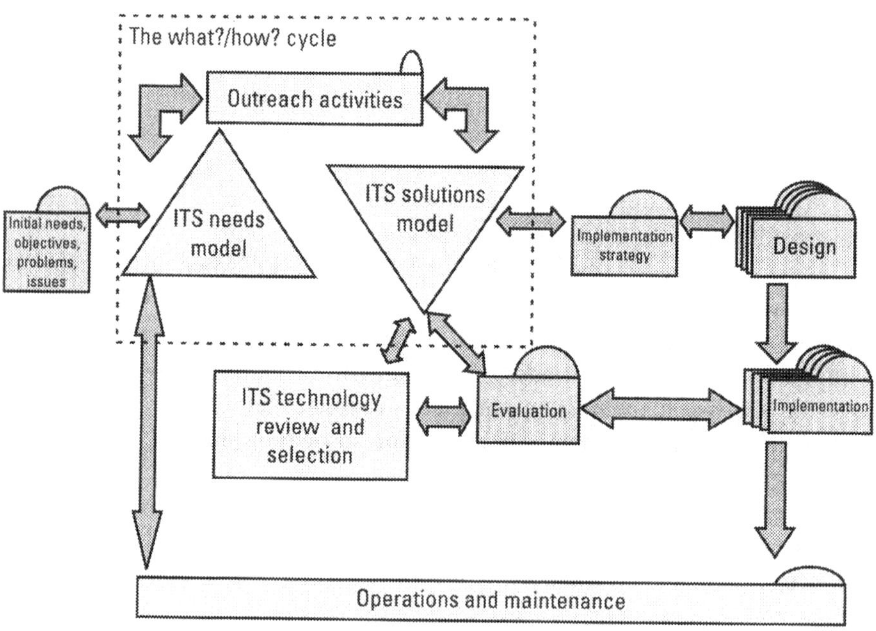


**Figure 4.1** The ITS Cooperative Development Methodology, high level.

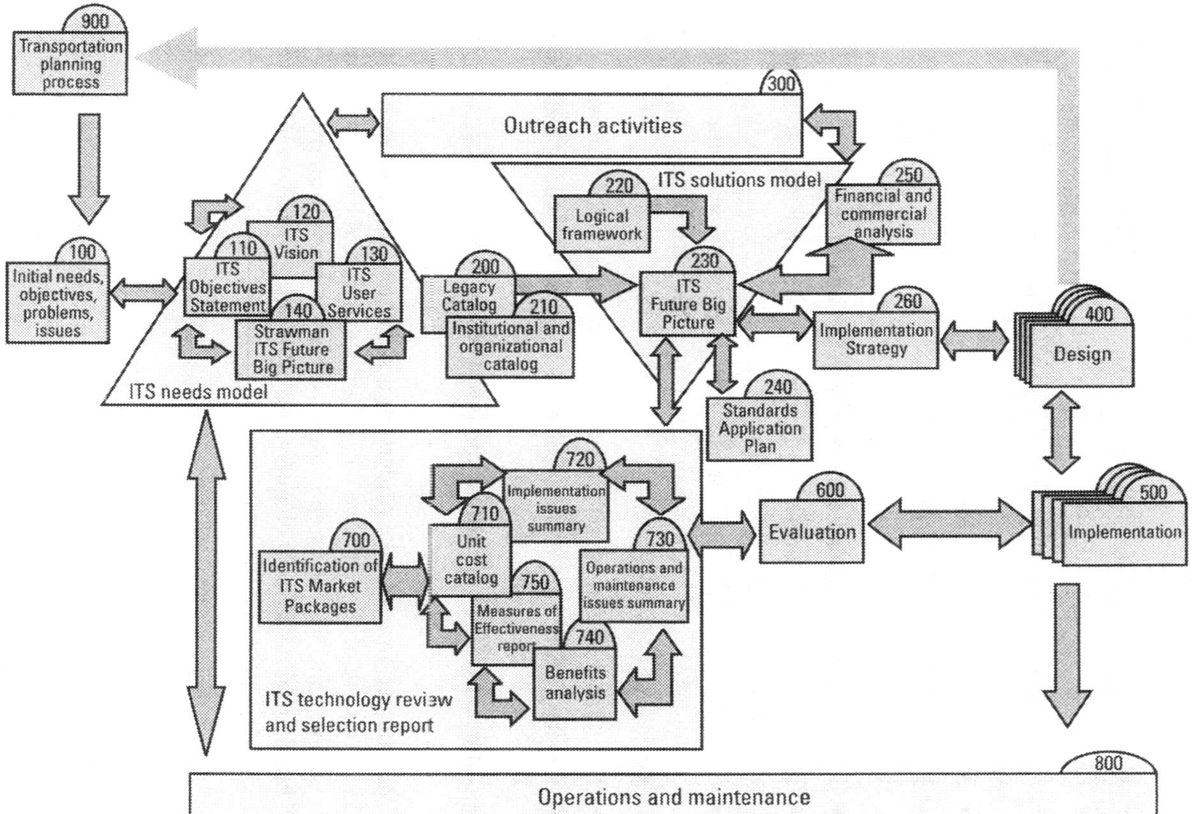


**Figure 4.2** The ITS Cooperative Development Methodo ogy, detail level.

- Invoke user-driven development, supported by clear communication of needs, objectives, technology capabilities, and effects;
- Highlight the direct correlation between *what* the user expects and *how* the system will work;
- Identify and confirm needs and objectives of stakeholder groups;
- Concisely encapsulate needs, objectives, issues, and problems;
- Support technical system analysts in defining technical solutions based on user needs;
- Sharpen the near-term focus;
- Provide a clear basis for evaluation;
- Facilitate ITS Market Package definition and detailed design;
- Identify infrastructure requirements and interaction;
- Describe the ITS Future Big Picture for the implementation within the context of an integrated ITS;
- Support full integration of the implementation with the National Architecture for ITS, make maximum practical use of existing systems;
- Explore ways forward;
- Define and illuminate possible paths from today to tomorrow;
- Engage ITS component suppliers in a dialog on what products and services will be available and when;
- Address the issues identified and described in Section 1.3.

This chapter aims to take readers on a swift tour through each of these steps. In the course of the tour, we will introduce each of the steps and explain what is involved and why the step is useful. This will provide an overview that we will build on in subsequent chapters, providing more detail on how to carry out the work required in each step.

## 4.2 ITS Needs Model

As discussed in Chapter 3, we want to support and take advantage of the what?/how? cycle to make our ITS planning and development as effective as possible. We make use of two models to do this. The first one is the *needs model;* as its name implies, it captures the needs that are to be addressed and satisfied by the proposed ITS. The needs model supports the *what* part of the what?/how? cycle. The second one, which is called the *solutions model,* captures the potential solutions that can be used to address the needs. This supports the *how* part of the what?/how? cycle.

The ITS needs model captures what you initially and finally want in a concise structured manner, making it easy for the user to explain requirements and for the system developer to understand the dynamics of the evolving requirements picture. It also accommodates changes and modifications, making it easy and desirable for the user to take a flexible approach and cooperate fully with the developer.

The ITS needs model we have used in our projects has the following components:

- 100: Initial needs, objectives, problems, and issues;
- 110: ITS Objectives Statement;
- 120: ITS Vision;
- 130: ITS User Services;
- 140: Strawman ITS Future Big Picture.

Each of these components is described in Section 4.2.1. The components in the ITS needs model have been numbered as the 100 series for easy identification. The steps are numbered in increments of 10 to accommodate the fact that we are always learning new things about developing and planning ITS and will probably add more steps as the process continues to evolve.

### 4.2.1 Step 100: Initial Needs, Objectives, Problems, and Issues

As discussed in Chapter 3, this is where the initially stated requirements for the users and stakeholders are identified, recorded, and discussed. This activity usually requires a search through existing documents such as current transportation plans for the area or region and any previous planning initiative deliverables. It may also include a series of interviews with key stakeholders to determine their views.

### 4.2.2 Step 110: ITS Objectives Statement

The ITS Objectives Statement is a concept derived directly from the system engineering world where it is commonly referred to as the ITS Objectives Statement. This was also the terminology utilized in the National Architecture for the ITS development program. It is quite simply a one or two paragraph statement capturing the high-level objectives of the proposed ITS development.

It is usually written in simple language and agreed on by all major stakeholders before being finalized.

Defining a simple Objectives Statement for the proposed ITS implementation might seem like a really trivial thing to do. However, the process of developing and agreeing on a simple high-level set of goals for the project involving the whole spectrum of users helps to define the work zone. We have also found that it maximizes the probability of a successful outcome to the project by giving the system developer a chance to satisfy all the key parties. Amazing though it may seem, many ITS projects start with different user groups holding conflicting and incompatible objectives for the project. Unless this is resolved and a reasonable consensus is reached at an early stage, there is little chance of a successful outcome that will satisfy all parties concerned.

To be honest, it is usually the case that the ITS Objectives Statement itself is fairly meaningless and does not contribute much to the overall project. It is the process of agreeing on the ITS Objectives Statement that made a significant contribution to the project in this case.

### 4.2.3 Step 120: ITS Vision

An ITS vision is a nontechnical, narrative description of what life will be like from the perspectives of the key stakeholders and users of the system once the whole system has been implemented. It is a future look at the transportation system for the region or area under study. The ITS vision can be developed by the system developer independently or in collaboration with a group of users or stakeholders. The power of the ITS vision lies in the externalization of the system developer's view of what the system will do. If this is agreed on with the users, then it forms a strong communications and marketing document for the entire system. Many system engineers consider this to be a waste of time and would rather launch into detailed, technical explanations about a proposed ITS system. However, we consider the vision to be vital due to its ability to communicate a view of the future ITS system in a nontechnical manner to a wide audience. Here again, the process of putting the vision together can be just as fruitful as the final product.

### 4.2.4 Step 130: ITS User Services

In order to support this initial focus on *what,* we have successfully utilized a tool known as the ITS User Service. This first saw the light of day in the U.S. National Architecture for ITS Development Program, and we have used it on many subsequent projects. It is a rather awkward piece of terminology but a very important tool. In simple terms, an ITS User Service is what the system has to provide in order to be considered successful by the client and the users of the system.

An ITS User Service is a clear, concise, unambiguous statement of user needs. In the National Architecture for ITS Development Program, 30 such ITS User Services were defined. However, we have had occasion, particularly in overseas projects, to modify and enhance this list. We do not believe that it is important to adhere to a prescribed list of ITS User Services; rather, we believe it is vital that the defined ITS User Services closely reflect the objectives, needs, issues, and problems of the specific transportation region or locale under study.

ITS User Services are used to support the process of collating and distilling objectives, needs, issues, and problems into a structured, easy-to-communicate format. Typically, the initial input from users will come in all shapes and forms. Objectives may be stated in terms of written reports. Problems can be defined during the course of personal interviews or meetings. Issues can be stated at any point in the requirement definition process. These all have to be analyzed, made compatible, filtered, and turned into ITS User Services. Once we have defined the ITS User Services, we will use them for two primary purposes. First, they will be used to feed back the user's input to confirm that we accurately captured what they told us. Second, they will form the basis for briefing the system developers on the parameters to be taken into consideration when outlining the proposed ITS solutions.

An important feature of the ITS User Service is that it allows us to focus the users on agreeing on what they want. This may seem like a trivial point, but we have witnessed many occasions in the early stages of ITS planning where key stakeholders and prospective users get caught up in long debates about appropriate ITS solutions to their problems. These debates usually feature a combination of discussion about what they want and their perceived understanding of what the ITS technologies can do. Unfortunately, in many cases, this perceived understanding of what ITS technologies can do varies widely from reality. It is often based on secondhand information or biased information about technology capabilities provided by salespeople. In our experience, this kind of debate is extremely counterproductive. To avoid it, we deliberately set out to ignore potential solutions until we have at least an initial definition of the problems, issues, needs, and objectives. This also has the useful affect of channeling all user energies toward agreeing on *what* among themselves before considering *how*.

It is also a very good idea at this point to define preliminary measures of effectiveness or performance parameters for each ITS User Service. These are the parameters that will be used to measure the performance or effectiveness of our proposed system and its components. It is generally the case that if you cannot define a Measure of Effectiveness to relate to your ITS User Service, then it is not a valid requirement. This may be viewed as the first step in the evaluation process.

### 4.2.5 Step 140: Strawman ITS Future Big Picture

This is one step you are unlikely to find in any formal system engineering texts. It has evolved from our direct experience of ITS planning and development initiatives. In the course of these initiatives, we have experienced a phenomenon we call *top-down tension.* Our customers get fed up with the top-down structured approach and want to get on and explore something more tangible. They also get perplexed about the abstract nature of the requirements and solutions models. We have found it very useful in these situations to go ahead and apply our best judgment and experience from previous implementations and produce a rough and ready approximation of what we think the final ITS Future Big Picture will look like. We refer to this as a Strawman, as it is designed to be a vehicle for debate and discussion and to be torn apart and redone in the light of new information about requirements.

We have found this to be an invaluable tool, because it supports the notion that people are much better at revising, enhancing, and modifying, than creating from scratch. The Strawman allows the system developer to incorporate the best experience, expertise, and intuition into a picture that can be used as both a requirements exploration and an expectation management tool.

## 4.3 ITS Solutions Model

This represents the other half of the "binary pair" of models supporting the what?/how? cycle. The solutions model provides the means to capture and evolve the *how* associated with ITS development and planning. It is the essence of how your needs can be satisfied by bringing ITS technologies together into a structured framework supporting the required interaction between components. The solutions model represents a finely balanced tool that on one hand contains enough detail to confirm that needs are being satisfied in a manner that ensures a balanced resource investment. On the other hand, it is pitched at a high enough level of abstraction to make subsequent changes to the proposed solution affordable.

Sections 4.3.1–4.3.7 discuss the main components of the ITS solutions model we have adopted on previous projects.

### 4.3.1 Step 200: Legacy Catalog

A Legacy Catalog is an inventory of all existing and planned systems and transportation initiatives within the sphere of influence of the proposed ITS. Any system or initiative that will interact with, or have influence on the ITS we are trying to plan and develop must be included. If the system or initiative is not under the direct planning and design control of the implementing agency and/or has gone beyond the point at which we can have any design or planning control, then it should be part of the Legacy Catalog. The Legacy Catalog is the primary instrument for ensuring that we have all the information available to embrace and protect sunk investment in prior systems. These are the "running horses" that cannot be stopped but must be saddled as part of the overall ITS development.

### 4.3.2 Step 210: Institutional/Organizational Catalog

Like the Legacy Catalog described in Section 4.3.1, this is an inventory listing of existing resources. In this case, the resources are the current or near-term planned institutional and organizational arrangements within the sphere of influence of the proposed ITS implementation. This is also an important input to the ITS Future Big Picture, particularly as it is very difficult to make any changes in existing institutional and organizational structures in the near term. Ideally, the technically and commercially optimum solutions would drive the institutional/organizational solution. In a more realistic scenario, the technical and commercial optimum will be compromised to fit existing arrangements or something very close to the existing arrangements.

### 4.3.3 Step 220: Logical Framework

This is the most abstract part of the process. It can be difficult to justify to technical and nontechnical stakeholders, as they have trouble seeing a meaningful output or deliverable. However, it is invaluable to the system developer and ultimately—albeit indirectly—provides substantial value to the stakeholders. The development of a logical framework is an extremely sensible step on the way to defining the ITS Future Big Picture because it defines the optimum grouping of processes and data flows. The logical framework can be developed independently of any consideration of hardware and software ensuring that the best and most effective groupings of work processes are defined. Starting from this logical optimum, it is then possible to take account of hardware and software constraints when moving on to develop the full ITS Future Big Picture.

The logical framework is a detailed description of the processing and data flows that have to be carried out and supported if all the ITS objectives are to be satisfied. It is the result of a detailed analysis carried out by the system developer. To develop the most appropriate logical framework, the system developer puts on a set of blinders and completely ignores any institutional or organizational arrangements that are currently in place. The system developer also pays no attention to the specific technologies that may be deployed at this stage. The whole idea is to develop a "perfect" logical framework that describes the processing to be carried out, identifies the most logical place to carry out the processing, and defines the data flows required to allow the whole framework to operate as a single entity.

To understand the logical framework, imagine several people at different desks in an office all carrying out separate tasks within an overall project. Each individual can autonomously carry out these tasks, but all employees need to swap information or data with one another to complete their tasks. A description of the work that is required by each individual to complete each task could be called a *process description.* The information or data to be shared between each pair (or each set) of individuals could be called a *data flow.* The data flow could be supported by a number of means—telephone calls, e-mail, faxes, memos, or even post-it notes.

Although we have not specified how the data flows are supported, we can define the data that has to be shared and the processing that each individual has to complete. In simple terms, a process is whatever the individual has to do to turn data in into data out.

Having defined the processes and the data flows, we have a logical framework. Chapter 10 provides additional information on the logical framework.

The development of a logical framework is a key step in identifying the synergies and similarities that exist between different parts of the overall system. It is at this point in the development process that the system developer can identify situations in which there is potential for duplication of effort in both processing and data collection. One of the prime objectives of the system developer is to ensure that the same data is collected and entered in the system only once but used many times.

This is also the point at which the developer begins to see the appropriate structure for the software required to support system operation. When the logical place to carry out the processing has been defined and the data flows to and from that process have been identified, then it becomes possible to visualize appropriate software modules. It is also possible to see what system interfaces may be standardized. Thus, the logical framework forms the launching pad for software development and the application of ITS and communications standards as well as providing the main vehicle for identifying synergy in the proposed system.

### 4.3.4 Step 230: ITS Future Big Picture

This is often referred to as the *layered framework* in ITS circles or the *physical architecture* by the system engineering profession. The ITS community uses the term layered framework, since the physical architecture component of the National Architecture for ITS is usually depicted as having three layers. We use the layers for one simple reason: it makes the various features of the ITS Future Big Picture easier to see and understand by enabling us to concentrate on one layer at a time.

The ITS Future Big Picture represents a multifaceted framework showing how all the proposed ITS technologies, products, and services will interact in the ultimate implementation. To develop a good ITS Future Big Picture, it is necessary to consider a range of issues spanning technical, commercial, and organizational aspects of the proposed implementation. Each issue has to be identified, evaluated, and balanced in order to define the best ITS Future Big Picture for a region or area. Therefore, another way of describing the process of developing the ITS Future Big Picture would be to call it a *multicriteria evaluation.*

The ITS Future Big Picture is important to the development of a successful ITS implementation, as it enables developers to see the whole picture and ensure that they get what they want, both now and in the future.

It is no coincidence that the ITS Future Big Picture is at the center of the ITS solutions model in Figure 4.2. It is the main product from this part of the process; the other steps feed into it.

This is the physical representation of the proposed ITS. It defines and illustrates the appropriate subsystems of which the overall ITS should be composed. The subsystems are defined by considering technical, organizational, and commercial issues and constraints. The interfaces to be supported so that each subsystem can cooperate as part of the overall system are also defined. Subsystem and interface definitions are technology-dependent in that the type of technology required is specified, but they are not technology-specific as design tradeoffs on the various options open have yet to be carried out. The layered framework is composed of three layers. A technical layer is developed by assembling a number of ITS Market Packages into the desired configuration to produce a technical solution composed of subsystems and communications interfaces. The picture is completed by two additional layers—an institutional/organizational layer defining the institutional arrangements and an organizational structure and a commercial layer identifying the commercial arrangements required to support the operation of the technical solution. It should be noted that the layers we have defined here—technical, institutional/organizational, and commercial—do not correspond to the layers defined in the National Architecture for ITS. We do not believe that it matters what layers are used as long as an ITS works. The development of a solution on all three layers requires a considerable amount of evolution and revision with inputs from several other steps in the process. The way in which these layer definitions work in practice is explained in more detail in Chapter 11.

### 4.3.5 Step 240: Standards Application Plan

This step in the process identifies the various ITS standards and other related standards that will be applied to the ITS implementation. The ITS Future Big Picture provides a clear understanding of the major subsystems and interfaces that are to be supported if the ITS is to function properly. The Standards Application Plan builds on this platform by assessing each subsystem and interface in terms of standardization requirements. Where possible, national or international standards are identified to support each interface. In cases where the national or international standards have yet to be developed, de facto standards are identified. The Standards Application Plan requires that a thorough review of the current state of ITS and related standards development be conducted.

### 4.3.6 Step 250: Financial and Commercial Analysis

This is a thorough identification and evaluation of the commercial and financial aspects of the proposed ITS Future Big Picture. It includes definition and quantification of the various revenue streams that will flow around the proposed ITS at key stages in its life cycle. This includes identification and assessment of capital and operating costs required to support the establishment and operation of the proposed ITS. An important part of this analysis is associated with the identification and evaluation of possible direct subscription payment methods for elements of the proposed ITS. If such mechanisms can be identified and proved to be commercially exploitable, then there may be a private-sector business opportunity available. This would enable the public-sector implementing agency to take advantage of private-sector leveraging to satisfy some objectives with private-sector funds.

At this point, it is also necessary to determine and evaluate the degree of public investment and subsidy. The financial analysis can take many shapes and forms depending on the culture and regulations constraining the public-sector implementing agency. However, a significant feature of any financial analysis will be the commercially oriented approach to rating business opportunities and evaluating possible risks associated with exploiting any opportunities discovered. Risk identification and evaluation will form a significant component of the financial analysis.

A financial analysis is a vital part of the ITS Future Big Picture as it provides the financial underpinning to support the proposed technical solution. This should show that the overall ITS is self-sustaining or alternatively quantify the degree of public expenditure required.

### 4.3.7 Step 260: Implementation Strategy

This is another key input to the development of the ITS Future Big Picture. The ITS Future Big Picture has to be technically feasible, commercially viable, and organizationally supportable. In addition, it must be possible to implement the ITS Future Big Picture. This is addressed by the Implementation Strategy, which considers today's starting point and tomorrow's ITS Future Big Picture and defines an appropriate sequence of events leading to full implementation. The Implementation Strategy takes account of a range of influence or factors that will affect the deployment, including the following:

- Local politics;
- The need for early winners;
- Technical feasibility;
- Risk;
- Commercial success.

## 4.4 Step 300: Outreach Activities

This encapsulates all the activities required to support and maintain an effective dialog among all the key participants in the project. These include both planning and execution activities such as the development of an outreach plan and the planning and coordination of stakeholder meetings and workshops throughout the planning and development process. The activities required to support the formation and evolution of an alliance or coalition to help with the planning and development of the ITS, and ultimately form the basis for the organizational solution, are also included here.

These activities are essential to the success of the development process because they support and enable the execution of many other steps in the process. Regular and meaningful interaction with key stakeholders is crucial.

## 4.5 Step 400: Design

This book focuses on the planning and development aspects of ITS, since we believe that the most pressing need is to move the starting point from today's design and build mentality to tomorrow's plan, develop, design, and build approach. However, we could not do a good job convincing readers about the merits of our approach if we simply ignored design, so we have provided some information here. We strongly recommend that readers treat this as background information and seek detailed guidance on design from other sources.

The ITS design activities should also include planning for later evaluation of the proposed ITS implementation.

## 4.6 Step 500: Implementation

After the Implementation Strategy has been defined and the detailed design work has been carried out, implementation can proceed. This can be performed in a few large steps, or more typically, in smaller, more manageable increments.

## 4.7 Step 600: Evaluation

Evaluation activities actually occur at many stages in the process, from the very beginning to the end. As discussed under ITS User Services, determination of Measures of Effectiveness is carried out very early in the process, thus starting the evaluation activities. Evaluation activities can occur pre- and post-implementation. In an incremental deployment with multiple stages, post-implementation evaluation for earlier stages will feed in at the beginning of the process for later stages of deployment.

## 4.8 ITS Technology Review and Selection

In this part of the process, the activities required for identifying and selecting the most appropriate technologies, products, and services to meet ITS needs and objectives are identified and evaluated. Sections 4.7.1–4.7.6 describe these activities.

## 4.9 Step 700: Identification of ITS Market Packages

This is where the ITS Market Packages are defined, on the basis of a thorough technology review that provides a status report on available technologies and products and their capabilities. ITS Market Packages are the basic building blocks of the ITS Future Big Picture, so we need to know what packages are available and assess what they can do and when they might be available.

### 4.9.1 Step 710: Unit Costs Catalog

This is the other half of the cost-benefit pair. In this step, unit costs for each ITS Market Package are defined and quantified. This requires that a thorough review of current market prices for ITS technologies and services is carried out and summarized in the catalog. The unit cost summary is important as it provides the cost elements required for the cost-benefit analysis used to justify the proposed implementation and to help in the selection of appropriate ITS Market Packages.

### 4.9.2 Step 720: Implementation Issues Summary

This is a practical assessment of the issues associated with implementing each of the ITS Market Packages. This is based on previous direct experience and on interviews with other implementing agencies and vendors.

It is important to evaluate the practical constraints associated with the implementation of each Market Package. Given the early state of development of some ITS technologies and products, this is likely to be an area where major risks lie.

### 4.9.3 Step 730: Operation and Maintenance Issues Summary

There is a separate operation and maintenance report created later in the process. This deals with how the proposed ITS will be run and maintained after implementation. However, operations and maintenance issues should be considered as an intrinsic part of the development and planning process, with technology and product selection taking account of these issues. Therefore, it is important to identify and evaluate the operations and maintenance issues associated with each ITS Market Package before the ITS Future Big Picture is defined. The operations and maintenance summary supports this by providing a list of issues with evaluation as input to ITS Market Package selection.

### 4.9.4 Step 740: Benefits Analysis

This is where the combined activities required to carry out an assessment of the likely benefits of the proposed ITS are summarized. Activities required to develop the benefits assessment include the following.

- Quantification of Measures of Effectiveness;
- Mathematical simulation modeling of the proposed implementation;
- Business modeling.

This is very important as it generates the material and documentation that will form the primary justification for implementation.

### 4.9.5 Step 750: Measures of Effectiveness Summary

To evaluate the cost benefit of the proposed deployment, it is necessary to be able to assess the effectiveness of each ITS Market Package, both individually and as a whole ITS Future Big Picture. This provides input to the evaluation process by defining parameters to be used as yardsticks for effectiveness and identifying typical values for each ITS Market Package.

## 4.10 Step 800: Operation and Maintenance Plan

These issues will have been addressed already in the Implementation Strategy and the design of the ITS. These steps are shown as a reminder that the life cycle of the ITS does not end at the implementation but at the end of the operational life of the system

Items covered in the operations and maintenance step include the following:

- Training;
- Servicing.

## 4.11 Step 900: Transportation Planning Process

As most readers know, there are highly developed transportation planning processes already in use in most developed countries. Unfortunately, in most countries that have already planned, developed, or deployed ITS, the relationship between ITS planning and the mature transportation planning process has been tenuous. This step in the process acknowledges the need for close integration between the transportation planning process linking objectives to conventional solutions and the process for ITS planning and development. Within the process diagram, this step can come before or after the other steps in the process, depending on the local transportation planning authority's position in the planning cycle. In some cases, ITS planning will precede the conventional transportation planning process due to the timing for updating the transportation plan. In other cases, the transportation plan will already exist but have taken no account of ITS capabilities, having been defined before ITS was considered for the region.

Now that we have made a quick tour of the ITS Cooperative Development Methodology, we can go into more detail. Chapters 5–19 present more detailed information and advice on how to carry out each step in the ITS Cooperative Development Methodology.

# 5

# Developing ITS User Services and the Strawman ITS Future Big Picture

## 5.1 Introduction

This chapter provides guidance on the execution of activities associated with step 100, "Initial Needs, Objectives, Problems, and Issues"; step 130, "ITS User Services"; and step 140, "Strawman ITS Future Big Picture" of the ITS Cooperative Development Methodology. These are all components of the ITS needs model, which is used to capture initial requirements and support the evolution of those requirements.

## 5.2 ITS User Services

ITS User Services have their roots in objectives and help to support the *what* part of the what?/how? cycle. So, how do we develop ITS User Services—by exploring requirements, of course!

During the course of many projects utilizing the ITS User Services concept, we have developed an approach to the development of ITS User Services that seems to be effective. First, it is important to recognize that there are three main parts to an ITS User Service, described as follows.

- *A label:* A convenient abbreviation that indicates what the ITS User Service is all about;

- *A description:* A textual description of what benefits or value the ITS User Service provides to the user;
- *Several "shall" statements:* The formal structured English stuff that system developers love to hear.

Our way of supporting the exploration of requirements with the stakeholders and the end users is to initiate and facilitate a dialog with and among stakeholders. Through this, we try to get them discussing with each other what they actually want to get out of the eventual system implementation. We call our facilitation tool the "so what?" analysis. In simple terms, it involves a group of stakeholders talking through various ITS User Service labels while we keep asking them "so what?" in terms of their particular group perspective. The sequence we typically follow is outlined in Sections 5.2.1–5.2.3.

### 5.2.1 Develop a Series of ITS User Service Labels

An ITS User Service label is a very brief five to ten word label indicating what the ITS User Service is. We have nearly always started by using the labels from the 30 ITS User Services as defined in the National Architecture for ITS as a checklist; these cover a wide range of ITS needs, objectives, problems, and issues. We then review stated needs from the user group and amend the ITS User Service list accordingly. It is important to note that it is not necessary to start with the 30 ITS User Services from the National Architecture for ITS, but they do make a good starting point for U.S. applications. For applications in other countries, it may be more appropriate to begin with the list of ITS User Services being developed by the ISO TC204 Working Group 1. Table 5.1 lists the labels for these ITS User Services.

In the course of one project it was suggested that it would be useful to introduce a rule for ITS User Service labeling: "The User Service label always starts with a verb." This rule helps to separate out ITS User Services from ITS Market Packages, which is an area where stakeholders and user groups seem to encounter difficulty and have a bit of confusion. Incidentally, we refer to it as the "Wagner rule" because it was suggested by our friend, Evelyn Wagner of PB Farradyne.

### 5.2.2 Conduct a "So What?" Analysis

We have found it very useful to carry out this analysis in small groups of six to ten people with the group being very user perspective specific. For example, on one of our projects, five stakeholder groups were identified as characterizing the entire user community. The five groups were consumers, transit

**Table 5.1**
ISO ITS User Services

| Service Category | Service Number | Service Label |
|---|---|---|
| Traveler information | 1 | Pretrip information |
| | 2 | On-trip driver information |
| | 3 | On-trip public transport information |
| | 4 | Personal information services |
| | 5 | Route guidance and navigation |
| Traffic management | 6 | Transportation planning support |
| | 7 | Traffic control |
| | 8 | Incident management |
| | 9 | Demand management |
| | 10 | Policing/enforcing traffic regulations |
| | 11 | Infrastructure maintenance management |
| Vehicle | 12 | Vision enhancement |
| | 13 | Automated vehicle operation |
| | 14 | Longitudinal collision avoidance |
| | 15 | Lateral collision avoidance |
| | 16 | Safety readiness |
| | 17 | Precrash restraint deployment |
| Commercial vehicle | 18 | Commercial vehicle preclearance |
| | 19 | Commercial vehicle administrative processes |
| | 20 | Automated roadside safety inspection |
| | 21 | Commercial vehicle on-board safety monitoring |
| | 22 | Commercial vehicle fleet management |
| Public transport | 23 | Public transport management |
| | 24 | Demand responsive transport management |
| | 25 | Shared transport management |
| Emergency | 26 | Emergency notification and personal security |
| | 27 | Emergency vehicle management |
| | 28 | Hazardous materials and incident notification |
| Electronic payment | 29 | Electronic financial transactions |
| Safety | 30 | Public travel security |
| | 31 | Safety enhancement for vulnerable road users |
| | 32 | Intelligent junctions |

representatives, commercial vehicle operators, infrastructure owners and operators, and a final group that cut across all the other four groups. We split those groups into the five components and then had each group conduct a "so what?" analysis on a subset of the ITS User Service labels.

We usually conduct the "so what" analysis utilizing a marker pen, a flip chart, and a facilitator. On some occasions we have automated the process using an LCD projector to project a large image for all to share and a word processor to capture the inputs and provide the prompts.

To conduct a "so what?" session, an ITS User Service label is selected and written on the flip chart and then the question "so what?" is asked of the group. The group is directed or steered a bit so that the response to "so what?" does not result in a "how." This is because we want to separate the *what* from the *how* in the process, with the *what* being represented and characterized by ITS User Services and the *how* by the ITS Market Packages developed later in the process.

This particular feat is easier said than done, but it may help to explain to the members of the group that if they feel inclined to even think about the word *how*, they should think instead of the phrase "by providing." This should improve the chances of obtaining a "so what?" answer, which consists of a set of benefits or useful utilities to the user rather than a potential solution.

At each stage, some information drops out that is valuable to the analyst later in understanding user needs, objectives, issues, or problems that are embedded in the ITS User Service label.

In terms of output from the "so what?" analysis, we really expect to get three separate components that make up a complete ITS User Service. The first component is confirmation that the ITS User Service label is useful in being a shorthand way of describing the ITS User Service. Second, we expect to be able to take the various layers of the "so what?" analysis to form the outline of an ITS User Service description. Third, we hope to use the raw material generated from the "so what?" analysis to create "shall" statements. The shall statements contain the succinct, structured information that help the system developer to start thinking about the appropriate technologies to address user needs.

Going back to the ITS User Service description, a key goal in developing the ITS User Service descriptions (the text that is used to describe the ITS User Service) is to get to the point where each user group can see itself in the ITS User Service description. The groups can see their needs, objectives, policies, and issues encapsulated, and the ITS User Service label relates specifically to their needs.

### 5.2.3 Lessons Learned

In the course of applying the "so what?" analysis, we have identified some lessons that we would like to pass on our readers.

The "so what?" analysis is an excellent tool for supporting a dialog between the user and the system developers and among the members of the user community. When this dialog emerges, you start to understand that there are more perspectives than you imagined even within what is supposed to be a homogenous grouping.

There seems, at first glance, to be a lot of work involved in the process. For example, if you start with one ITS User Service label and you come up with five "so what?" responses and then go on and say "so what?" to those responses, you may create another five each. You can see that the amount of work grows exponentially. It requires quite a lot of resources to support the entire process. The good news is that the process does tend to close again as repetition starts to creep in after level two or level three. This repetition can be easily dealt with in a fairly mechanical way. Consequently, we have been able to attain output from the process that justifies the investment in resources and would, therefore, recommend the use of the "so what?" analysis as a way of supporting the definition of ITS User Services.

We have also learned that the "so what?" analysis provides a way of making sure that the first cut of ITS User Services created for the project are fairly representative of the users' demands. The process also develops a high degree of ownership.

The whole question of ownership is something that has been somewhat neglected in the development of ITS. Perhaps the assumption is that engineers and system designers know a lot more than the user. Unfortunately, this is usually not the case, particularly when you are talking about the users' problems, issues, and objectives. In fact, the users are the experts in these areas, while the system developers and designers need to be experts in how the technology can be used to address user needs. Ownership transfer is key to the process—not just coming up with and agreeing on a set of user needs or a solution but making sure that the ownership for those user needs and the solution lies firmly with the user community.

Another interesting feature of the "so what?" analysis is that the typical end result after several layers will be what we refer to as an *irreducible* or *goal.* These constitute a fairly small number of relatively high-level objectives such as the following:

- Save money;
- Protect the environment,
- Save lives;
- Improve the economy;
- Improve personal wealth.

An interesting insight from the use of the "so what?" analysis is that sometimes users can be rather impatient and a little frustrated about the top-down nature of the technique approach. We have seen this exhibited in numerous projects and have come to refer to it as *top-down tension.* This is where users get impatient and frustrated and want to leap off and do something, anything, as long as it means getting some results and getting something to show for the effort.

In applying the "so what?" analysis, we have experienced the situation in which some users in the user groups simply want to install devices without going through a rigorous process that they perceive to be analysis paralysis. They believe that we are producing another report that goes on the shelf and that there is not going to be any action. Accordingly, it is important to counter the top-down tension, designing into your approach some way of relieving that tension and showing users that the effort will yield some meaningful results that will, in turn, lead to tangible action.

The existence of top-down tension illustrates the continuing influence of the "what?/how?" cycle: You have to start the cycle somewhere, preferably by asking users what they want, but often at an early stage, depending on the user group, you will find it necessary to explain *how* you can actually satisfy user needs. Frequently, a degree of skepticism builds up about the continued analysis of *what* people want when they receive no explanation of how you plan to satisfy their needs.

We have also learned that it is important that the "what?/how?" cycle rotates fairly rapidly. Once users have some insight into the technological possibilities, their needs change. They start to climb the learning curve about ITS technological capabilities, and they get into much more meaningful dialog because there is a real, tangible end in sight.

Finally, always bear in mind when using the "so what?" analysis that rigorously going through the layers and sticking to the rules is not essential, since the process itself does not have a great deal of value. The value is in the dialog that it supports between the users and the system developers and among the users themselves. In fact, the "so what?" analysis should be considered a fairly flexible tool to get this critical dialog underway.

## 5.3 Strawman ITS Future Big Picture

Like many things in life, beginnings are important in the development cycle. This step in the process encapsulates the activities required to identify, explore, evaluate, and confirm the user requirements that the proposed ITS will have to address and satisfy.

In an ideal world, a comprehensive group of potential users of the system would get together behind closed doors and emerge after a while with a definitive set of objectives, problems, needs, and issues to be addressed by a proposed system. The consultant or system developer would then take this and come back some time later with a fully operational system that would perform the required jobs. Subsequently, the user group and the client would check off all the things the system does against their original set of objectives, problems, issues, and needs; note that they had all been addressed; and then cheerfully sign off on the deployment—everyone would be happy. Unfortunately, this never happens due to a number of factors. First, the initial user group may not include all key users and participants. Identifying key users and participants, or *stakeholders,* is an important part of successful system development, but it is also one of the most difficult parts for system developers to handle on their own.

In practice, you rarely start off with a complete set of stakeholders, and, in fact, we believe that beginning this way should not be a goal. The set of stakeholders tends to snowball when you first try to identify all the stakeholders that you believe you need and then get them together for a meeting. From that meeting there will probably be suggestions regarding people who were omitted or not invited who should have been invited. It is wise to take full advantage of this powerful mechanism. It is better to have many minds—local minds, in fact—telling you who should attend rather than trying to pre-empt that decision and think it all out yourself.

Another way of identifying stakeholders would be for the system developer to sit down with a small number of well-informed system users and identify who they think should be involved. This is probably the way most system user groups are initially defined, as some users usually have a clear idea of the local situation. They would pinpoint the main players and the stakeholders whose objectives would have to be satisfied for the system to be considered successful.

In our experience, this is a good starting point, but it never provides the full list of key stakeholders. This is mainly because user perceptions of the system capabilities are vague and the full impact of the system is not yet understood.

The identification of stakeholders for an ITS deployment is a particularly critical aspect of the whole exercise. If you do not identify and invite people to come along who would be impacted by the ITS, or you forget about some key group, it is possible that at some point in the future, that group will cause more difficulty than if it had been involved from the beginning. Therefore, it is actually worthwhile to spend a little bit of time and effort in doing a fully comprehensive analysis of the likely effects of your proposed ITS. Unfortunately, this gives us another chicken-and-egg situation, because it would suggest that we would have to complete the architecture before fully identifying all the likely effects on the region. This is, of course, impossible, as you want your stakeholder group to work with you in defining the architecture.

Accordingly, it is necessary to make an initial assessment of the impacts of the ITS and the architecture. This is where a Strawman ITS Future Big Picture or candidate architecture solution can be very valuable in actually taking a leap ahead to the end of the process. By using your skills and judgment to tap into experiences from previous implementations and previous architecture development programs, you can develop a solution that could be used as a vehicle for communicating with the stakeholders. We call this the Strawman ITS Future Big Picture, since you will put it on the table and ask the users to attempt to pull it apart like a straw man.

As the needs of each project vary considerably, it is not possible for us to provide you with a recipe for developing a Strawman ITS Future Big Picture. Sometimes we have utilized a proposed institutional and organizational framework; sometimes we have used a proposed technical solution. It really depends on the needs of your particular user group and the consultant's or developer's own previous experience. However, we can offer some guidance by highlighting and explaining the essential ingredients of a Strawman ITS Future Big Picture.

### 5.3.1 Based on Needs

Carry out an initial review of the needs, objectives, issues, and problems that the stakeholder group or lead customer has expounded. Make sure that the Strawman ITS Future Big Picture clearly addresses these requirements.

### 5.3.2 Credible

Although you may end up throwing it out when the stakeholders have finished with it, the Strawman ITS Future Big Picture has to be credible to be useful. There is no point in incorporating highly advanced technologies that cannot be explained to the stakeholders. Similarly, it is a waste of time to suggest an approach that obviously could not be afforded or implemented.

### 5.3.3 Drawn From the Consultants' or Developers' Experience

This is the key to a good Strawman ITS Future Big Picture. Drawing on the knowledge the consultant or developer has won from earlier development projects, it is possible to produce a highly effective Strawman ITS Future Big

Picture. This is a balancing act though, since you do not want to fall into the trap of using an earlier off-the-shelf solution without a full check that it will meet your needs. There is a real danger here that you will like the Strawman ITS Future Big Picture so much, you will not want to go through the other steps.

### 5.3.4 Very Rough and Obviously Incomplete

This is one way to avoid the danger of blind adoption of the Strawman ITS Future Big Picture. Making sure that the Strawman ITS Future Big Picture is obviously incomplete and not able to be implemented helps to make the point that it is after all only a good guess at the most appropriate solution.

### 5.3.5 Simple to Explain and Understand

Remember that the main objective in developing the Strawman ITS Future Big Picture is to provide a communications tool to facilitate the discovery and understanding of the real requirements. It is important that the consultant or system developer keeps this in mind and avoids the urge to look smart through over sophistication.

### 5.3.6 Graphically Oriented

Building on the previous point, we have found that it is very helpful to describe, define, and communicate the Strawman ITS Future Big Picture in mainly graphical ways with little reliance on text descriptions.

### 5.3.7 Disposable

For the Strawman ITS Future Big Picture to do the job, it has to be disposable. The consultant or developer must ensure that the resources invested in the development of the Strawman ITS Future Big Picture are small enough for it to be viewed as disposable by the whole development team.

# 6

# Developing and Applying an ITS Objectives Statement and ITS Vision

## 6.1 Introduction

This chapter deals with step 110, "ITS Objectives Statement," and step 120, "ITS Vision" in the ITS Cooperative Development Methodology. The development of an ITS Objectives Statement and an ITS Vision for a project are very important parts of the overall development program for an ITS Future Big Picture. They set the scene for requirements definition and provide good communications tools.

### 6.1.1 ITS Objectives Statement

Starting with the ITS Objectives Statement, we will describe how to go about developing these parts of the ITS Cooperative Development Methodology. The only way we know to develop a robust ITS Objectives Statement is to get all the major stakeholders in a room and not let them out until they have agreed upon some sort of ITS Objectives Statement. Even if the consensus and compromise process results in a high-level statement with a degree of ambiguity remaining, its purpose will have been served. This should be carried out right at the beginning of any project as it provides the first step in the long consensus and convergence process.

A facilitated workshop usually works best. This may be a single plenary session, or it may incorporate breakout sessions. Either way, it is necessary to end up with a single statement regarding the mission in developing the system.

You may also want to develop Strawman ITS objectives as the basis for discussion and revision during the workshop. The discussions during the workshop will also help to frame the space within which the ITS User Services will be developed later.

Our main word of advice for developers, consultants, and implementers is to prepare for this workshop thoroughly and deploy your most experienced staff. This is very much a scene-setter. You should also be prepared to revise the ITS Objectives Statement in light of insight gained during the development of the other parts of the needs model (ITS Vision, ITS User Services, and Strawman ITS Future Big Picture).

### 6.1.2 ITS Vision

One of the primary reasons for developing the ITS Vision is to provide a means for the client and the system developer to externalize their own internal visions for the future. This enables their respective views of the future to be synchronized to form a common understanding of what will really be expected from the system in the future. Typically, the client group will have a "user" view of the final system, based on what services and functions it expects the system to provide or what transportation problems and policy objectives it expects to have addressed. In many cases, this is a "bottom-up" view tightly focused on specific applications.

On the other hand, the system developer usually has a clear "top-down" picture of how each required element might work with others as a single integrated system. The vision forms part of the overall process of removing ambiguity from requirements. This ensures that the requirements defined and the proposed solutions being identified for the region are lined up with what the customer really wants and expects as well as what the system developer has envisioned as the final solution. Of course, another important reason for doing this is to allow the client, typically a group composed of local authority and local government representatives, to develop a nontechnical description of the future. They are then able to share it, not just with the system developer, but with a wider audience beyond the core group of people who are steering or directing the development of the ITS project.

The ITS Vision is a primary component of the needs model, which in turn is part of the two-stage iterative *needs/solutions* cycle for recursive ITS Future Big Picture development.

The ITS Vision can be used as a sales document to be taken to prospective private-sector investors or to be taken to private-sector companies that are already participating in the region by providing information services or ITS services or products. The ITS Vision for the local region can then be shared with a particular group, and an understanding can be developed of how the public and private sector might work together. Synergy can be identified, and perhaps public funding and public investment in ITS can be leveraged with parallel private-sector investment. This could never be achieved if an ITS Vision or an ITS Future Big Picture were not created in the first place. This is because it would not be possible for the local government people to externalize and share their vision with other interested groups such as the private sector or the general public.

With reference to the general public, another important use of the ITS Vision is to provide the basis for public consultation, public information brochures and leaflets, public information meetings, consultative documents, and consultation meetings. By these means, the local authority can share with the general public the status and the background related to the project and allow the general public to produce input that will shape the development of the ITS Future Big Picture as it goes along.

A very important aspect is to make sure that the various stakeholders, including the general public, are actually involved in the loop from the beginning and kept actively engaged in the process. This is to ensure that a resistance to the proposals does not develop and cause problems when the proposals are due to be implemented.

Another use for the ITS Vision that is sometimes overlooked is to support the consensus formation process within the client group. Regional and national ITS Future Big Pictures for ITS developments, in particular, tend to be directed by a steering committee composed of a diverse group of professionals with multiple perspectives and perhaps an uneven, or inconsistent, perception of the capabilities and likely impacts of ITS in their region. Thus, the ITS Vision becomes an information-sharing tool and a key component in the overall ITS needs model.

From a nontechnical, nonarchitecturally aware point of view or from the perspective of the users and the groups of people who are not familiar with ITS, this is probably one of the most important parts of the needs model. From a system engineering perspective or a system development perspective, of course, that may not be the case. The technically oriented engineers or developers are typically more interested in specific technical details and concise engineering design requirements and definitions and sometimes overlook the importance of the ITS Vision for the general audience and for the client.

This is just one of many indications that we are dealing with multiple viewpoints, attempting to derive a single agreed upon set of requirements to which we can apply a wide range of potential ITS solutions. There is an amazing potential for problems to arise due to ambiguity and misunderstanding with a consequently high risk factor associated with ITS developments.

Probably the best way to illustrate the essentials of an effective ITS Vision is to provide an actual example. The one we have chosen was actually developed for a real client but describes the future in a yet-to-be-built city. In the remainder of this chapter, we will describe the ITS Vision that was developed. We will also provide you with commentary on some key aspects of the ITS Vision that may be useful to those developing an ITS Vision for their own region or state. Technologies and applications are not separated or structured in the ITS Vision. Instead, the user perspective is used as the focus and the technologies and applications are described in the order in which they would be encountered in the final implementation.

What follows is a nontechnical description of what the final ITS Future Big Picture or architecture of an ITS might look like for a region or city. It was written from multiple perspectives to illustrate the ITS from the perspectives of all the major stakeholders and all the people who are likely to be impacted by the ITS in their daily lives. The main idea in developing the ITS Vision is to allow all stakeholder groups to get a good idea of their relative roles and responsibilities or how the ITS might impact them in the future. The ITS Vision forms an important part of the needs model for ITS development and is intended to be a nontechnical, straightforward, exciting, and illustrative way to actually explain to people what the system might be like.

## 6.2 ITS Vision Example: Putrajaya

This ITS Vision provides several user perspectives of a hypothetical ITS implementation in Putrajaya. These are intended to illustrate the potential impact that ITS might have on everyday life. The system described is a vision of how the proposed implementation might appear; it is not intended to be a hard and fast system design. This would be developed during the course of a formal development and design process.

The following user perspectives—specially selected based on an early assessment of the perceived problems and the important stakeholder groups in the region—have been adopted for the Putrajaya ITS Vision:

- Ambulance driver;
- Bus driver;
- Business traveler;
- Commercial vehicle operator;
- Commuter;
- Local authority traffic engineer;

- Policeman;
- Taxi driver;
- Tourist.

These user perspectives are described in Sections 6.2.1–6.2.9.

### 6.2.1 Ambulance Driver

Azizul was proud of the job he did as an ambulance driver; he had a real sense of doing a worthwhile job that had a direct impact on people's lives. In fact, in some cases, he was responsible for saving lives. This made him very happy.

The one part of his job that he found to be very frustrating related to the problems of getting to an emergency scene through the traffic and locating the precise place where he was required.

Many times in the past, he had wasted valuable minutes picking his way through traffic jams and searching for an obscure address. He well knew that there was a direct relationship between the delay in getting to an emergency scene and the chances of survival of the accident or illness victims. This was particularly true for heart attacks and road traffic accidents.

Azizul was very pleased that his managers back at the emergency response center had recognized this problem and were now utilizing the new Putrajaya Intelligent Transportation System (PJITS) to track and dispatch ambulances. The installation work had started last month, and most of the ambulance fleet had now been fitted with on-board vehicle tracking and driver information terminals. Azizul's ambulance had been equipped only last week, and he had just completed the training course.

The new system not only tracked Azizul's ambulance through the road network but also provided him with route guidance advice on how to get to the emergency scene. He was confident that the people back in the emergency response center were now making good use of all the vehicle location information at their disposal and with the help of the new emergency vehicle management system were deploying ambulances even more efficiently than before.

*(The point being made here is that training was provided as part of the ITS implementation. Many potential ITS users may be afraid of new technology; this helps to reassure them that someone will help them to use it and that it is okay to be unfamiliar with new technology.)*

He remembered that during the training course he had been told that the emergency response center (ERC) had direct communication with the PJITS, so that the tracking and location information could be relayed from PJITS to the ERC. This link also enabled the ERC to request emergency priority for any ambulance on a particular route. The PJITS then coordinated with the traffic control system that adjusted the necessary signal timings as the ambulance approached each junction. Azizul was very impressed by this feature as it amazed him that the system could give his ambulance priority without disrupting the entire downtown traffic pattern.

He had heard it said that the system had already been directly responsible for saving at least one life by getting the ambulance involved to and from the emergency scene in record time. As far as he was concerned, that one life justified the whole investment in the system, and he knew there would be many more lives saved over the coming years. "Progress indeed," he thought.

### 6.2.2 Bus Driver

Nor had noticed that the people on his bus seemed to be in a much happier frame of mind these days. "Maybe it's due to the recent introduction of the PJITS system," he thought to himself. The new system was having a considerable impact on the efficiency of the transport system in the entire Putrajaya region and along the Multimedia Supercorridor to Kuala Lumpur and the new Kuala Lumpur International Airport (KLIA). In particular, the public transport system had been greatly affected. All his regular passengers now had smart cards to pay their bus fares, and this saved Nor the problem of dealing with all those small coins and enabled passengers to get onboard his bus more quickly.

He liked the fact that boarding was now quicker as he hated to get behind schedule. The smart cards could also be used in local shops and supermarkets, so his passengers were able to buy other things beside bus fares and could load the smart card with more money while they shopped. In addition, the larger shops had pretrip information terminals that people could use to find out the predicted arrival time for the next bus home and investigate different routes and interchange possibilities. He was convinced that this type of facility had made his passengers more relaxed as they could now rely on the bus service.

He was also less concerned now about getting stuck in traffic congestion and being late to arrive at the next bus stop. For a start, this did not happen so often now as the traffic control part of the PJITS was able to manage the city streets and expressways. On the rare occasions when things did go wrong, due to an expectedly busy day on the roads, construction work, or some incident or other, he knew that the PJITS would inform passengers about the delay and give him priority at traffic signals. This would give him the opportunity to make up some time.

The smart bus stops, installed as part of the PJITS implementation, would tell waiting passengers how many minutes they would have to wait for the next bus. It would also offer information on alternatives such as LRT or taxi. If any of the passengers wanted to opt for a taxi, it would even allow them to call one to the bus stop by inserting their smart cards into the smart bus stop terminal. This would identify them as bona fide travelers and deduct the taxi fare from their cards. (No more fumbling for change or opening purses or wallets in public.)

*(There is a large emphasis on describing positive, tangible benefits to the traveler. The real benefit is decsribed as well as the direct effect of the technology application.)*

Nor had noticed that more and more, passengers would frequently appear at his bus stops just before he arrived. "This must be an effect of the smart bus stops and the information terminals in the shops," he thought. This had been an interesting effect of the new system. When potential bus passengers arrived at the smart bus stop and determined that the bus would not arrive for another 10 minutes, they decided to do some shopping in adjacent stores while they waited. This had pleased the local shopkeepers as this was good for business, and many of them had installed terminals inside their shops, so that people could shop until the last second before the bus arrived. The bus information was reliable enough to allow this, and some shops even had an announcement system that advised shoppers of the imminent arrival of the bus.

Another reason for the increased happiness of bus passengers and drivers was the comforting thought that the new PJITS was continually monitoring the condition of the major components on the bus, ensuring that problems would be swiftly rectified. This also meant that the emissions from the bus were being constantly monitored, so air quality was not being damaged in the new city of Putrajaya. Nor knew this particularly well as his driver display unit was constantly advising him of the appropriate driving speed for optimum engine performance and for schedule adherence. It helped him to give his passengers a nice smooth ride with predictable journey times.

### 6.2.3 Business Traveler

Aslan had decided to get up a little earlier this morning to gain time to consult the PJITS before breakfast. He had come to rely on the new system to provide him with current information on the status of the transportation system and inform him of the various options available.

*(This section indicates that the new ITS has matured to the point where people feel they can rely on it. There is also a strong relationship between the features and services of the ITS and real business benefit of winning the job.)*

This was a big day for him. His company, a large systems and software company based in Putrajaya, was bidding for a multimillion Ringgit contract to supply a turnkey system to a major financial institution. Today was to be the final presentation to the client's board of directors at company headquarters in Singapore. As the leader of the bid team, Aslan knew that if he and his team performed well today, they would clinch the deal. It was important that he and his team arrive on time and be well prepared. In fact, this presentation was so important that his team was already in place in Singapore, preparing for the presentation. He had not traveled with them due to an important commitment on another project. As a result, the pressure was on him to make sure he arrived in good time.

He inserted his smart card into the PJITS in-home terminal. This identified him as a subscriber to the travel information service and, after asking him some security questions, presented him with the welcome screen. As Aslan had used the system several times before, the system had started to learn about his favorite modes of travel and had presented the appropriate menus as a welcome screen. Information on his subscription status and travel and information format preferences was stored on his smart card.

"Hmmm, I see that the system thinks there may be some delay in getting to the airport today; maybe I should investigate alternative modes and routes," he thought. He quickly asked the system to display times, costs, and reliability indices for the journey from his home in Putrajaya to KLIA. He also informed the system that his ultimate destination today would be downtown Singapore and asked that it set up a link to the Singapore ITS for information retrieval and reservation purposes.

The information was retrieved promptly and it informed Aslan that the road network would be operating at reduced capacity today due to construction work and predicted traffic volumes. Consequently the reliability index for private car, bus, and taxi journey times was only at 55%. This is not a good number if your prompt arrival is vital, and it convinced Aslan that today was a good day to go multimodal. He would leave his car at home in the garage and head for the public transport system. *(The ability of the user to configure the ITS to his or her own needs and requirements is stressed here.)*

He called up the public transport information menu and asked for information on options for getting from Putrajaya to KLIA. He placed an emphasis on journey time reliability, so the system automatically ruled out bus-only and taxi-only options. After further investigation he concluded, with the help of the system, that a bus/rail link option was best and asked the system to supply schedule, fare, and stop/station location information.

The system did so promptly, printed out the information, and advised him that it had noted that the total return fare for the journey would be RM25 but that he only had RM10 left as a balance on his smart card. Did he want to transfer funds to the smart card from his checking account? He said yes to this option but specified his corporate Visa account as the source of the funds transfer, as this was company business. The system transferred RM100 to his smart card as he requested.

"Right!" he thought, "That takes care of the journey to the airport; what about the other end?" He asked the PJITS about Singapore travel conditions. It said that the Singapore transport network was running normally, so Aslan requested that a prereservation be made for a local taxi to collect him from the airport and take him downtown. This was confirmed, and the system reminded him to check the status of the taxi from an in-flight information terminal just prior to arrival in Singapore. He also decided that he felt lucky today and asked the system to display a range of four-star restaurants in downtown Singapore that had reservations available. He made a quick choice and reserved a celebration lunch for his team.

"Great," he thought as he logged off the system, "I have 30 minutes to have my breakfast and get down to the bus stop." He had his breakfast, feeling relaxed and confident that he would get to his presentation on time.

As he finished his breakfast, he decided that he would take his personal travel information terminal with him today. This was a combined cellular telephone and information terminal that would enable him to stay in touch with his office and monitor the changing conditions on the transport network. He slipped it into his pocket, downed the last of his coffee, and headed for the bus stop, which was only 150 meters from his house. When he got there, the smart bus stop display was confirming what he already knew—that the bus would be along in 2.5 minutes. "Good," he thought, "just enough time to call my team in Singapore and check that they are all ready."

He was just finishing his call when the bus arrived. He waved his smart card at the reading plate next to the driver, and the system confirmed that his final destination was the airport today but that he had chosen to interchange onto the LRT. It deducted the maximum fare for the route and reminded him that he would have to use his card again to check out as he left the bus in order to get the appropriate credit. Its onboard system also confirmed that his destination had been logged and would be announced on arrival.

This was always a relief to Aslan as he did not use the bus system very often and did not always know exactly where to get off. Sure enough, just as the bus was approaching the LRT station, the variable message sign above the driver flashed up the destination, and a pleasant synthesized voice confirmed that the bus was about to arrive at the LRT.

Aslan held his smart card up to the reader next to the exit door of the bus; the system confirmed that it had credited him the difference in fares, and he got off. As he was proceeding to the LRT, he checked on his personal travel information terminal and confirmed that his flight on Malaysia Airlines to

Singapore was going to be on time. The system also indicated that he had been allocated seat 2A, which was his favorite seat as it allowed him to get off the plane ahead of the rush.

The rest of the journey to the airport proceeded smoothly. The LRT was equipped with the same type of display and ticketing equipment as the bus service so the standard of passenger information and service was just as good. *(Another example of a direct tangible benefit to the user.)*

On arrival at the airport, he used his smart card to get "fastrack" check-in at the Malaysia Airlines desk. This speeded up the check-in process by electronically transferring his details and preference from the smart card to the airline central reservations system. There would be no paper tickets to fold up and stuff into his wallet.

The flight left on schedule, and shortly before arrival in Singapore, he confirmed the status of the taxi he had prereserved, utilizing the travel information terminal on the back of the seat in front of him. He noted that his taxi would be number 12 and would be waiting for him at the arrivals area. He was a little anxious to note that the information system was carrying a bulletin stating that there had been a fairly major traffic incident on the road link from the airport into Singapore. A fuel tanker had overturned, and several roads had been closed off as a precaution.

The police and emergency services were dealing with the situation, but response time and overall coordination was not that good. Aslan thought, "If only these chaps in Singapore had the same foresight as the PJITS people in adopting a fully automated incident management system, then I wouldn't be so concerned about possible delays." He consulted the terminal and concluded that it would still be okay to take the taxi, but that it was very important for the taxi to take a bypass route avoiding the incident area. He made a note to ask the taxi driver what route he proposed to take.

Aslan was relieved that Malaysia Airlines ran true to form that morning as the plane arrived at the gate exactly on time. He was even more pleased to see taxi number 12 waiting for him at the curbside arrivals area.

As soon as he got in, he asked the driver about the choice of route. The driver said that he knew all about the incident and would be using the bypass route under guidance from his in-taxi information system. He also asked Aslan if he would like to pay for the taxi fare by smart card as this taxi was equipped to do this. "Trust the Singapore taxi operators to be up to speed on modern customer service," Aslan thought.

So his plan worked to perfection, and Aslan arrived at the meeting venue in good time. His team, under his leadership, gave a great performance and won the contract. The only slight problem was that he had to use the PJITS system to change his return reservations. This was due to the start of the presentation being delayed one hour by the client. It turned out that the managing director of the financial institution had not been aware of the tanker incident and had been stuck in a traffic jam for an hour. "Maybe we should have tried to sell him some ITS as well," Aslan joked with his team over the celebration lunch.

*(This reinforces the notion that the client is smarter than everybody else because he is deploying and benefiting from ITS.)*

### 6.2.4 Commercial Vehicle Operator

Winston arrived at the depot a little early this morning. He wanted to make sure that his boss, Norman, the fleet manager, had renewed the license on his goods vehicle as he had promised. It had expired the night before, and Winston did not relish the prospect of having to stay in the depot all day if the license was not there. Worse still, he may have to go to the licensing office and line up himself to get the license.

He was pleasantly surprised when he arrived at the depot and noticed that his vehicle license had been renewed overnight. He was even more surprised to note that the new license was in the form of an electronic tag. "Progress indeed," he thought. He realized that this should not have come as a surprise really. Norman had been introducing new technology at a fast pace and was full of new ideas and implementations.

Only last month, he had arranged for the installation of an in-vehicle information unit for Winston's truck. This had been part of the PJITS initiative and had provided Winston with real-time instructions on which delivery to make next and advice on how to get there. Norman now had a fleet management terminal in his office and was able to send dispatching instructions to his entire fleet of 100 vehicles. The data was relayed to the PJITS control center, which then transmitted it to the in-vehicle equipment. Once Norman had made his decisions with the help of the management system, the instructions were relayed to the in-vehicle units by the PJITS control center.

Norman was really pleased with the performance of this system as it had shown a reduction in fuel consumption of more than 15% across the whole fleet as well as improving his on-time delivery rate to 95%. He was really saving money in not having to pay those "just-in-time" delivery penalties.

Winston, after some initial cynicism, was also quite pleased with the results of the system. It was helping him to avoid being stuck in traffic jams by displaying congestion information and displaying alternative routes. It also enabled him to earn more of a bonus by making more on-time deliveries every day.

He was also rather comforted by the fact that the on-board unit, as well as monitoring his driving speed and total miles traveled each day, was keeping a close eye on key safety components on his truck. He was reassured to think that brakes, steering, tire pressure, load stability, and equilibrium were all being checked continuously and reported both to him and Norman back at base. Norman was using this data to produce monitoring reports, which he passed on to the maintenance department, so they could make better plans for routine maintenance and parts delivery.

Winston noticed that Norman was already at his desk and went over to see what the new licensing arrangement was all about. "Hello Winston, I see you've noticed the new electronic license then?" Norman said as he saw Winston approaching. "I suppose you'll want an explanation, so sit down and I'll tell you all about it."

Norman then explained that the electronic license had been linked to the equipment already on Winston's truck and that the unit held electronic proof that the license fee had been paid for the next year. "It's really great," Norman said. "We don't have to go to the license office ever again. We just dial in to its central ordering system and read the electronic unit numbers, and the agency deducts the license fee from our company's bank account. They even send us an electronic reminder for each unit a few days before it's due for renewal. I was able to renew licenses for 20 of our vehicles last night, it's so easy. What's even better is that the license office has given us a discount of 10% off every license as we choose to use the electronic system. They say it saves them administrative staff time so they pass some of the savings on to us." *(This illustrates seamless, integrated operation of regional ITS.)*

"We've also become part of the probe vehicle fleet for the PJITS. The same electronic unit, which holds the license information, will also allow your truck to be located at key points on the road network. PJITS will use that information to determine your average journey time around the various roads on the network. This will be used to update the traffic information they supply to the other subscribers and us. They will also be able to supply me with vehicle tracking information, which will make our fleet management even more effective."

"The other good news is that you can also use it to go through the Lorong Plustag at the toll booths on the expressway. No more stops at the toll booths, Winston. You can drive past all those vehicles waiting to pay by cash and not stop. Your truck will be identified automatically and our account with PLUS will be debited."

Winston left Norman's office with the intention of remembering not to be surprised at the new ideas springing into his life, "I suppose this is the way life around this company is going to be from now on."

### 6.2.5 Commuter

Kadijah was as keen to get to work promptly today as any other day. She prided herself on her excellent attendance record and first-class time keeping. This had become a lot easier recently thanks to the PJITS. She lived in Putrajaya and had to commute to downtown Kuala Lumpur every day to work. Her secret in keeping good time was to always allow an extra 30 minutes for the journey. That way, if the bus ran on schedule, she would be 20 minutes early; if not then she would always arrive just in time. Although this added up to an extra three hours per week (she worked on Saturdays as well), she did not mind the sacrifice. (This is an example of information availability providing additional choice and flexibility to the traveler.)

Recently, this had all changed thanks to the PJITS. Kadijah's in-home information terminal gave her access to the most up-to-date bus schedules and details of how well the service was performing on any given day. She had used this for three weeks and was gaining confidence in the reliability of the information.

Kadijah now only allowed an extra 10 minutes just in case, which saved her more than two hours per week. She had put the time to good use by enrolling in a distance learning course at the local Putrajaya Technical Institute. She had enrolled in this new "smart school" and was learning computer science by carrying out her studies using the same home terminal that provided the public transport information. The course was taking up more than the two hours she was saving on travel time, but she did not mind as it would eventually enable her to get a better job.

She had not created any extra hours in the day; it was just that some of the hours were now more usable. Instead of waiting hopefully at the bus stop, she could make better use of that time at home or in the office. Improved journey time, reliability, and better access to current transport information had really improved the quality of her life.

Even when she had started her journey, there was now current information available at the smart bus stops on when the next bus would be along and what alternatives were available. On some rainy days, she had used this information to good effect by deciding to share a taxi with the other women standing at the bus stop. They had all entered their smart cards into the bus stop terminal and summoned a local taxi. This was a little more expensive then taking the bus, but it took them right to the doors of their respective offices.

On other days, she had reviewed the current transport situation and decided to delay her journey to work. The system had predicted that if she delayed her departure by 20 minutes, she would still get to Kuala Lumpur at the same time as if she had left now. This was because of the variation in traffic conditions. Kadijah also had an agreement with her boss that she could vary her start time by up to one hour each day as long as she worked the required number of hours over a two-week period. He referred to it as *flexitime.*

### 6.2.6 Local Authority Traffic Engineer

One of the biggest headaches a traffic engineer has is trying to figure out just what all those drivers out there are actually doing. Where are they coming from? Where are they going to? What routes are they choosing? How many of them are lost? These are the most interesting questions to consider when trying to design a traffic management set-up and when trying to manage the operation of a city traffic network.

In this respect, Lisa faced just the same problems as any other traffic engineer working in any other major city in the world; Putrajaya had similar difficulties and challenges. However, Putrajaya was different—it was an exciting place to practice the traffic engineering profession. The implementation of the PJITS had provided a new set of state-of-the-art tools for the traffic engineer. Thanks to the advanced traffic management features of the new system and close coordination with City Hall in Kuala Lumpur, both city and regional traffic were now being handled with unparalleled effectiveness.

What really excited Lisa, though, was the volume of high-quality data about current traffic flows, journey times, and travel demand patterns that were now available from PJITS. She had been delighted when she was selected to be a member of the transportation research group within the PJITS control center because she had known that she would have access to all this lovely data. It had been a dream of hers, since her days in college, to be able to devise new traffic and transportation management strategies that would make life better for everyone in the city.

Getting the right data had always proved to be the problem. Throughout her professional career, Lisa had put up with coarse pictures of traffic flows and demand patterns. She knew that these had to be treated with care as they were five to ten years old and based on 5–10% samples of total traffic. She did not complain about this as she remembered from her younger days how much effort went into organizing and carrying out those manual traffic counts and roadside origin/destination interviews. All things considered, she was lucky to have any data at all.

The real-time traffic flow data from the traffic signal control system detectors had been supplemented by the range of traffic and vehicle sensors operating in the PJITS. These provided speed, journey time, origin, and destination data. Additionally, the integrated ticketing system part of PJITS was generating a comprehensive view of public transport demand.

This data was all being collected, analyzed, integrated, and stored by the PJITS management information system. The transportation research group had access to all this data and the tools to manage and analyze it.

There were currently 10 research projects under way at the center. These were investigating and devising various management strategies including the optimization of the total transportation network. This was being considered through demand management and public transport fare structuring techniques, the use of probe car data to enhance traffic modeling and demand forecasting, and the management of goods flow from ports and airports to manufacturing sites.

This latter task was Lisa's own pet project as she had come up with the initial concept. The idea they were working on involved the use of the PJITS to increase the reliability of goods vehicle journey times and directly influence the flow of raw materials into Malaysia and the flow of manufactured export goods out. It made her particularly proud to think that her work might even have a direct effect on the Malaysian economy by increasing international competitiveness. She also liked the fact that she got to work with both university and government economists on the project. This has widened her horizons and given her a better appreciation of the economic impacts of transportation.

"This new PJITS system has certainly made a big difference in my life," she thought. "Now things that we could only dream of before are becoming possible; there are no data constraints to get in the way."

This user perspective makes the point that transportation planners don't have enough good data or present information and that an ITS is a two-way street with data flowing in and out. The data is valuable to traffic engineers, transportation planners, and perhaps also to private sector companies seeking good business sites.

### 6.2.7 Policeman

Khairun was very concerned when he received the message on his in-vehicle information terminal. It was from the emergency response center, telling him that he should proceed to the federal highway immediately as a tanker had overturned and was blocking the road.

He had attended one such incident before and remembered that it had brought the entire highway network to a complete standstill for over eight hours. Apart from the fact that he was required to work all night on that occasion, the other bad memory he had concerned the difficulties the police and fire team had encountered in determining what the vehicle was carrying and how to deal with it.

*(This describes how bad things were before ITS in order to highlight the effects of ITS on this stakeholder group.)*

All this ran through his mind as he sped to the scene of the incident. On his way he noticed that the PJITS system was working well and that his vehicle was being identified as an emergency vehicle in urgent mode. Consequently, he was being given priority at signalized intersections, which was an immense help in getting there quickly. The location of the incident had now been posted on the PJITS, so this was now displayed on the map display of his in-vehicle unit. The PJITS was suggesting a slightly longer route in terms of distance, as it avoided busier traffic areas and would give him a predicted journey time of 15 minutes, with a forecasted journey time reliability of 98%. He knew that the fire and other emergency vehicles making their way to the scene would also be making good use of this PJITS facility.

On his approach to the scene he noted that the VMS at key decision points on the network were advising drivers of the problem and suggesting alternative routes. A press of a key on his terminal gave him a map display showing the current level of congestion on the surrounding road network. The problem was growing quickly.

He also noted on his terminal that a number of other traffic officers were being deployed to monitor the traffic situation on surrounding roads. He knew from the training he had received on the PJITS that it had access to hundreds of contingency plans for traffic rerouting and critical point monitoring. This was probably what was happening right before his eyes. Police were being deployed to look after roads that had not been so important before the incident but had now become critical links in the diversion strategy.

When he got to the scene of the incident, he was pleased to note that no one had been injured, so the ambulance that had been dispatched to the scene was now just standing by in case of problems with the spillage. The fire department had just arrived and was currently interrogating the onboard electronics on the overturned tanker. They were able to do this from a safe distance as the ruggedized communications system on the truck was still functioning.

The fire chief, John, was monitoring the terminal onboard his vehicle and noting that the tanker's load consisted of nylon salt. He had not encountered this substance before, so he requested handling instructions from the PJITS. While these were being prepared, he viewed another menu which showed the contents of the tanker's black box system. This showed the speed, direction of travel, acceleration, braking, steering, and engine management functions on the truck during the last two minutes prior to the incident. This was being prepared for transmission to police headquarters for safe storage and later analysis to establish the cause of the problem.

Meanwhile, the handling instructions came back from PJITS. The system had direct links with the national hazardous materials database and relayed handling instructions and precautions directly from this source. John was glad to note that the substance was only mildly caustic and that the main precaution required would be eye protection. He immediately transmitted this information to his fire crew, who received the information in the form of a display projected onto the inside of their safety visors.

In the meantime, Khairun was receiving this information on his terminal and was preparing to carry out his job tonight. This was to liaise with the PJITS information and control center and manage the local traffic congestion in the immediate vicinity of the incident. He knew that affected roads in the wider area were already being managed and that his role was to deal with the traffic that had been close to the incident and was now jammed up.

After brief consultation with PJITS and the fire chief, using his terminal, they had all agreed that it would be at least two hours before the road could be reopened. Khairun figured that the best bet for local traffic management would be a temporary contraflow arrangement to take the stuck traffic along one lane of the opposing roadway. Before the introduction of PJITS, this would have been too dangerous to even consider for both motorists and police. Now however, it would be possible to execute this with minimum danger.

This was thanks to the variable message lane control signals along the expressway, combined with remotely operated emergency crossing in the central median. Khairun and his other police colleagues would only have to deal with one or two difficult vehicles, while the majority would be guided by the signals.

PJITS started to alter the VMS displays in sequence starting 5 km away from the incident scene. These displays instructed drivers on the opposite roadway to slow down gradually over the 5 km. As drivers approached the scene, they could see the signs saying 70 km/h; then the next sign would say 60 km/h, then 50 km/h, and so on. The signs also warned drivers to move out of the fast lane over the course of the 5 km.

Once the new flow pattern had been established, Khairun contacted his emergency dispatch center and requested permission to open the emergency crossover gates on the central reserve. The center checked with PJITS to confirm that the traffic pattern had stabilized and then issued Khairun with the security code necessary to open the gate. Subsequently, Khairun accessed the remote gate control from his terminal, and the gates opened on cue.

The rest of the operation was now routine. The officials diverted traffic through the gate and manually directed the one or two difficult vehicles. Consequently, the traffic started to flow again.

In the meantime, the firemen had started the clear up operation, and the crane had arrived to lift the tanker out of the way. Khairun noted that since all required resources seemed to have arrived in good time, maybe he would not have to work all night.

Sure enough, the scene was cleared within two hours, and even better, the traffic buildup had been effectively managed to minimize the network recovery time. The combination of temporary contraflow and area wide traffic control had kept the whole situation under control. Khairun cheerfully set about reversing the contraflow process and then went home to have his supper.

On another occasion, Khairun had cause to use the PJITS for a completely different reason. He had been informed by dispatch that a driver, having committed a series of traffic violations, was now the subject of a high-speed chase. He was directed to join the chase as the errant driver and vehicle passed his station. Once he had spotted the runaway and joined the chase, Khairun had engaged the intelligent vehicle system (IVS) fitted to his vehicle. This was the same system that was fitted to all patrol vehicles and enabled Khairun to give chase with a higher degree of safety. The system automatically slowed and redirected his patrol car if he was in danger of getting too close to the target vehicle or the other pursuing patrol cars. It maintained an *exclusion zone* around his patrol car by controlling steering, brakes, and accelerator. It also provided warning when there was an obstacle ahead and managed the interface to the traffic signal control system, enabling pre-emption to be granted to the chase convoy as it progressed through the region. Khairun had used the system to good effect, as it enabled him and his colleagues to append the errant driver with no casualties or damage to property.

### 6.2.8 Taxi Diver

Business had never been better for Sri. At first he had been rather cynical when the salesman told him that this new equipment would be good for business. The deal was that by buying the central equipment and equipping all of his taxis, he could tap into a brand new source of business—the PJITS. It appeared that the PJITS included major facilities for public transport management and that the system operators had included taxis under this definition. Consequently, every PJITS information point, whether at home, in the office, or at major public locations such as shopping centers, had a taxi reservation facility. In addition to all the real-time information about buses, LRT, and trains, the traveling public could also opt to hire a taxi.

What was even more interesting to Sri was the fact that people could use their new smart cards to place the call and pay for the taxi. This meant less work and a safer job for his taxi drivers as they did not have to handle so much cash.

More importantly, it meant that he could send his taxis to where they were really needed. The PJITS provided him with demand summaries, which tipped him off about major demand hot spots as well as providing him with the pre-reservation facility. He was beginning to see that taking part in the PJITS was going to be one of his better business decisions. The cost of installing the equipment and even the monthly subscriptions to the information service were more than covered by the increased revenue generated by the extra volume and increased utilization levels of his taxis.

Even his drivers were happier as they could now make their bonuses with less work and less cruising around looking for fares.

### 6.2.9 Tourists

Bruce and Sheila had been planning this trip for some time. It had always been Bruce's ambition to visit Malaysia and sample some real Asian culture. Sheila was also looking forward to the trip. She had heard that apart from the natural beauty of the country, the friendliness of the people and the quality of service made a trip really worthwhile.

After all, it was not such a big deal in the age of jet travel. Ten hours on the plane from their native Sydney would take them directly to KLIA and their dream holiday.

They had planned the trip meticulously, starting with a visit to their local travel agency, where they had collected brochures on hotels, air flight and other information. Sheila had been particularly interested in the brochure explaining the PJITS system, which was now fully operational in the Putrajaya region.

It had attracted her attention at first because it had puzzled her that there should be a brochure on what looked like traffic control in the office of her local travel agency. On closer inspection, she noted that it was describing much more than a traffic control system and that it could have direct relevance to their trip. It turned out that the PJITS could provide them with a range of services that would make their trip easier. She pointed this out to Bruce, but he was engrossed in the magazines describing the sights and culture of Malaysia.

When it was time to collect the tickets from the travel agent, Bruce was working late, trying to make up some time in advance of his holiday, so Sheila went to collect the tickets on her own. At the travel agency, she inquired about PJITS and the travel agent explained the range of services that could be provided for a small, all-inclusive subscription. The travel agent explained that the service included the issue of airline tickets in electronic form on a smart card and that the same smart card would be used to access the rest of the services. The agent then went on to give Sheila a short training course in the use of the system and entered some security information onto the card. They also made a separate card for Bruce.

Sheila explained all this to Bruce that evening, but he was still distracted by the Malaysia tourist materials and looked as if he really was not paying much attention. They had been married long enough for Sheila to have understood that this was the way Bruce was. He would not really get interested in the travel arrangements until the morning that they left. She would probably have to do most of the packing as well.

When the departure day finally arrived, sure enough Bruce started to take an interest in all the travel arrangements. He was rather surprised to realize that they were going to use the smart cards instead of airline tickets. "I knew he wasn't really listening when I explained this to him," thought Sheila. He was even more surprised when they got to the airport, and he discovered that they could "fastrack" on the check-in as they had the smart card. They were flying Malaysia Airlines (MAS), and as the airline was participating in PJITS, it had installed smart card readers at a few of the check-in stations. The airline was, of course, planning to use the smart card readers for other locally based services.

Ten minutes later, they were checked in and on their way to the airline lounge area. Sheila then sprang her next surprise by asking Bruce to check the flying time and route that their flight would take today. "C'mon Sheila, I would have to go back to the MAS desk to do that; we've just come from there!" he replied. Sheila then informed him—"yet again" she thought to herself—that his smart card would allow him to access the travel information terminal located in the lounge. He could get all the information he needed there, with no need to go back to the MAS desk.

Bruce, looking a little sheepish by this stage, proceeded to do this and was slowly navigating his way around the screen menus when Sheila appeared at his side. "How did the system know to ask my date of birth and mother's family name, and how did it know the correct answers?" he asked her. "That's simple," she replied, "the smart card holds security information that I provided to the travel agent. It's just checking that you really are who you say you are."

Sheila noted that Bruce was plodding his way through the menus and making a big job out of it. "Let me show you how to do it," she said, pushing him gently to one side and taking charge of the terminal. Bruce was completely amazed at the speed at which she found her way around the menu-driven system and requested the desired information. At that point, she had to confess that this was not all natural talent, since the travel agent had given her a training session and had set up both their smart cards with information that told the system to display information in her preferred format. "No wonder she's so familiar with the system—she had specified the data format herself," Bruce thought, making a mental note to keep a closer eye on Sheila in future; she was getting ahead of him.

They boarded the flight, and it passed without incident as all the best flights do. Shortly before arrival, Sheila used the in-flight information terminal in the seat in front of her to check the latest arrival time estimate and make a reservation on the high-speed rail link from KLIA to Putrajaya. The system confirmed the reservation, and as she had already specified their hotel as the final destination on today's journey, it asked if she would like to reserve a taxi from the railway station. She did not know enough about where the railway station was in relation to the hotel to make that decision, so she asked the system to display a map of Putrjaya highlighting both railway station and hotel. This showed her immediately that it would be a 25-minute walk to the hotel. This would be a nice walk under some circumstances but not when arriving in an unfamiliar city with a lot of luggage. She went back to the earlier screen and accepted the offer of the taxi. The system told her that taxi number 42 would be waiting for them at the railway station.

The flight touched down approximately 45 minutes late due to head winds encountered en route. Bruce was a little concerned about this as he was convinced that they would miss their train and taxi connections. However, the train service operated at 30-minute intervals, so they just caught a later one. He was surprised that the system had automatically updated their reservations to the later train and had even allocated seats for them.

He was even more surprised to find that taxi number 42 was waiting for them at the railway station. They did not know that the original taxi number 42 had been redispatched to another call when the taxi dispatcher had received information from the PJITS about the flight delay. This taxi 42 had simply changed its VMS and turned up a few minutes ago.

They were just about to arrive at the hotel when Bruce then got that sinking feeling. "Oh no!, we didn't change any money. I don't have any Ringitts and I'm sure this taxi driver will be really annoyed if we try to use dollars or, worse still, a credit card!" As they pulled up to the hotel, Sheila said, "Relax Bruce, I'll take care of this," and handed the driver her smart card. "Thank you, Madam," the driver replied as he swiped the card over his smart card reader.

Bruce had learned not to be surprised anymore. He had also learned to pay more attention when Sheila was trying to explain something to him.

Once inside the hotel, Bruce decided that it was his turn to show Sheila that he knew a thing or two about travel information. While Sheila was in the shower, he quickly accessed the in-room information system using that smart card again. Finding his way around the menus, he quickly found details of a nice Malaysian restaurant, and even better, he managed to make a reservation and book a taxi. Sheila was very impressed.

During the rest of their stay, apart from fantastic tourist sights and rich culture, they were also impressed by the range of transportation and "yellow pages" information available to them at many locations in and around Putrjaya. For a few days they even rented a small portable information terminal that they used to guide them around and give them background information on sights and landmarks.

Sheila, of course, managed to make use of another feature of the smart card during her stay. She had quickly realized that many of the larger local shops would accept payment for goods and services using the smart card. She had made good use of this as the currency conversion was done automatically and on favorable terms. Her use of this facility had been made easier as the hotel had provided specially modified telephones in all their rooms. This enabled Sheila to place the smart card on the telephone and dial in to her bank in Sydney. She could then transfer funds from her account to the smart card. Sheila thought this was great: "It's like having an ATM in your hotel room."

When they finally returned to Australia after the trip, Bruce had to admit that he was just as impressed with the PJITS system as he had been with the sights and cultural diversity of Malaysia. "High praise indeed," thought Sheila.

## 6.3 Essential Elements of an ITS Vision

Continuing with the theme of the vision for ITS, this section summarizes the key elements that should be contained in a successful ITS vision. First, it is necessary to use local names to identify the key stakeholders whose perspectives are going to be described. It is also important to have a preconceived idea of all the stakeholders and what their perspective might be. For example, the taxi driver, bus driver, commuter, traffic engineer, and other stakeholders would have to be configured and tailored to the local context. Second, it is important to describe the benefits of the ITS and the various features and functions to each user group from each perspective by just describing in everyday terms how their lives would be affected by the ITS. Taking the Putrajaya vision as an example, you can see that the careful use of humor and the careful use of complementary material on the local context helps to bring the thing to life. It also helps to make the rather technical content more acceptable and more palatable to the local client.

Another important element is the incorporation of legacy systems descriptions into the vision. For example, systems that exist already or infrastructure that may be already on the ground or planned should be described as part of the overall context when describing how the new system might operate and effect the lives of people in the area. A particularly important element of the ITS vision is the style in which the document is written—that is, actually writing the vision as if it is through the eyes of the user. Using the first person and using interactive dialog provides a more easily understandable way of providing the information. The use of local names is also pretty important when describing the perspectives of the various stakeholders. One nice little tip in this respect is to make use of names of the various client groups and stakeholders as you know those are definitely local names, and it is also quite entertaining to clients to see their names used in a document like that.

When describing the impacts and benefits to the stakeholder group, it is very important, if possible, to mention tangible benefits in terms of time savings, savings in money, or increased business opportunities that would be available as a result of the implementation of the ITS. It is also important to write the vision in such a way that each stakeholder group starts to comprehend, understand, and identify its respective roles and responsibilities in the ITS Future Big Picture. The groups can see that there is a place for them in the future, understand what they may have to do to support the ITS, and start to make plans and align themselves with what the future might be. In addition, of course, having seen a version of someone's idea of the future, they can then be stimulated to come up with their own ideas and shape their own input. They can crystallize their own ideas about things they might like to see changed in the proposals to better reflect the way they would like to see the future.

Another feature of this particular example of the vision is the focus from the perspective of the user. The document is not structured in terms of technology groups or application groups but in fact is focused as a sequence of events that link all the various applications and technologies into a seamless overall system. The focus is on how that system interacts with, and provides benefits and services to, the user group.

# 7

# ITS Technology Selection and Review

## 7.1 Introduction

This chapter addresses the activities encompassed in steps 700 to 750 of the ITS Cooperative Development Methodology as indicated in Figure 7.1. In addition, it aims to provide system developers with some concise guidance on how to go about the activities involved in this part of the development process. We also want to build on the introduction to ITS technologies provided in Chapter 2 by talking a bit more about ITS technologies, but this time in terms of technology bundles or ITS Market Packages. We address this by describing a few of the most popular ITS Market Packages.

Our final goal in this chapter is to remind the system developer of the overall objective—needs-driven technology selection. By mapping identified ITS User Services to ITS Market Packages, we can support the cooperative development of the system through the "what?/how?" cycle.

Furthermore, this chapter provides some guidance on how to go about the steps collected under the box entitled "ITS Technology Review and Selection Report" in Figure 4.2. All the steps lead to the development of this report, which in turn support the "what?/how?" cycle by providing the input required to explain the proposed solutions to the users. This is not intended to be a recipe that you follow rigidly until you achieve the desired results. Rather, the information provided here should be considered as illustrative and used as a model for your own approach. For example, the entire box is divided into a number of steps, each with its own deliverable or report, reflecting the way we might approach it. You could shuffle the steps around into different report combinations depending on the needs of the customer.

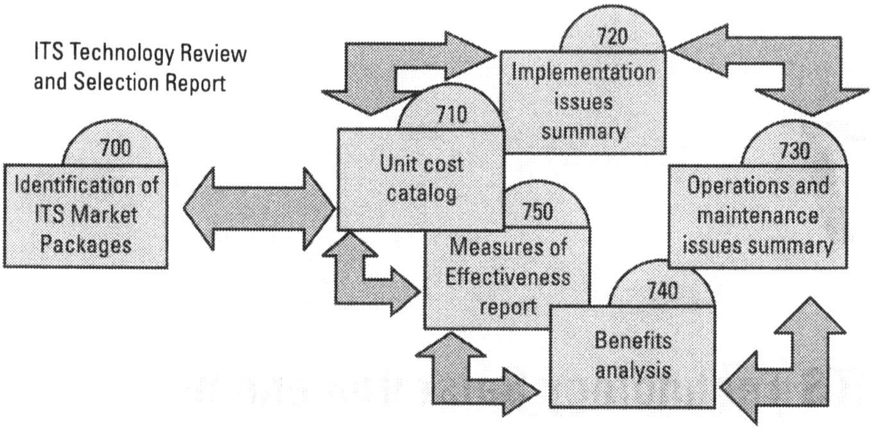


**Figure 7.1** ITS Cooperative Development Methodology steps 700 to 750.

There is a great deal of relevant information for this series of steps in the National Architecture for ITS, particularly in the "Implementation Strategy" document. In this document, the National Architecture for ITS is characterized using ITS Market Packages to provide a deployment view of the architecture or ITS Future Big Picture. A total of 53 ITS Market Packages are defined and described in this document, and while they are constrained to technology bundles relevant to the National Architecture for ITS, these ITS Market Packages could form a good starting point for your ITS Market Package definition. In fact, if your ITS planning and development initiative is in the United States, it is highly recommended that you start with these ITS Market Packages, since federal funding for your eventual implementation may depend on your conformance with or compatibility to the National Architecture for ITS. Naturally, you will expand and enhance the ITS Market Package list in the course of your development and planning activities, especially if you have identified and defined ITS User Services that are not part of the National Architecture for ITS User Service list.

Before we go any further, let us remind you what an ITS Market Package is. In Chapter 2, we defined an ITS Market Packages as follows:

> A group of technologies that have been *bundled* together to meet a defined need within a market area.

They represent the groupings of technologies that comprise the products and services available in the market. ITS Market Packages are, from a system planning perspective, the basic building blocks of ITS Future Big Pictures.

They provide the raw material to be used to explain the "how?" to the stakeholders in support of the "what?/how?" cycle. Note that ITS Market Package descriptions are technology-specific, as they describe the types of enabling technologies that are bundled together to address the need or provide the service. They are not technology-dependent because they do not specify the particular design variants or brands of technology. For example, an ITS Market Package might specify a traffic signal controller, but it would not define the brand of controller or the detailed functionality. This level of detail is deliberately left until later in the design process to keep the "what?/how?" cycle as light as possible.

As we move from system planning to detailed system design (between steps 260 and 400 on the ITS Cooperative Development Methodology), ITS Market Packages are reconfigured and transformed into ITS Equipment Packages. These are technology-specific and embody detailed design choices and technology tradeoffs.

In Chapter 2 we introduced you to the world of ITS by describing some real-world applications and a range of ITS enabling technologies. We promised that we would come back to the subject of ITS enabling technologies in this chapter.

This time we are going to address ITS enabling technologies as bundles and describe them in terms of ITS Market Packages. Remember we said that ITS enabling technologies rarely exist in the stakeholder marketplace on their own. They have to be bundled together to provide a service and have a measurable benefit, so they would be hard to sell individually. We have to come clean here and say that this is not strictly true! Although ITS enabling technologies do not do anything for stakeholders, there is actually a system integrator market in ITS enabling technologies. Many vendors specialize in the manufacture and supply of equipment that is destined to be incorporated, by another company, into an ITS Market Package. Their target market is the system integrator and developer rather than the stakeholder in an ITS development and planning initiative. It is therefore the developer's responsibility to have a current awareness of these enabling technologies and translate and combine them into ITS Market Packages for the stakeholders.

At this point in the system development process the developer or integrator has a primary role as the "bridge" between what is required and how it can be achieved. Considerable skill is required to translate the highly technical and detailed descriptions of technologies, products, and services to meaningful ITS Market Package descriptions. The designer may well get into the position where it is necessary to cross the boundary between the ITS Future Big Picture and design (ITS Market Packages to ITS Equipment Packages), in order to effectively communicate the technology features and capabilities to the stakeholder group. For example, when describing ITS Market Packages involving the use of traffic sensor technology, it may help to show an example of a specific sensor type. This might imply that the design choice or sensor type has already been made and ITS Equipment Packages have been identified and defined. We believe that it is worth running this risk for the sake of effective communications. The situation can be managed by explaining the fact that the illustration is serving purely as an example.

The developer will also act as a bridge in terms of "retro-system engineering" previous designs and deployments back to higher level ITS Market Package descriptions. We know that this is going to seem strange: Why remove detail from designs already in place? Isn't this re-inventing the wheel?

Well, if we were communicating with a stakeholder group full of system engineers and technologists, then perhaps it would be. However, we are trying to communicate technology capabilities to a mixed audience. Our primary goal is to provide just enough information to make broad choices that can be clearly linked back to the needs, issues, problems, and objectives captured as ITS User Services. We are trying to avoid quantum leaps in the development process, straight to final design solutions, until we fully understand the real requirements. Therefore, we need to use the designs and implementations around us as illustrations but get to the underlying "usefulness" of these designs rather than adopt them wholesale at this point. We need to get to the fundamental solution that will satisfy the requirements but avoid branching off into a specific design variant until we all know enough to be sure that it is the most appropriate choice.

Consider the following illustration. Suppose that the requirements for the project point to the need for a global communications capability. The stakeholders have identified that they need two-way voice and data communications with any of their staff at any location in the world. If we leap straight to a design solution, we define the need for a cellular telephone system that will work anywhere on the globe. This is going to be a technical challenge and could be expensive. Incidentally, Motorola is launching just such a service, called Iridium. Having leapt straight to this design solution, we did not explore alternatives that might also satisfy the requirements. If we had stayed technology-dependent but not technology-specific, it would have left the way open to identify a design alternative.

In this case, the design alternative would be to make use of multiple cellular telephone systems around the world. These would be integrated, or made to operate seamlessly, through the use of a smart card interface. All staff members would be issued a smart card that contained their account information and favorite telephone numbers. This would be inserted in a cellular telephone rented in the particular country visited. Hence, the global communications capability would be achieved through a different combination of technologies. It all boils down to keeping your options open as long as possible, while supporting a convergence on the agreed solution.

At the moment, the ITS Market Package information situation is messy. It should get better as the methodology is adopted, and the market responds with clearer information, projecting products and technologies as ITS Market Packages. Currently, a lot of effort is required to translate vendor information into ITS Market Packages.

So what kind of ITS enabling technologies bundles or ITS Market Packages are we seeing in today's market? Well, it would take an additional book to provide you with a comprehensive picture of the entire spectrum of ITS Market Packages and technologies available today. This would also detract from our main purpose of providing you with the ideas and concepts, rather than a recipe book. Accordingly, we have focused on a small number of relatively popular ITS Market Packages.

The ITS Market Packages described in Section 7.2 are fairly popular at the moment. Note that many people do not consider these to be ITS Market Packages; they consider them to be deployments, applications, or designs. However, you will note from the way we describe them that further design work, including technology tradeoffs and selection, would have to be carried out before any of the ITS Market Packages could be specified, procured, and deployed. We run the risk of confusing the issue here by talking about popular ITS Market Packages as a way of bringing ITS Market Packages to life with some examples you can relate to. What we are actually describing is the ITS Market Package equivalent of some detailed designs that are currently being deployed. Note also that our ITS Market Package examples do not correspond with those described in the National Architecture for ITS documentation.

## 7.2 Review of Popular ITS Market Packages

Sections 7.2.1–7.2.4 describe popular ITS Market Packages.

### 7.2.1 Freeway Management

These Market Packages bundle the following group of enabling technologies:

- *Sensors:* Traffic sensors to sense current traffic speeds, volumes, and spacing.

- *Surveillance:* CCTV cameras to provide images of current traffic conditions. These video images are very important in supporting traffic center operators to make decisions during incidents. (The cameras could also be regarded as a form of traffic sensor.)
- *Communications:* From the traffic sensors to control center and from control center to roadside driver information displays.
- *Control center:* Where data is collected, processed into information, and presented to the traffic or transportation manager for decision support.
- *Driver information displays:* Roadside VMS carrying informational messages.
- *Traffic control devices:* Ramp metering traffic signals, roadside VMS carrying mandatory driver instructions, moveable barriers to control traffic flow.

This ITS Market Package, with the addition of emergency vehicle dispatching elements and appropriate institutional/organizational procedures, could also provide incident management capabilities. Any unusual event on the road network that has the potential to affect traffic conditions is referred to as an incident. Incidents are unexpected events that cause nonrecurring traffic congestion as opposed to the recurring traffic congestion that happens every day because of higher demand for use of the road network than the network is designed to accommodate.

This package builds on the freeway management package by providing additional features and services to support the incident management process. Typically the incident management process has the following steps:

- *Detection of the incident:* Sensing that something unusual is happening on the road network;
- *Verification of the incident:* Confirmation that a real incident has occurred that requires a response from the traffic manager;
- *Response to the incident:* Deciding on the appropriate resources to deal with the incident and dispatching those resources efficiently;
- *Clearance of the incident:* This is the physical process of clearing any disabled vehicles involved in the incident and any associated debris from the road network, then restoring normal capacity.
- *Traffic management* in the zone of influence of the incident.

During and after the incident, traffic management is required within the zone of influence of the incident. This helps traffic to adapt to the conditions prevailing during the incident and return to normal conditions as quickly as possible after the incident has been cleared. Note that this post-incident period often has the largest effect on traffic as the network may take a considerable time to recover.

Incident management would be supported through the addition of enhanced communications links from the control center to other agencies responsible for incident and emergency management such as hazardous materials agencies, fire, police, and ambulance. Expanded freeway and traffic management features may also be added to facilitate traffic management during and after the incident.

### 7.2.2 Urban Traffic Control

Urban traffic control (UTC) constitutes the use of traffic signals to control the flow of traffic on urban surface streets and arterials. This category has similar elements to freeway management but utilizes the familiar red, amber, and green traffic signal aspects to provide time-sharing for traffic at road intersections.

There are a few variants available for this ITS Market Package. The simplest one is independent standalone traffic signals controlled by an on-street traffic signal controller. Timings for the signals are based on historical traffic counts and remain constant until the signal controller is reprogrammed at periodic intervals.

The next variant employs traffic sensors on each approach to the intersection under signal control. The sensors detect the presence of vehicles and the length of the queue on each approach, and the local signal controller adjusts the timing to suit the prevailing traffic conditions. This is known as vehicle-actuated or adaptive signal control.

Another variant links each of the local controllers to a central control room, where the timings at each intersection are coordinated to allow traffic to flow smoothly on an area-wide basis. These are known as UTC systems or area traffic control (ATC) systems.

### 7.2.3 Electronic Payment for Toll Collection

The design equivalents of this ITS Market Package have been a very popular application in the United States and Europe over the past five to ten years. The basic concept behind this ITS Market Package is the automatic identification of suitably equipped vehicles as they pass particular points along the highway.

The vehicle identity is matched to an account held by the driver, and money is deducted corresponding to the appropriate toll due for the use of the highway facility.

This ITS Market Package consists of the following components:

- *In-vehicle unit:* Some form of transponder or communications device that supports dedicated short-range communications (DSRC) between the vehicle and the roadside reader infrastructure;
- *Roadside reader equipment:* Equipment installed at the roadside with capability of communicating with the unit onboard the vehicle;
- *Communications network:* Connecting the roadside reader equipment with the back office system located in a central facility;
- *Back-office accounting system:* Manages the necessary accounting and audit processes from a central facility, ensuring that the correct users are charged the correct fee; includes account establishment and maintenance and billing activities;
- *Enforcement system:* Some method of detecting violations and enforcing the prevailing regulations. This is often accomplished at the design solution level through the use of video or still cameras to record an image of the violating vehicle and the license plate. The license number is then linked to the appropriate vehicle owner through access to the local department of motor vehicles database.

### 7.2.4 Autonomous Vehicle Navigation

This is a very technical name for something you may have seen. The design equivalents of this ITS Market Package are in rental cars already. Hertz calls it NeverLost, and Avis calls it the Avis Satellite Navigation System. Both systems, or ITS Market Packages, have similar technology bundles. Inside the vehicle, there is a driver display and input unit. In the trunk, there is a computer, a CD-ROM drive, and a vehicle location system. For the design deployed by Hertz and Avis, the vehicle location system uses a combination of dead reckoning, map matching, and GPS. Dead reckoning is a well-established technique that makes use of a compass and an odometer (distance-traveled measuring device) to determine the vehicle's current location. Assuming that the vehicle departed from a point of known location, bearings and distances traveled can be used to calculate current position.

Map matching makes use of sophisticated algorithms to continually compare maneuvers made by the vehicle with patterns on a digital map database.

For example, if the onboard compass and wheel sensors detect that the vehicle has made a left turn, traveled 500 meters, made a right turn, and traveled 1 kilometer and has just completed another right turn, the algorithm would search the digital map database for the exact location on the road at which the vehicle would be, having just completed the series of maneuvers. Having found the corresponding location on the network, the software would correct the vehicle location determined by the dead-reckoning technique to this more accurate estimate of the vehicle's current location.

There are many other ITS Market Packages in use around the world. We hope that this small selection provides the basis for understanding the concept. You have probably already figured out that many of these packages have overlaps. Several of them make use of a communications link from the vehicle to the control center, for example. This is known as ITS Market Package synergy and has to be taken into account when determining the technically optimum deployment sequence. The existence of this synergy is one of the primary reasons for developing an ITS Future Big Picture in the first place. Taking an integrated approach to the planning and development of ITS enables the synergy to be exploited, money to be saved, and effectiveness to be increased.

## 7.3 What ITS Market Packages Might We See in Tomorrow's Market?

We have included this section to remind you that ITS Market Packages are the fundamental building blocks of the ITS Future Big Picture. Therefore, it is desirable that you identify ITS Market Packages for the future as well as for short-term deployment. We have found that it is a good idea not to constrain your thinking to near-term deployable technologies only. To satisfy the ultimate requirements over a 20-year time frame, for example, you cannot avoid including future building blocks that require further development, testing, or maturing.

Sections 7.3.1–7.3.3 present ITS Market Packages that we believe will be feasible in the next five years or so.

### 7.3.1 Automatic Lane-Keeping

Automatic lane-keeping technology consists of equipment on the vehicle, the roadside, or both that enable the vehicle to sense the appropriate transverse position on the roadway and actuate the vehicle steering system to comply with this position.

### 7.3.2 Adaptive Cruise Control

This technology consists of equipment on the vehicle, the roadside, or both that enable the vehicle to sense the appropriate distance from the vehicle in front and actuate the vehicle accelerator transmission and braking system to comply with this position.

### 7.3.3 In-Vehicle Signing

Instead of all those large variable message signs at the side of the road, we believe that in the medium term, an ITS Market Package will be available to support in-vehicle signing. This will consist of an in-vehicle display unit, possibly integrated into the dashboard, and a short-range communications link from the roadside signs and beacons to the vehicle. The roadside signs would transmit their information for display inside the vehicle enabling a view of the sign to persist and be available to the driver long before the desired exit and long after the sign is little more than a dot in the rear view mirror. It might also be possible to extend this ITS Market Package to support in-vehicle VMS, eliminating the need for roadside VMS.

Now that we have set the scene, let's take a look at the part of the ITS Cooperative Development Methodology that encompasses the review of ITS technologies and the identification and selection of ITS Market Packages. This is addressed in steps 700 to 750 in the diagram as illustrated in Figure 7.2.

## 7.4 Step 700: Identification of ITS Market Packages

This is where the ITS Market Packages are defined, based on a thorough technology review that provides a status report on what technologies and products are available and what their capabilities are. ITS Market Packages are the basic building blocks of the ITS Future Big Picture, so we need to know what blocks are available and assess what they can do and when they might be available.

So where do ITS Market Packages come from? There are two primary sources. The first is the National Architecture for ITS and the ITS Market Packages that have been defined by the National Architecture for ITS development teams. The second is the system developer or ITS technology specialist, who synthesizes new ITS Market Packages based on knowledge of technologies and how they fit together to provide a service or satisfy a need.

While the National Architecture for ITS provides an excellent source for ITS Market Packages, we encourage you to develop your own set, perhaps taking the National Architecture for ITS Market Packages as a starting point but

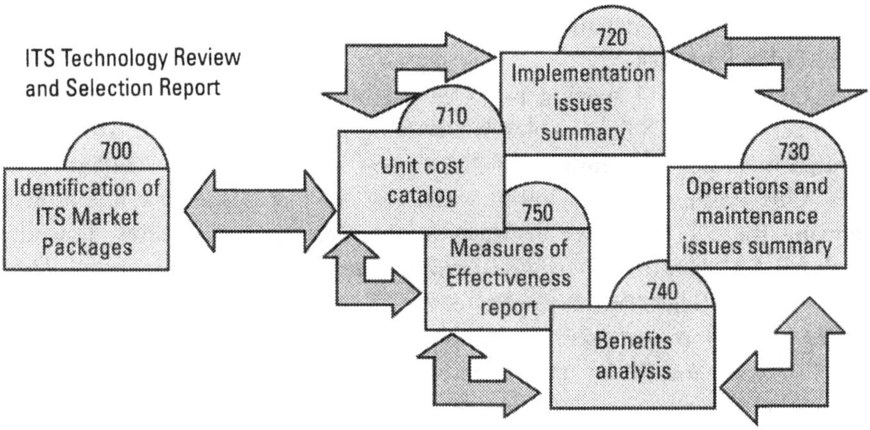


**Figure 7.2** ITS Cooperative Development Methodology steps 700 to 750.

building on them from your own knowledge of the current ITS market. This requires that you either have a good state-of-the-art knowledge of the ITS technologies available in the market, or you have to carry out a technology review.

The first job involved in carrying out the technology review and selection is to find some ITS Market Packages. When you go about this task, you have to bear in mind that the ITS equipment supply industry is still relatively immature. In many cases you will identify pieces of a ITS Market Package rather than complete ones. In an ideal mature situation, the various vendors and suppliers would have a very good picture of user needs in the market and what ITS Market Packages would prevail. This is because their understanding of the market would lead them to develop comprehensive bundles of technologies that provide a service to a well-defined sector of the market or a specific group of users to satisfy clearly defined ITS User Services.

Most vendors at this stage of the market have either insufficient exposure to the users to get a real picture of needs or believe that the user and customers have not yet homed in on what they really need and want. Consequently, the momentum is currently behind the provision of a fragmented range of smaller units, either partial ITS Market Packages or Equipment Packages.

Note that in most markets suppliers and developers pushing technology at the customers drive the early stages. Later markets seem to be driven by customer-stated needs, which tend to drive the products and services that are available.

So what are we trying to say here? We are preparing you for the fact that there may not be too many ITS Market Packages available off the shelf. The consultant or developer may have to assemble smaller acquisition units into ITS Market Packages.

One very good starting point would be to make use of the National Architecture for ITS deliverables as a starting point. In particular, the implementation strategy document provides a good initial list of ITS Market Packages. One point to bear in mind when you look at this is that the aim of showing these ITS Market Packages in the document is to show how the National Architecture can be characterized in terms of deployment units. They start with the architecture and split it into ITS Market Packages. If you are proposing to develop your own architecture then you will start with the ITS Market Packages and build the architecture. Either way is perfectly valid. Remember that these are all tools to be used in any way that gives you the best advantage.

You also have at least a preliminary idea about what the stakeholders need and want based on your initial attempt at the needs model. Accordingly, another good exercise is to scan the market for products and services that might fit the needs exhibited. Even if you decide to use the National Architecture for ITS deliverables as your starting point, it would be a good idea to carry out a review of current ITS technologies. This is because the ITS market is very dynamic, and new ITS Market Packages may have emerged since the architecture deliverables were published. That is another reason why we have sidestepped the issue about providing a comprehensive catalog of ITS Market Packages in this book.

No matter what approach you take to the identification, description, and definition of ITS Market Packages for your development and planning initiative, bear in mind, at all times, the main objective in putting ITS Market Packages together in the first place. What you are trying to do is to enlighten the stakeholder group as effectively as possible about what technology possibilities and capabilities are out in the current and future marketplace. This is to facilitate the confirmation of their needs, issues, problems, and objectives, enabling them to evolve their requirements with your cooperation as they learn more about possible "hows." We believe that this will be very frustrating to system developers and technology specialists who feel that they have a strong grasp of the appropriate technologies to be deployed for a particular purpose. However, we need to remember that we are not attempting to design or procure at this point in the development cycle. In fact, we are trying to avoid detailed design and the activities associated with procurement and deployment at this stage in the planning and development of the system. Remember that when we first talked about the "what?/how?" cycle we mentioned that the key to success in supporting the cycle is not to invest large resources or commitment to a particular approach while the cycle is in live-action mode. If we invested time and money in design and procurement at this point, we would violate that principle and get far too rigid too early.

ITS Market Packages turn out to be a delicate balancing act for developers and technologists. On one hand, you need to provide enough detail to explain to the stakeholders how you plan to address their stated requirements. On the other hand, you have to resist the temptation to design the solution at this point and keep the loop light on resources. The resource expenditure and commitment to a single solution have to be managed to enable the developer to accommodate and even encourage the stakeholder group to evolve their initial requirements and take ownership of the eventual solution.

Developers and technical specialists may well find themselves in the situation where they know a lot more about the details of the technology bundle or ITS Market Packages than is explained in the ITS Market Package description. You will probably know more about the design variants that are possible and the practical aspects of implementing the technologies. However, we have found it much more fruitful to resist the temptation to lead the stakeholder group blindly down a particular solution path simply because we believe that it is the best one. We have tried this, and it has always come back to bite us later when the design approach does not exactly match the evolved stakeholder requirements and they have not built full ownership in the chosen solution.

ITS Market Packages are a means to an end. They are a tool to be utilized to help you to support the "what?/how?" cycle. While they are only as good as the results they help you to achieve, we have to confess that we like them a lot as they have helped us carry out the cooperative development approach in many contexts.

## 7.5 Step 710: Unit Cost Catalog

This is the other half of the cost-benefit pair. In this step, unit costs for each ITS Market Package are defined and quantified. This requires that a thorough review of current market prices for ITS technologies and services is carried out and summarized in the catalog. The unit cost summary is important as it provides the cost elements required to carry out the cost-benefit analysis used to justify the proposed implementation and help in the selection of appropriate ITS Market Packages.

In addition to identifying and describing the ITS Market Package, it is essential to estimate the cost of the ITS Market Package. We have used the term estimate, since it will not be possible to determine an exact cost. Further detailed design tradeoffs and choices have to be made before exact costs and benefits can be calculated.

Sections 7.5.1–7.5.7 describe the elements that should be included in the estimate of the unit cost for the ITS Market Package.

### 7.5.1 Capital

How much does it cost to initially purchase the ITS Market Package? This includes some investigation into the relationship between volume and cost. It could be very important, if you were buying something like an automatic vehicle location (AVL) system, to understand the kind of discounts that may be available if you decide to equip 100,000 vehicles in a deployment stage rather than 1,000, or to equip 100 miles of road with a freeway management system instead of 10 miles. These will provide important inputs into the deployment sequencing or implementation strategy work to be carried out later.

### 7.5.2 Customization

In addition to the basic cost of buying the equipment, how much more money will be required to adapt or customize it to the exact needs defined in the needs model? It is important at this point to keep value for money in mind on behalf of the users and stakeholders. Often a fairly minor change in the needs model will enable the customization costs to be minimized or even zeroed. This provides a vital input to the ongoing needs exploration activities providing good material to support the users' understanding of the relationship between what they are asking for and what it will cost. We have participated in many planning and development projects where the needs and the costs have somehow become separated resulting in either very expensive implementations or disappointed stakeholders.

### 7.5.3 Implementation and Setup

How much will it cost to install, test, and set up the ITS Market Package as well as buy it? For example, for some types of in-vehicle equipment, it will cost a significant amount per vehicle to retrofit the equipment into the vehicle. For on-road equipment, factors such as the duration of installation may be important, as the road may have to be closed while the equipment is installed. This aspect of the ITS Market Package is closely related to the maturity of the technologies incorporated into the ITS Market Package. Newer technologies may provide better cost-to-performance ratios but have a higher implementation and setup cost or a longer installation period as installers may be less familiar with the equipment.

### 7.5.4 Operations and Maintenance

How much has it cost to operate and maintain in other implementations, or how much is the predicted estimate? For example, if you decide to purchase a surveillance system ITS Market Package utilizing leased communication lines, then the cost of operating the communications will be rather significant. Similarly, if you choose to install cheaper equipment with a shorter operational life cycle it could turn out to be more expensive over the whole life of the system. This raises another interesting point: How many users have a clear idea of how long they plan their system to last? We suspect that many do not even ask this question and do not look beyond the initial inauguration of the system. We believe that whole-life costing of ITS implementations should be used as the basis for selection of approaches.

A good example here would be the comparison of inductive loop-based traffic detection for travel times versus the use of toll tags and probe car data. While the loop-based system might provide the lowest capital cost, the tag system could provide the lower whole-life cost, since the need to maintain inductive loops on the pavement surface has been avoided.

The question about the desired lifetime of the system needs to be answered as part of requirements analysis. This raises another interesting point. As ITS technologies become established in the market, then unit costs will inevitably fall. The emergence of new technologies seems to be taking place at an ever-faster rate, providing the distinct possibility that a better or more efficient approach may be available. Perhaps the combination of these two factors mitigates toward the adoption of shorter term procurement strategies, incorporating the use of cheaper technology options, with shorter payback periods.

On the other hand, development, design, and deployment costs are a significant component of any ITS deployment, so it may be better to take a longer term approach, based on longer payback periods, enabling the costs of design and deployment to be amortized over a longer period.

The bottom line to all this is that the decision on the strategy to be adopted should be conscious and not by default. The various options should be fully explored by the developer and the stakeholder group, leading to a deliberate choice of procurement strategy.

### 7.5.5 Current and Future Year Projections

As the particular ITS Market Package may not be implemented in the first phase of deployment, it is useful to have an assessment of the likely future costs of the ITS Market Package. Of course, this is only going to be approximate, but it may uncover potential for a substantial increase or decrease over time, thus influencing the proposed deployment sequence. This is particularly important for potentially high-volume items such as toll tags and other in-vehicle units. As witnessed in consumer electronic markets, higher sales volumes can have a significant impact on unit costs.

### 7.5.6 Integration Costs

Under this heading, you have to address the following issue:

> Has this ITS Market Package already been deployed with your other proposed ITS Market Packages, or will your deployment involve substantial new effort in developing interfaces and system integration?

This issue is closely related to the current state of ITS standards development. If a standard has already been developed supporting the interfaces between the ITS Market Package and others, then there is no problem. However, many interfaces have no standard available as yet, so it is necessary to look carefully at how this should be addressed. When reviewing the possibilities, you should look at the situation today and tomorrow. With regard to today, there may well be de facto standards that will support the integration of the ITS Market Package. However, it is important to ensure that the de facto standard will also provide support for migration to the ITS standard under development.

### 7.5.7 Step 720: Implementation Issues Summary

This is a practical assessment of the issues associated with implementing each of the ITS Market Packages. It is based on previous direct experience and on interviews with other implementing agencies and vendors. It is important to evaluate the practical constraints associated with the implementation of each ITS Market Package. Given the early state of development of some ITS technologies and products, this is likely to be an area where major risks lie. It is important to note here that as ITS Market Packages do not have the design detail required to implement or design ITS, it may only be possible to get a rough idea of implementation issues. A more detailed evaluation would have to be conducted as part of the detailed design process. At this stage, the most important point is to frame the questions and issues rather than obtain detailed answers.

### 7.5.7.1 Implementation Issues

Some implementation issues that could be considered here include the following.

- Is there an available contractor who can install it successfully?
- What is the degree of installation difficulty?
- How many lanes will have to be closed down and for how long?
- Are reliable installation duration estimates available?
- How quickly can it be made to work?
- Will the traveling public see a long installation effort and get fed up with the delays negating the positive impacts when they finally arrive?
- Has it worked before in the field?
- How much does it cost to install?
- How quickly can it be implemented—are there significant constraints such as fixed installation resource capacity or production capabilities?
- Is the ITS Market Package designed to conform to recognized standards or is it a standalone ITS Market Package? How flexible is the ITS Market Package in accommodating emerging and future standards?

This activity may draw on the standards review work carried out in step 240 of the ITS Cooperative Development Methodology.

## 7.6 Step 730: Operations and Maintenance Issues Summary

There is a separate operation and maintenance report created later in the process. This deals with how the proposed ITS will be run and maintained after implementation. However, operations and maintenance issues should be considered as an intrinsic part of the development and planning process, with technologies and product selection taking account of these issues. Therefore, it is important to identify and evaluate the operations and maintenance issues associated with each ITS Market Package before the ITS Future Big Picture is defined. The operations and maintenance summary supports this by providing a list of issues with evaluation as input to ITS Market Package selection. Issues addressed should include the following:

- Operation and maintenance costs;
- Staff availability to operate the proposed ITS Market Package;
- Ability of existing staff to operate and maintain the ITS Market Package;
- Staff training required;
- Whether there is a need for the user agencies to recruit additional staff;
- Ability of the new staff to fit in the existing culture—or need for cultural change to accommodate the ITS Market Package;
- Suitability of balance between capital cost and operating cost to organizational requirements;

Although whole-life costing is routinely applied in the transportation community, it seems to us that the guidance suggested by such analyses is often routinely and deliberately ignored. It is sometimes expedient to opt for the solution that provides the lowest initial investment but not the lowest whole-life cost. This enables the "I've started so I'll have to finish" effect to operate, locking in future investment. It can also be used as a way to get around initial resistance to implementation.

Another aspect to consider as part of the operations and maintenance analysis is the intended operational life of the ITS Market Package. It may be the case that you expect the ITS Market Package to be superseded or significantly modified within a short time frame. If this were the case, then the whole-life cost analysis would be skewed in favor of lower initial investment and lower longevity. To give you a specific example, say you choose to deliver en route driver information by way of roadside VMS, and you have a choice of cheaper shorter life solutions and more expensive longer life solutions. A preliminary look at whole-life costing may show that the more expensive longer life approach would provide better value for money. However, this assumes that you will keep using the VMS over their full design life. If your sweep of the ITS technology market shows that in-vehicle information systems will penetrate the local market within the design life of the VMS, you would have to assume that the VMS would become redundant before the end of the design life.

## 7.7 Step 740: Benefits Analysis

In this step, the combined activities required to carry out an assessment of the likely benefits of the proposed ITS are summarized. Activities required to develop the benefits assessment include the following:

- Quantification of Measures of Effectiveness;
- Mathematical simulation modeling of the proposed implementation;
- Business modeling.

This is very important as it generates the material and documentation that will form the primary justification for implementation.

We have already stated a few times in the book that ITS Market Packages build into ITS Future Big Pictures and that ITS Future Big Pictures do not get implemented, designs do. So how can you possibly figure out benefits for something that does not get implemented?

Well, it is really all about balance. On one hand, we do not want to invest the resources required to carry out detailed design at this stage in the "what?/how?" cycle. On the other hand, we need sufficient detail to make choices between ITS Market Package options.

An important way to evaluate the potential benefits of the proposed implementation is to demonstrate that the original needs, problems, issues, and objectives of the key stakeholders can be addressed by the proposed solution. This is carried out by mapping the proposed ITS Market Packages back to the ITS User Services defined earlier in the development process.

The first job is to ensure that all ITS User Services are addressed by at least one ITS Market Package. Then it gets more difficult, as you have to decide what ITS Market Package works best in addressing each ITS User Service. We do not have a recipe for success in this task but would recommend that you apply some form of rating system, based on a combination of qualitative and quantitative evaluation parameters.

When and if you get to the point where it is difficult to make a decision based on a simple rating system, then it may be feasible to engage some form of mathematical simulation model to produce benefit assessments with additional resolution. These models can typically take account of network topology, traffic flows, and driver behavior in determining likely benefits. This will be covered in more detail in Chapter 19.

## 7.8 Step 750: Measures of Effectiveness Report

To evaluate the cost benefit of the proposed deployment, it is necessary to be able to assess the effectiveness of each ITS Market Package, both individually and as a whole ITS Future Big Picture. This provides input to the evaluation process by defining parameters to be used as yardsticks for effectiveness and identifying typical values for each ITS Market Package. This involves asking "so what?," "what does it do to meet my objectives?" and "has it been applied to a similar set of requirements somewhere before?"

Do not forget that ITS Market Packages do not yet have the design detail and technology specific tradeoffs that come during the design process. It may only be possible to get an indication rather than specifics at this point. When ITS Market Packages have been subjected to detailed design and have become technology-specific, they are known as ITS Equipment Packages. We need to introduce these because of the following limitations of ITS Market Packages.

- They do not define the units you will purchase.
- They cut across subsystems.
- They do not obey deployment-sequencing boundaries.
- They are blind to procurement requirements.

Therefore, we need another building block type called an ITS Equipment Package. Defining ITS Equipment Packages involves taking the ITS Market Packages and reconfiguring them for purchase and deployment by translating your needs to realistic solutions.

This raises the question of why we bother with ITS Market Packages, why not go straight to Equipment Packages? Remember that the primary objective of the methodology is to support the "what?/how?" cycle. This is achieved by carrying out needs analysis while simultaneously defining solutions. The proposed solutions are then presented back to the users to support revisions to the initial "what?" ITS Market Packages provide just enough detail for the users to appreciate the proposed solutions and see how they might want to change their minds about needs, in light of the new information they have learned about the technology capabilities.

Equipment Packages are part of detailed design. They take account of subsystem boundaries and procurement approaches. They are technology-specific, quoting vendors, products, and service names and descriptions.

ITS Market Packages are bundles of enabling technologies that provide a service defined in the ITS User Services and that have measurable benefit.

ITS Market Packages also support the exploration of ITS Market Package sequencing, or deployment strategies that take account of technology availability, political, commercial, and institutional/organizational factors.

The system developer has to walk a tightrope at this point. ITS Market Packages presented to the user must be based on achievable solutions.

Developers really have to carry out an initial Equipment Package review, then translate these into ITS Market Package building blocks for presentation to the users.

One way to look at it is to think of ITS Market Packages as "how?" stage 1 and ITS Equipment Packages as "how?" stage 2.

# 8

# Incorporating Legacy and Protecting Sunk Investment

## 8.1 Introduction

This chapter deals with the activities summarized in step 200 of the ITS Cooperative Development Methodology, "Legacy Catalog." This includes the issues involved in identifying and assessing systems and initiatives over which we will have no design control, but that will affect or influence the system we are responsible for developing. Such systems or initiatives may currently be in place or planned and may require varying degrees of integration with the planned system in order to satisfy requirements.

We start by describing and defining what is meant by legacy, then move on to discuss some of the issues that should be addressed when incorporating legacy. We finish the chapter by developing a list of suggested activities designed to address the issues and provide an effective approach to managing and incorporating legacy within an ITS planning and development project.

## 8.2 What Is Legacy?

The term *legacy* is a convenient abbreviation for a fairly wide set of concepts. Therefore, it is worthwhile to take some time to explain exactly what we mean by the term when it is used in the context of ITS development and planning. When we talk about legacy in everyday life, we usually mean the things our predecessors or ancestors have left for us. This is usually in the form of a sum of money bequeathed, but it could also be something in which our ancestors invested their time and resources for us to benefit from later. In the case of system development projects, legacy means something similar with some additional twists specifically for ITS.

We use the term legacy to describe anything we encounter during the system development process that will have an effect on our system or will influence our system or interact with it. Legacy is usually considered in system engineering to take the form of previously planned or implemented systems, or systems being developed in parallel to ours by independent implementers and developers.

Therefore, the definition of legacy in an ITS project is inter-related to the definition of the scope of the project. One way to think about legacy is to imagine that we could draw a circle around our project, defining everything inside the circle as under our design control and everything outside of the circle as outside of our control. Then anything on the outside of the circle that needs to interact with things inside the circle would be legacy. Also, anything outside that will have an effect or influence on how things work inside the circle must also be considered as legacy. The circle defines the boundary of our direct responsibilities and actions within the development project. It defines what we are being paid to do.

However, we cannot provide a complete solution if we restrict our attention to things inside the circle. Our development work takes place within the context of a wider set of systems, projects, and initiatives that are not directly our responsibility and not within the scope of our project yet which must be taken into account if we are to develop a comprehensive solution.

System engineers sometimes illustrate this with what is known as a *context diagram.* This is the circle we have described with connecting lines drawn from the circle to items outside that require interaction with, or have influence on, our system. Figure 8.1 illustrates this concept with a sample ITS context diagram.

The context diagram also represents the first step in defining the logical view of the ITS Future Big Picture for an ITS. We will explain more about this in Chapter 10, where we talk about the development of the logical view of the ITS Future Big Picture.

Let's talk about the things outside the circle. So far, we have been a bit fuzzy about defining these. In a non-ITS information system development project, these things would be defined as the hardware and software that compose other existing and planned information systems. For ITS, we have found it valuable to extend this definition to include both information systems and other transportation initiatives such as new road construction or other conventional transportation projects such as new bus stations or new rail facilities.

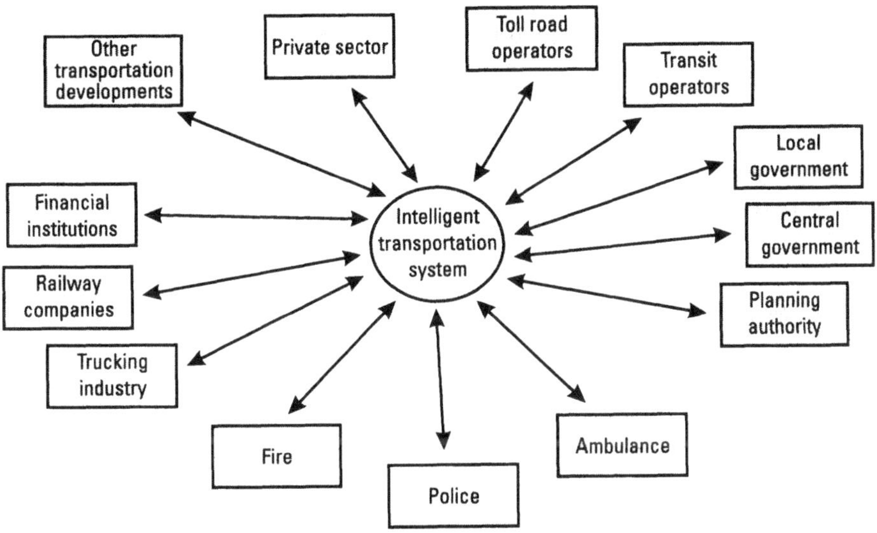


**Figure 8.1** Sample ITS context diagram.

These are not information systems implementations, but they are likely to have a significant impact on the transportation patterns and trends within a region. Consequently, they will influence and affect the proposed ITS. This is, of course, a two-way street, as our planned ITS may well influence and affect them also.

Such systems and initiatives may already be completed or in place. They may be in the planning process or have funds committed. Either way, the course of these systems and initiatives is assumed to be outside of our control or outside the planning horizon for our system development.

We also consider that there is another important dimension to legacy with respect to ITS planning and development. This is the "people" dimension. Most legacy we have encountered has multiple resources associated with it. In addition to hardware, software, asphalt, steel, or concrete, these resources include people. This includes the experience and learning that people have acquired in the operation and use of the legacy and the way in which people relate to each other within existing organizational frameworks evolved to support the legacy.

It can also include the less tangible "ownership" that people feel for the legacy. Having invested time and effort in learning how to use the legacy and perhaps even having been involved in the development and implementation of the legacy, the people involved may have developed a strong affinity with that legacy, whether it be an information system or an infrastructure project.

The incorporation and accommodation of the people dimension in legacy is just as important as the other more tangible aspects. In our experience, failure to understand and manage the people dimension leads to what we can best describe as the "not invented here" syndrome, in which good technical system proposals are rejected due to resentment and resistance from people that are critical to the success of the project.

## 8.3 A Few Examples of Legacy

In order to bring our description of legacy to life, Sections 8.3.1–8.3.4 present examples we have encountered in the course of applying the ITS Cooperative Development Methodology in a number of locations. There are, of course, many more examples of legacy, and this is not intended to be a comprehensive list. However it should help to clarify what we mean by the use of the term legacy in the ITS context.

### 8.3.1 An Existing Traffic Control System

This is a very commonly encountered legacy system in ITS. Traffic signals have been around for a number of years, have proven benefits, and are well understood. Consequently, many metropolitan areas have extensive existing traffic signal control systems with varying degrees of automation and communication.

The more sophisticated systems may well have sensor subsystems and communications networks that may be candidates for incorporation into new proposals. Some systems also have a command-and-control or traffic management center that could be utilized to house a new center for the proposed system. From a people perspective, the organization or team that operates and maintains the traffic control system would also be considered as part of the legacy.

### 8.3.2 An Existing Freeway Management System

This is similar to the traffic signal control system. Freeway management systems may well have components that can be included in the new system operation, or they may have to be interfaced as part of a geographical expansion of coverage. Typically, such systems include sensors, driver information displays, communications networks, and control centers. The legacy may cover an adjacent area or region, in which case links may have to be established or may have been installed within the region under consideration. Here again, the team of people running the current system would be part of the legacy. Additionally, any relationships that have been established between that team and other transportation agencies would be good candidates for inclusion in legacy.

### 8.3.3 An Existing Communications Network

In addition to existing subsystems and components, legacy can also consist of communications networks or systems. These could be existing local or wide area networks utilizing copper wire or fiber optics. They could also be short- or long-range wireless communications utilizing radio frequency or infrared communications media. The legacy might also include the communication protocols being used to process the data signals. From the people perspective, the communications network operating organization, whether it be public or private, should be considered as part of the legacy.

### 8.3.4 A Proposed Light Rail System

Such proposed new infrastructure can take many years to develop, plan, design, and implement. Given the complexity of major infrastructure and the need for extensive public consultation, the gestation period for such an implementation can span many years. Although it may not be on the ground today, it may have to be considered as part of the ITS Future Big Picture. Information and management systems associated with the new infrastructure may have to be interfaced as part of the new proposals, and the question of degree of integration between systems needs to be addressed. Electronic payment systems associated with new transportation infrastructure may all hold high potential for synergy and integration. From a people perspective, the transit agency operating the light rail system could be considered to be part of the legacy.

## 8.4 Why Is Legacy Important?

Legacy is a very important aspect of ITS development for one simple reason: It is very unlikely that you will start the development and planning of your ITS from a "greenfield" situation. In other words, you very rarely start off with a blank sheet of paper and have the luxury of developing an ITS in isolation. In most cases, you will have to take full account of what has been done in the past and what is happening now beyond your span of influence.

In many cases the work done in the past may have resulted in capital expenditures that dwarf your current budget for ITS deployment. Even if they do not, the people responsible for those earlier initiatives may be your clients for the current project. In fact, if the legacy initiatives have been at all successful, the people responsible may well have ascended to higher levels in the client organization and become key decision makers regarding your project. If that is the case then they will likely be the champions for your project as well, so you need to nurture their legacy and avoid "shooting the early adopters." They are the pioneers who placed their faith and energies in new technology and implemented at the start of the curve.

There may be a temptation to avoid unnecessary complications by scrapping everything and starting again on a new streamlined integrated system design that solves all the interface problems. However, it is usually the case that legacy systems and initiatives are producing high levels of benefits and are perceived as a success. Therefore, it is very important to assess them and embrace as much of them as possible.

The protection of sunk investment in the transportation network is also a valuable plank in any justification for spending money on a new system. People like to have their earlier decisions vindicated, praised, and fully engaged as growth platforms for new initiatives. Even more important is the need to ensure that every last scrap of previous funding, sometimes referred to as sunk investment, is utilized and exploited in the course of developing and implementing the new one.

Finally, the identification and proper management of legacy is one of the key elements in a user-driven approach to ITS planning and development. It would be very difficult to plan and develop an ITS that meets all the needs and objectives and addresses all of the problems, if legacy is not taken into account. Although not within the scope of the project, there may well be features and attributes of legacy that will have substantial impact on your planning and development work.

## 8.5 Issues Associated With Legacy in ITS

As you have probably gathered by now, legacy is a complicated subject. Sections 8.5.1–8.5.8 discuss a few issues associated with legacy in ITS that we have encountered and had to manage in the course of previous projects. We are not suggesting that you will encounter these on every project, but it is just as well to know and understand these as potential issues.

### 8.5.1 Comprehensive Legacy Identification

The development and maintenance of a comprehensive inventory of the legacy to be taken into account is vital. In the case of information systems legacy, we are looking for information that will help us to decide if it should be incorporated or removed, the best way to incorporate the legacy, and the likely effects of the legacy on the overall system plan. Therefore, we are looking for information about the data input, data output, and operating characteristics of the legacy.

In the case of infrastructure or transportation project legacy, we are mainly looking for information on how the legacy will affect the transportation trends and patterns within the region being addressed by our proposed system. We might also be looking for information that would support the definition of ways in which the planned ITS could be extended or modified to work in closer harmony with the legacy.

### 8.5.2 Likely Effects or Consequences of Legacy

An analysis and evaluation of the effects and consequences is required as the basis for decisions on legacy management. These effects might include impacts on the budget required for the new system or requirements for additional funding to upgrade or modify legacy for incorporation. It might also include review and consequent modification to the initial requirements identified and defined with the client group earlier in the process.

### 8.5.3 Smooth Transitions

We have found that the one thing to avoid at all costs in the planning and development of an ITS is sudden, dramatic change. The creation of a situation in which a quantum leap is required, whether in terms of technology, organization, or people, typically results in a high level of associated trauma. Whether the decision is to abandon or incorporate legacy, the best approach usually involves the identification and definition of a smooth transition from the current situation to the desired one. Disruption in the normal flow and organizational structure of the organization responsible for planning and implementing the ITS can only lead to delays and make the process far more difficult.

A decision to incorporate or abandon legacy may have a significant impact on the proposed Implementation Strategy for the ITS. It may be necessary to adopt an interim system configuration that allows legacy to continue to be utilized for a period, supporting a smooth transition to a new system.

### 8.5.4 Building on Success

A good ITS planning and development approach will build on previous successes by incorporating as much of the legacy as possible. There may be instances where previous implementations have obviously failed for one reason or another, and maybe it will be unavoidable to propose their removal. However, even this type of legacy should be evaluated carefully to ascertain if any aspects, features, infrastructure, or organizational aspects can be salvaged for use in the proposed system. We have found that effort invested in identifying, then building on successful aspects of legacy pays large dividends in terms of acceptance of our later proposal and the evolution of consensus regarding the ITS Future Big Picture as it emerges. We suppose that this is a facet of human nature rather than some specific aspect of ITS.

### 8.5.5 Dead Ends

Sometimes legacy will be identified that simply has no future due to technical, organizational, or commercial reasons. Ideally, this should be identified and slated for abandonment or removal. In some cases, it may be necessary to develop detailed documentation explaining the reasons for removal and the nature of the "dead end" phased out. Sometimes this may require extensive discussion with the client and stakeholder group in order to attain agreement and ownership in the decision to abandon the legacy. On other occasions, it may have been a decision anticipated and expected on behalf of the clients and stakeholders. In fact, we have worked on some projects where the original motivation for initiating the planning and development of a new ITS was to ensure the removal of an existing unwanted legacy.

### 8.5.6 Legacy "Ownership"

In addition to cataloging legacy such as hardware, software, asphalt, concrete, and steel, the people dimension should also be considered. In addition to identifying the legacy that you are faced with, it is important to understand the history of the legacy, including the people involved in the development, planning, and implementation of that legacy. This fits with what we said earlier about the people dimension and the importance of accommodating the people. This also helps with the development of your understanding of the key players and people involved in the development of your system.

### 8.5.7 "Immovable Objects"

This describes an "ownership" effect. There may well be elements of legacy that, from a system engineering or technical perspective, should be abandoned as soon as possible. However, due to the people dimension described in Section 8.2, the removal of the legacy does not form part of an acceptable solution. It is necessary to recognize this type of situation and take appropriate action. This might include the development of a reasoned justification for the removal of the legacy, to be used in an effort to win people over. It could also involve the development of a phased strategy for transitioning away from the legacy over an acceptable time period.

It could, of course, also include the modification or alteration of your system planning and development approach to incorporate the legacy into the new system. In this case, it would be very important to ensure that the new constraints imposed as a result of legacy incorporation are fully identified and explained to the client group. This may well result in changes to the originally identified and defined requirements for the project.

### 8.5.8 Paradigm Shift From "One Shot" to "Cyclical Renewal"

We have encountered many clients with a "single shot" paradigm when it comes to planning, development, design, and procurement of systems and the implementation of transportation projects. Unfortunately, this paradigm does not fit ITS. Whether you like it or not, the decision to plan, develop, and implement the first ITS project or phase puts you on the way to a regular cycle of renewal, upgrade, and replacement. We suspect that this is true for almost everything a transportation agency or authority acquires or procures. It seems to be particularly relevant to ITS, perhaps because of the swift pace of technology development, the shorter project cycles, and the high rate of return that can be achieved with respect to benefits attained.

## 8.6 Managing Legacy

We have more or less concluded that it is rarely an acceptable approach to completely abandon all previous systems and start again with a clean sheet. In most cases, ITS developers will be faced with the task of managing some form of legacy as part of the development of the new system. Indeed, incorporation of and interaction with legacy could be considered to be part of the requirements for the new system. So how do we work with legacy and either build it into our new proposals or successfully remove it from the plans? Sections 8.6.1–8.6.3 describe a few steps that we have found useful in the past.

### 8.6.1 Catalog Legacy, Build an Inventory

The first thing to do when you have been assigned a task or responsibility for a project is to find out what you have inherited. This involves the development of some kind of inventory or list of items. It is useful to bear in mind when putting this together that the list is not just about technical detail. It is also about institutional/organizational and commercial aspects of what has gone on in the past and what is coming down the pipeline at the same time as your project. Consistent with our strategy throughout the book, we do not intend to provide you with a rigid checklist or recipe for success in legacy management; rather we intend to identify the main issues and provide you with some thoughts on a practical way to address these in your development approach.

Starting with the development of the inventory, we want you to understand that this is not just a data collection exercise. You will probably be developing this at an early stage in the development cycle when relationships between the developer, the key client contact, and the major stakeholders of the system are still in the formative stages. Accordingly, in addition to collecting the data you need, you are also afforded the opportunity to have some "quality time" with these people. In fact, this part of the development cycle along with requirements definition hold the major opportunities for the system developer to legitimately ask the stupid or naive questions that are essential catalysts in yielding the input required for good system design.

We have seen some companies relegate this task to very junior staff, missing the opportunity for exploiting client and stakeholder interface opportunities. Obviously junior staff needs to be involved, not just to gain experience but to play an appropriate role in the cost-effective execution of the work. However, it is also essential, in our opinion, that experienced staff plays a significant part in these activities. We would not dare to presume to tell experienced consultants and system integrators how to manage their business, so let's just say that a balanced staffing structure, appropriate staff assignments, and a management approach is highly desirable in the course of addressing legacy.

We view the legacy activities in the development cycle as a close relative of the requirements definition activities such as ITS Objectives Statement, vision development, and the development of ITS User Services. The Strawman ITS Future Big Picture can also be a useful tool in developing and understanding the legacy you have to manage. In many cases, presenting a Strawman solution will elicit very meaningful responses from key individuals. Utilizing these elements of the ITS Cooperative Development Methodology in close coordination has also proven to be an effective approach, as they are mutually supportive.

It is also important not to forget the iterative nature of the development process. The main reason you are developing the ITS Future Big Picture before moving on to detailed design is to have the opportunity to revise and modify plans to take account of new information about needs, before major investment is committed.

The first task in the development of an inventory of legacy is to identify and assimilate existing sources of information. In most cases there is extensive documentation regarding implemented and planned systems and infrastructure. The client organization is probably a good starting point in the gathering of this information since it will probably have an important role in the implementation and operation of transportation systems within the region under consideration. Using a combination of direct interviews with key staff and literature search techniques, it should be possible to develop a first version of the legacy inventory.

It is unlikely that the legacy inventory will be complete at this point, as there will probably be agencies other than the client organization involved and influential in transportation provision in the region. The initial interviews and literature search may well have identified some of these agencies and pointed out the need for more information. This can rapidly turn into a very large-scale and time-consuming operation as the number of information points grows. We have found it useful to hold a requirements identification and definition workshop as part of the overall outreach program for our system development at this stage in the process. Legacy identification and description can be considered as part of requirements analysis activities, since the need to embrace legacy and protect sunk investment is usually an important requirement in ITS planning and development.

### 8.6.2 So What Kind of Information Should You Be Seeking?

In an effort to get as much information as possible, we loosely target our activities around a core set of information but simply use this as a starting point to support information gathering. A typical core set of legacy data might contain the following items:

- Maps showing existing and planned road networks;
- Maps showing existing and planned transit networks;
- Planned transportation infrastructure projects;
- Planned or existing information or ITS systems;
- Data dictionary showing data inputs and outputs from existing and planned information systems;
- Transportation master plans or other planning documents describing transportation demands and planned approaches to satisfying or managing demand;
- Policy statements from key transportation agencies or organizations.

This list is intended for use as a starting point or as a tool to get you thinking about the actual needs of your project. We do not believe that it is possible to develop a "recipe" that would work for all situations. If we could, then there would not be much need for system integrators and ITS consultants! The guiding principle should be to "beat the bushes" for information and to not refuse any piece of information you are offered even if it seems irrelevant at the time.

### 8.6.3 Sources of Information

There will be many potential sources of information on what transportation-related systems, infrastructure, and initiatives the new system will have to incorporate or work with. Sections 8.6.3.1–8.6.3.3 describe some of these.

#### 8.6.3.1 Direct Interviews With Client Organization Staff

In many cases, we have found this to be the best starting point. Either the client staff has access to all the required information, or it can tell you who else to talk to. This activity also involves getting around the region, identifying the major players, and asking them what they have and what they plan. It takes a lot of effort, but it is probably the best way to get comprehensive information about legacy. Not only are you building a good picture of legacy, you are also building relationships with many of the key transportation stakeholders in the region.

Incidentally, we have found it very valuable to enlist the help of staff based in local offices for this work. It may be that the local staff is not comprised of ITS specialists and maybe not even transportation professionals, and this may lead you to believe that staff members have nothing to contribute to the legacy picture. However, we have found that local staff nearly always has something to add to your picture of the local scene. We have noted that just about anyone who has a drivers license considers his or herself qualified to express an opinion on transportation systems within a region anyway!

Seriously though, we have found it very worthwhile to engage the help of locally based staff, even those not directly involved in transportation. We touch on so many aspects of life that it is difficult to live in a region and not have some knowledge of the transportation network.

#### 8.6.3.2 Previously Commissioned Reports and Studies

It is highly likely that any significant legacy will have been the subject of a design report, feasibility study, or some other document describing it. Your client and perhaps other key stakeholders should have knowledge of the previously commissioned reports and how to access them. These might include transportation master plans for the region. They might also include the overall transportation plan for the region. This has proven in the past to be a good source of information regarding the legacy that is not yet implemented but has funding committed or is in advanced development, planning, or design.

#### 8.6.3.3 Questionnaire Surveys

One potential method of reaching a wider audience of organizations and agencies involved in providing or operating transportation facilities within a region is to deliver a questionnaire type survey to be completed by target agencies. This approach is very effective in making the best use of limited resources for information and data collection, but it has some drawbacks. The major disadvantage of this approach is the difficulty in getting interviewees to respond either on time, or at all. You almost have to adopt *Readers Digest*-style marketing tactics and offer them some sort of incentive like the chance to win a free vacation or car if you are to get any sort of reasonable response rate. We have witnessed the considerable expenditure of time and resources on this activity in previous projects. An unsupported questionnaire just does not seem to work.

To get around the drawbacks, we have found it useful to adopt a hybrid approach to questionnaire surveys. One hybrid involves the combination of the questionnaire approach with a plenary meeting of the target audience. This forum can be used to explain and discuss the information you are trying to obtain and also to encourage the agencies to fill in the questionnaire on the spot. You can improve your response rate dramatically by asking the interviewees to fill in the questionnaire before they leave. On one project in which we were involved, the team responsible for conducting the survey positioned themselves at all exits from the room and collected responses as interviewees left!

The meeting approach also provides the opportunity for interaction with the agencies involved, enabling questions to be raised and answered regarding the questionnaire.

Another hybrid involves the combination of the questionnaire with a direct personal interview with the target interviewee. This can be in person or by telephone. We have successfully applied a combination of both telephone and direct interviews in sequence starting with a series of telephone interviews to get an initial idea about legacy and who is worth talking with in more detail. A general questionnaire is used to support the telephone interview, structure the flow, and record the results. The results from the initial telephone interviews are then evaluated and used as the basis for the development of a second wave of direct personal interviews with targeted interviewees. These second-wave interviews are supported with a more detailed questionnaire customized using the results of the telephone interviews. This has the advantage of keeping costs down in the first wave, yet giving you access to good information for focusing the higher level of resources required on the second wave of direct personal interviews.

## 8.7 Evaluating Legacy—Does it Go or Does it Stay?

Having gathered the information required in developing the legacy inventory, the next job is to use the information to support the assessment and evaluation of the legacy. This boils down to what superficially seems like a simple decision: Can the legacy systems or initiatives be incorporated and accommodated in the planning and development of our ITS, or will they have to be abandoned? This gets a little bit more complicated when you remember that legacy also includes transportation issues within the region as well as systems and initiatives that are in the pipeline and existing "on-the-ground" implementations. The incorporation and accommodation of legacy is also not a binary or black and white decision as there are degrees to which the legacy can be incorporated and different levels of system integration.

The following is a list of factors that you might want to take into account when making the legacy assessment and evaluation.

- How old is the legacy?
- How much money—capital expenditure, operating, and upgrade costs—has been invested?
- Is it technically feasible to incorporate the legacy technology?
- Have advances in standardization made the legacy obsolete or incompatible?
- What will be the approximate cost of incorporating the legacy?
- Define the adaptations that would have to be made to the proposed ITS Future Big Picture to accommodate the legacy.
- Does this represent a reasonable cost compared to the benefits of preserving the legacy?
- What is the foreseeable working life of the legacy relative to this implementation's life cycle?
- Who operates the legacy?
- Who will operate and maintain the legacy in the ITS Future Big Picture?
- Is the operator part of the stakeholder group? If not, then should he or she be?

- Does the client or stakeholder group value the legacy?
- If the legacy is at the planning stage then can the development be synchronized with this project?

Sections 8.7.1–8.7.8 offer some specific advice on actions to take in the legacy evaluation activity.

### 8.7.1 Review Emerging ITS Standards

Bearing in mind that we are trying to develop a Future Big Picture for ITS within a region, encompassing legacy and providing flexibility for future expansion, deciding how to apply emerging ITS standards is an important issue. When assessing the value of legacy, it is necessary to consider how well emerging and future ITS standards align with existing systems and infrastructure. This is a deep issue with many variables involved in the decision.

One set of issues relates to the maturity of the standards affecting the legacy system. If it looks as though the emerging standard is close to attaining universal acceptance, then a potential change to accommodate it would be more worthwhile than if it is still evolving. This is the reason why we have a separate step in the Cooperative Development Methodology that deals exclusively with assessing the current status of standards development and developing a plan describing how to apply standards to the system under development. This is described in Chapter 12.

When considering all the issues, it is important to remember that the primary objective is to protect sunk investment, while embracing legacy. Balancing this is the desire to reduce future risk by taking advantage of new standards to guarantee interoperability and encourage open systems. In the situation where some part of your legacy does not align with or support an emerging standard, then a value judgment needs to be made. Does the risk reduction justify the loss of sunk investment? Will abandoning existing, perhaps proprietary, systems with closed interfaces and tied-in suppliers and the ensuing loss of sunk investment be justified by the increased flexibility to purchase from multiple suppliers?

Do not forget that the loss of sunk investment may well involve more than the actual dollars spent on the development and deployment of previous systems. It may include the loss of familiarity with the operations and maintenance aspects of the existing system. It may also include a change in the availability of staff trained to use the system. Therefore, a simple analysis of benefits and capital investment is not enough.

What we refer to as the "organizational momentum" behind the legacy needs to be identified and evaluated. This also involves the identification of the key players within the implementing organizations, making sure that they are onboard as part of our outreach program.

### 8.7.2 Assess Value and Benefits Already Attained Versus the Size of the Sunk Investment

A more elegant way of describing this is *payback analysis.* This should provide some input into the legacy go or no go decision but will not be the sole deciding factor. The analysis requires the determination of the total benefits obtained from the legacy system or initiative and the total amount of sunk investment made in terms of both capital and operations. This analysis should also give some indication of the remaining life of legacy systems and their usefulness in providing benefits to the traveler and the implementing organization. If there is a proposal to abandon legacy of some form, then this payback analysis may play an important part in the justification process.

### 8.7.3 Review Technology Capabilities and Capacities

Can the legacy system provide the capabilities required to satisfy the ITS User Services as defined in the requirements stages of the process? Remember that requirements are set for the short term and the long term so the full range of requirements should be considered when evaluating the suitability of legacy.

With regard to legacy communications networks, the carrying capacity needs to be assessed as well as the capabilities of the legacy. A data-loading analysis to determine present and future carrying capacity requirements for the communications system should provide you with a good indication of whether legacy communications networks can be viable in the future. The age and condition of the network or the switching equipment may also be an issue. This is not a simple task, as advances in communications technology such as new and more efficient protocols are making it possible to squeeze much more data along existing wirelines than had previously been believed possible.

Items like database management systems and information processors also need to have a specialist analysis carried out to determine the potential for expandability and scalability to meet predicted future needs. Another variable in this respect is the ever-decreasing cost of both processing power and data storage space. It may well be a cost-effective strategy to exploit the legacy to the fullest for a couple of years, then invest in new processing and data storage.

### 8.7.4 Identify Issues, Problems, and Opportunities

In addition to simply compiling a catalog of legacy to be taken into account, we have found it useful to develop a list of issues, problems, and opportunities associated with the legacy. Developing this as you go along provides an ideal basis for further analysis and evaluation, should it be necessary. The kinds of things that might be included here include an assessment of existing communication systems with points on whether existing infrastructure or protocols form a constraint for the future. It could also include information on the likely effects of planned transportation infrastructure and identify problems or opportunities that the planned implementation might pose for your system.

### 8.7.5 Identify Opportunities for Fast-Track Action to Synchronize Systems and Initiatives

At an early stage in the evaluation and assessment of legacy, it may be possible to identify early, or fast-track actions that can be taken to make the legacy fit better with the proposed system. If the legacy system is in fact in the development or planning stage rather than existing, then it may be possible to make some late changes to the legacy. Another approach may be to suggest some immediate changes to the evolving ITS Future Big Picture and start discussion and consideration of these proposed changes while the legacy evaluation work is ongoing. This allows time for full discussion of the likely impacts of incorporating the legacy into the proposals and gives the stakeholder group an early indication of how legacy is being managed and accommodated.

### 8.7.6 Use the Legacy Inventory as Input to the Development of the ITS Future Big Picture

All this consideration and evaluation of legacy provides an input into the development of the ITS Future Big Picture. It may well provide key input that significantly affects the shape of the ITS Future Big Picture. If the legacy analysis shows that a change in the proposed ITS Future Big Picture would help to embrace legacy and protect sunk investment then that change should be seriously considered. Remember that one of the main reasons for developing the ITS Future Big Picture, or architecture, on the way to system design, is to provide practical possibilities for changing your mind. The evolving ITS Future Big Picture may well go through many such modifications and alterations on the way to completion. In fact, there may be something wrong with your development process if it does not.

You should also bear in mind that there is a danger of pouring good money after bad in attempting to incorporate legacy that should really be abandoned. It may be better to face the problem of communicating this to the client now, rather than have him or her find out in the years and dollars to come.

### 8.7.7 Inform the Client Group That Legacy Has To Be Abandoned

When an assessment of legacy indicates that something just cannot be incorporated into the ITS Future Big Picture, we have found that communicating this information to client and stakeholder groups requires some care. Even if the system or infrastructure to be replaced or abandoned has obviously served the client well and is due for upgrade or replacement, the way in which this is handled is important.

The way in which the information is conveyed is probably as important as the information itself. We have found the following process to be useful:

1. Identify the owners or champions of the legacy;
2. Explain to them the reasons for the decision;
3. Try to convince them to support the decision and take ownership in the new proposals;
4. Explore the history and background of the legacy with them;
5. Prepare a succinct presentation explaining the benefits of abandoning the legacy;
6. Review this with the legacy champions;
7. Deliver this to the rest of the stakeholder group with the active participation of the champions.

This seems to work because of the close involvement from the people who were responsible for the legacy. In most cases, the legacy will be viewed positively, and the champions will be seen as successful. Therefore, getting them involved in the decision to abandon the legacy has a double effect. Concentrating on the job of convincing the champions that the legacy has no role in the future enables you to focus resources on this tough problem. When this has been achieved, it should be relatively straightforward to make the case to the other stakeholders.

We have stated several times throughout the book that you should not confuse the development of the ITS Future Big Picture with the subsequent job of design. In discussing the issues associated with legacy and the protection of sunk investment we have come perilously close to the (blurry) dividing line between the two. In fact, it may be necessary to continue into the detailed design of a system before the final decision on legacy can be made.

As before, our best advice is to remember that we are trying to develop an ITS Future Big Picture through making major decisions about what should be included in the system and how the different elements should relate to each other. If there is not enough detail to make a decision at this point in the development of the system, then leave the legacy in the picture until later. We would advise against carrying on into the detailed design purely to establish a legacy judgment, as there are many other factors to be considered in the development of the ITS Future Big Picture that may change or override our choice on legacy. The creation of the ITS Future Big Picture for ITS in a region typically has to support the evaluation of multiple selection criteria based on the requirements of a diverse set of stakeholders.

### 8.7.8 Try to Incorporate Legacy Over the Long Term

As the Future Big Picture is being developed to meet requirements over a long term of perhaps 15–20 years, it may be possible to define a transition or migration path for incorporating legacy. This is one of the advantages of developing the Future Big Picture since it supports the definition of a transitional strategy from where we are today to where we want to be tomorrow. Slow transitions are much easier to support than abrupt changes in equipment and procedures and provide the implementing agency with much needed time to get comfortable with changes and persuade stakeholders to adopt the new system.

This overlaps with the identification and development of an Implementation Strategy for the whole system as discussed in Chapter 14, as the strategy for the whole system would have to take account of constraints imposed by the legacy management strategy.

The transition strategy might initially allow for a lower level of integration than is ultimately desired. There are three levels of integration that we define for transition purposes:

- *Level 1:* Nonconflict or peaceful coexistence. Systems do not have to communicate but are operated as independent, nonconflicting units.
- *Level 2:* Data sharing and cooperation. Operation is still independent, but data and information is shared and acted on accordingly.
- *Level 3:* Mutual codependency. Processes in one system are carried out purely to support another.

Level 1 or 2 integration may be tolerated for a number of years as the legacy system is phased out or modified to support level 3. The legacy system may be left as a standalone independent system or may have minimal data flows supported to and from the new system. It may also be possible to introduce a third-party temporary subsystem to support data exchange between legacy and new systems.

It may also be desirable to leave selected ITS User Services or requirements unfulfilled in the short term as a way of accommodating legacy. This would require a review of initially stated requirements such as needs, objectives, issues, and problems with the client or stakeholder group.

## 8.8 Summary

This is all we have to say about legacy and the protection of sunk investment. The art lies in saving what you can but also recognizing the time to stand up and make the decision to abandon. If it comes to that, then careful management and communication of this message is essential. Of course, the whole reason you have to deal with legacy in the first place is that you very rarely have an opportunity to start from a "greenfield" situation. The starting point for every ITS development project with which we have been involved has never been the "ideal starting point," and we do not believe it ever will be. We are reminded of the story about English tourists Fred and Doris paying their first visit to the Irish Republic. They are touring in their rental car with the aim of getting to Dublin. However, they get completely lost. After a protracted discussion about stopping to ask for directions, Fred capitulates and stops next to a farm. He calls out to the Irish farmer sitting on the wall next to the farmyard, "I say, old chap, can you tell us how to get to Dublin?" The farmer scratches his head for a few minutes and then provides his considered reply, "To be sure, if I was going to Dublin, I would not start from here!"

With regard to legacy the same is true in most cases. To get to the ITS Future Big Picture, ideally we would not start from today's situation.

# 9

# Institutional/Organizational Issues

## 9.1 Introduction

This chapter addresses the issues and activities associated with step 210, "Institutional and Organizational Catalog," in the ITS Cooperative Development Methodology. We have defined a complete ITS Future Big Picture as being composed of three layers as shown in Figure 9.1. This chapter takes a look at how we should go about defining and developing that second layer.

This is definitely the "people" part of ITS planning and development often turns out to be one of the most complicated and perhaps the most difficult area in any ITS development project. It is also the area in which the development of a Future Big Picture can be very valuable. It addresses the way organizations currently operate and the way people in those agencies do their jobs. It helps in the selection of the most appropriate organizational structure.

We start the chapter by defining institutional and organizational issues. This tends to be fairly abstract, so we have included specific examples of institutional/organizational issues that we have encountered in the course of some previous projects. Section 9.2 describes these briefly and discusses why institutional/organizational issues are important.

In order to bring the institutional/organizational issues to life with some specific application context, this chapter begins by exploring the institutional and organizational issues through specific ITS application scenarios. Finally, we conclude the chapter by addressing the question of how to manage these institutional/organizational issues effectively based on some activities we have found to be valuable in addressing and managing these issues.

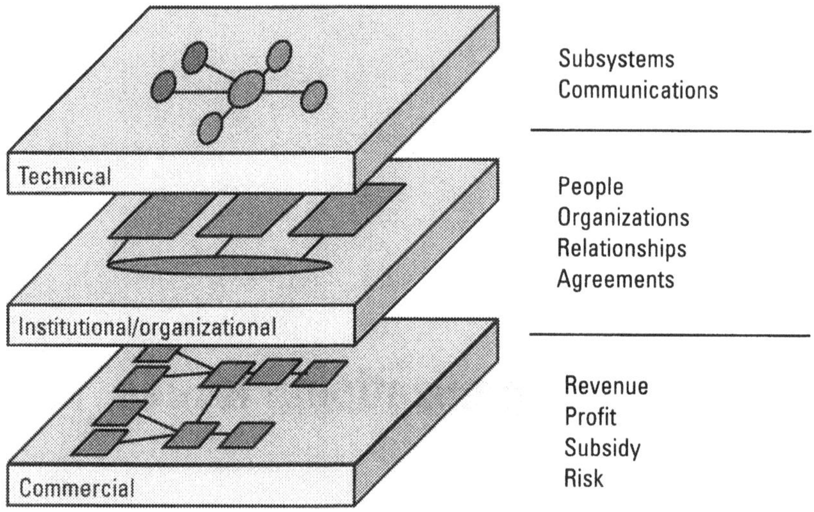


**Figure 9.1** The three layers of the ITS Future Big Picture.

## 9.2 What Are Institutional/Organizational Issues?

Let us start off by exploring and defining what we mean by institutional/organizational issues. The term *institutional/organizational* is used to describe the relationships between organizations and people that support the proposed technical solution. As the name suggests, they also encompass the barriers and constraints imposed by institutional structures such as legislation and legal requirements. Institutional/organizational issues are the "people" and organizational issues associated with the implementation and operation of an ITS and include the following:

- Nontechnical issues, problems, constraints, and barriers;
- *People-based* arrangements that are required to support the technical and commercial arrangements;
- One-on-one personal interactions and organizational communications;
- Identification and definition of, and agreement on, mutual goals and strategies;
- Coordination of organizational objectives and goals;
- Formal arrangements such as memoranda of understanding that provide the framework for cooperation;
- Legislative constraints and issues.

Since the whole subject of ITS is fairly new and most current institutional/organizational arrangements were not evolved to suit the needs of ITS, there are typically two ways to look at institutional/organizational issues. One perspective is in terms of the needs of ITS, in particular the technical solution being proposed. The other perspective is in terms of what support can be provided by the existing arrangements—in other words, what can be done to support ITS without significant change to the existing arrangements. It is usually necessary to develop both perspectives, as it is not always possible to develop new institutional/organizational arrangements or to modify existing ones to provide ideal support for the proposed technical solution. In fact, it may be necessary to revise or amend the proposed technical solution to fit within existing arrangements.

The decision about which perspective to use depends on the local context and the degree to which compromises in requirements satisfaction can be accommodated or are acceptable. We will focus on the first perspective, looking at the ideal institutional/organizational arrangements to fully support the proposed technical solution. This should provide a good platform for exploring and explaining the issues and a starting point for developing the second perspective.

Ideal institutional/organizational arrangements can be determined by analyzing the logical view developed as part of the technical solution. Figure 9.2 shows a diagrammatic representation of a technical solution. As explained and discussed in Chapter 11, the bubbles represent processes or actions to be performed, while the lines or arrows represent data flows between processes. These data flows carry the essential data from one process to the next. A process is defined as the action required to convert data in to data out.

If we now consider the institutional/organizational solution we can draw a parallel diagram as shown in Figure 9.3.

The bubbles or processes are replaced by organizations, and the data flows are replaced by relationships between organizations. The institutional/organizational component of the Future Big Picture is composed of many individual organizations, conducting multiple parallel activities. Their own specific policy objectives and ITS Objectives Statement drive organizations. They are also directed by their own management and operated by their own staff.

The individual organizational activities are coordinated and synchronized through connecting relationships between individual organizations. This begins to suggest the importance of the relationships in the role of "gluing" the individual components together at the institutional/organizational level, rather like the role communications systems play at the technical level.

Sections 9.2.1 and 9.2.2 take a closer look at the boxes (organizations) and the arrows (relationships), respectively.

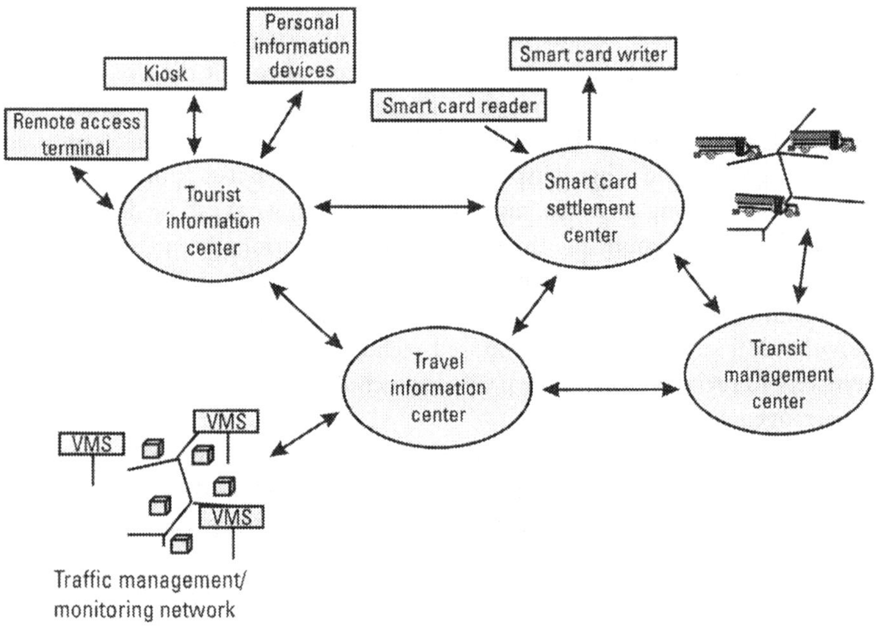


**Figure 9.2** Technical solution, logical view.

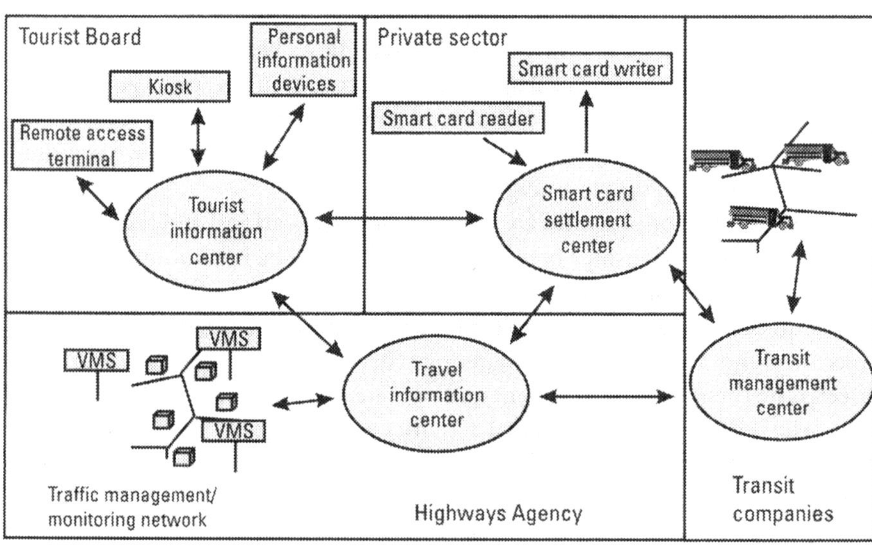


**Figure 9.3** Institutional/organizational framework for ITS.

### 9.2.1 Inside the Boxes (the Organizations)

In the transportation context, it is highly likely that the organization will be focused on a particular facet of the regional transportation system. For example, a transit agency would be focused on the planning and operation of buses or perhaps light rapid transit, while a city traffic engineers' department would be focused on the operation of urban surface street control for maximizing throughput of passenger cars. There may also be private-sector organizations with cultures and objectives that are very different from the public-sector organizations. Each organization may be carrying out one or more of the processes identified in the logical view.

### 9.2.2 Arrows (Relationships Between Organizations)

The relationships between the individual organizations may be specially formed to support the flow of data and information required for the technical solution to operate. Alternatively, it may be possible to use existing relationships, arrangements, and agreements to support the technical solution. There are multiple parameters that can be used to characterize a relationship between organizations, including the following:

- *Nature of the relationship:* Ad-hoc or systematic;
- *Formality:* Formal or informal;
- *Frequency of use:* Frequent or infrequent;
- *Criticality:* Subjective measure of how important the effectiveness of the relationship is to the operation of the whole system;
- *Level:* What staff levels in the respective organizations participate in the relationship.

A careful analysis of the logical view of the ITS Future Big Picture can yield a great deal of information about what the ideal institutional/organizational arrangements should be.

So far, we have been a bit general and abstract about institutional/organizational issues. Let's start to talk about some specifics.

We have encountered a wide range of issues in the course of a number of ITS planning and development projects that we have categorized as institutional/organizational. These include the following:

- What is ITS, and what options are open?

- What is the likely impact of ITS on users?
- What are the legislative issues?
- What is the liability?
- What are the quality issues?
- Who operates the system?
- Who owns the system?
- What data or information needs to be shared?
- What data will not be shared and why?
- How is data shared between agencies?
- Who pays for the system implementation, and how is it funded?
- What needs to be funded?
- What potential funding sources are available?
- Who pays for the maintenance and upkeep of the system?
- Who provides the subsidy or takes the profit?
- Is private-sector funding a possibility?
- What organizational structure best supports the new system?
- Who controls the system and determines the effects on the transportation network?
- Who will be the implementers?
- What are the roles of the implementers?
- What is the role of the market?
- Does the ITS fit with regional and national transportation planning processes?
- Whose responsibility is information security?
- Who owns the intellectual property rights?

Each of these issues is discussed in Sections 9.3–9.26.

## 9.3 What Is ITS and What Options Are Open?

This may seem to be a strange issue to include; however, it relates back to one of our earlier comments about the transportation community's lack of awareness of information and communications technology capabilities. This turns into an institutional/organizational issue when organizations do not take the actions to plan or implement ITS, because they are not aware of the possibilities. This issue also touches on the fact that the skill set required to support many ITS functions and operations is completely different to that found in traditional transportation. Therefore, it becomes necessary to affect organizational change through increased awareness of ITS capabilities and through the training of existing staff and perhaps the acquisition of new staff. Organizational structures and business and procurement processes may also have to be changed to support the new ITS paradigm.

## 9.4 What Is the Likely Impact of ITS on Users?

This covers a wide range of related individual issues, including social equity concerns. Will the proposed system benefit the community in a balanced, even-handed way, or will it benefit some and provide a "dis-benefit" to others? For example, if part of the proposed ITS implementation called for the introduction of congestion pricing, under which road users would pay a fee to use roads during busy periods, would this provide an unfair advantage to the well-off who can afford to pay the fee and a disadvantage to the less well-off who cannot? Would the introduction of a technology-based system provide benefits only to a certain demographic group in the community? For example, would the introduction of a common stored-value ticketing system for transit fares advantage the younger members of the community, as they are comfortable with new technology and change, while disadvantaging the older members of the community? Would it be the other way around?

These all fall into the category of undesirable, maybe even unexpected, side effects.

## 9.5 What Are the Legislative Issues?

ITS is new and requires a whole new approach to the management of transportation, including much more inter-agency and public-private coordination and cooperation. Consequently, there may be legal constraints, rules and regulations that were useful for traditional transportation but that act as barriers to ITS planning, development, and implementation.

The organizations involved in the planning, development, implementation, or deployment of ITS may not be able to legally do this under the terms of their governing regulations and legislation. This is related to the established goals, objectives, and mission of the organization.

An example that comes to mind involves the rules and regulations governing the procurement of goods and services by central and local government agencies. In many cases, these have been developed to support asphalt-, steel-, and concrete-type procurements, preventing cooperation or collusion between bidders and regulating the flow of information from the procuring agency to bidders. In some cases, these regulations act as barriers to ITS procurement by preventing and constraining the type of cooperation required to successfully procure the ITS. For example, these rules might discourage a consortium of prospective bidders for a project from developing a concept and then approaching the government agency to negotiate procurement. The agency would be obliged to take the concept and its associated intellectual property and make it public knowledge. Thus, the beneficial effects of risk sharing and cost reduction are not realized due to legislative issues.

## 9.6 What Is the Liability?

What happens if the system fails and causes damage or injury to someone? The transportation community is very good at assessing and managing liability associated with conventional transportation infrastructure but may be unfamiliar with the needs of ITS. This is especially true when the system is composed of multiple subsystems, operated by different organizations. In those cases, it may be unclear where liability or responsibility lies. An oft-quoted example is the application of intelligent vehicle technologies. If a system within the vehicle, responsible for control of essential safety functions such as distance-keeping, or lane-keeping, fails, then who is to blame? Is it the driver for buying the system in the first place, the automobile manufacturer for selling it to the driver, or the road operator for failing to provide a safe driving environment?

These issues should be addressed by making sure that the liability elements are completely identified and fully understood. It may be necessary to make use of specialist legal services to conduct an assessment of the likely impact of the liability elements.

## 9.7 What Are the Quality Issues?

This is closely related to liability in that it requires the identification of one or more organizations to take responsibility for a particular aspect of the ITS operation. If several organizations are each responsible for the quality of parts of the ITS, then it will be necessary to address the question of who takes overall responsibility for ensuring the quality of the whole system or the final results produced by the system. This covers the monitoring and management of the quality of the data being collected and provided as input to the system as well as the quality of the information and services being provided to the traveler as output from the system. Each data flow in the logical view of the technical solution should have some quality management arrangements associated with it.

## 9.8 Who Operates the System?

It is worthwhile taking a bit of time to discuss what we mean when we talk about operation of the system. This category covers the following issues:

- Employing and managing operational staff;
- Providing day-to-day management and direction of system operation;
- Defining short-term objectives and strategies for system operation;
- Performing all the tasks required to support effective system operation;
- Serving the end users of the system.

When we talk about *the system* we are actually referring to a discrete subsystem within the overall ITS Future Big Picture. For example, when considering incident management, who decides on the strategies to be adopted for region-wide traffic control, which effectively controls the traffic during and after the incident? Who employs the staff operating the system and taking the actions that have an effect on the transportation network? Also, who is making the decisions that shape the short-term operational policies rather than the higher level organizational objectives that are already agreed upon? The decision regarding which organization is best suited to operate specific subsystems can be pretty complicated and intertwined with many of the other institutional/organizational issues listed here.

## 9.9 Who Owns the System?

This is often closely related to who pays, but not always. The implementing organization that pays for the system may be acting on behalf of other organizations or may only have the authority to develop and implement the system before handing over to others for operations and maintenance. This raises the question of the difference between "who owns" and "who operates." We differentiate between these closely related issues, because it is a distinct possibility that the owner and the operator are two different entities within the context of ITS. Sections 9.9.1 and 9.9.2 discuss the characteristics of the system owner and operator, respectively.

### 9.9.1 System Owner

The system owner must perform the following functions:

- Pay for the system to be planned, developed, and implemented;
- Define the high-level objectives to be satisfied by the implementation and operation of the system;
- Define the target end users or customer base for the system.

### 9.9.2 System Operator

The system operator must perform the following tasks:

- Employ and manage operational staff;
- Define short-term operational objectives and strategies;
- Perform all the tasks required to support effective system operation;
- Maintain the system;
- Serve the end users of the system.

There is, of course, another organizational model in which the owner and operator are one and the same organization. Either model will work depending on the specific local circumstances. One of the major influences on the decision to combine or split the owner and operator role will be the existence or not of experience and expertise in the operation of information technology-based systems within the system owner's organization. Another factor may be whether the owner's organization has the freedom to legally assume the role of system operator, if this is not currently part of the organization's standing orders or authority.

In addition to the decision about who does what, there is an overarching decision on how the various responsible agencies will interact with each other. The overall structure for cooperation can take several forms, listed as follows:

- *Centralized operation:* An existing organization or agency takes on the lead role and operates the system in a centralized fashion on behalf of other organizations. Alternatively, the various organizations involved may decide to create a new organization with representation from each participating organization. In either case, the single entity acts as the leader in terms of system operation.

- *Decentralized operation:* Each organization decides to maintain its current jurisdictional boundaries and keep control over its current responsibilities. Some form of data, information sharing, and collaborative agreement is struck in order to support regional coordination and synchronization. This typically involves the adoption of a peer-to-peer "network of equals" approach, rather than defining and nominating a leader.

## 9.10 What Data or Information Needs To Be Shared?

The logical view of the ITS Future Big Picture is again useful at this point in the process. If it has been developed correctly, the logical view will define and describe the processing that needs to be carried out within the system. It will also indicate precisely what data needs to be shared between logical processes in the ITS Future Big Picture. This provides an excellent basis for identifying roles and responsibilities for each agency and stakeholder involved in the ITS development. This will of course be the basis for an "ideal" institutional/organizational structure as it defines the arrangements that would produce maximum efficiency and synergy and takes no account of current organizational arrangements. However, it does provide a great starting point.

Typically, a single organization will be responsible for each process, so it should be fairly easy to determine who should share the data with whom as well as the data content, format and structure to be shared. If it turns out that it is not practical for a single agency to take responsibility for a single process, we would recommend that the logical view be revised to take this new information into account.

## 9.11 What Data Will Not Be Shared and Why?

There may of course be data that the participating organizations decide will not be shared. In order to support the development of effective agreements, it is important to identify and define the data not being shared as well as the data that will be shared. It can also be very helpful to catalog the reasons for not sharing the data, particularly if the decision is based on the current institutional or legislative framework.

## 9.12 How Is Data Shared Between Agencies?

It is not enough to be able to define the data and ultimately the information that needs to flow around the proposed system. There must also be an organizational framework of agreements and understandings between agencies to enable the data to be exchanged. The terms and conditions under which the data will be shared needs to be defined and described to the mutual satisfaction of the sharing agencies, and any internal barriers to striking such agreements need to be identified and addressed. There must also be a definition of the proposed media channels and protocols to be used to support the necessary flow of information. These are defined in the technical layer of the ITS Future Big Picture.

In many cases, especially where the organizations involved are all in the public sector, it is possible to develop or "strike" an information exchange bargain or agreement. This involves the definition of a *quid quo pro* arrangement for swapping data. In many cases, organizations working in the public sector in transportation will have existing agreements on sharing responsibilities and perhaps on exchange of data and information. These could form the basis for agreements to support the proposed ITS, or they could form constraints to the agreements required.

## 9.13 Who Pays for the System Implementation and How Is it Funded?

There are a whole set of issues associated with the question of who pays and what funding mechanism is to be adopted. These are described in Sections 9.14–9.18.

## 9.14 What Needs To Be Funded?

If you are going to ask someone to pay for something, then you have to be able to describe clearly and concisely what it is they will be paying for and what the key characteristics and benefits of the proposal will be.

## 9.15 What Potential Funding Sources Are Available?

In many cases, especially in the public-sector funding arena, the number of potential funding sources available is sensitive to the detailed nature of the proposal. For example, if the proposed ITS is primarily aimed at increasing the efficiency of the transportation network, but it can also be shown to have a significant effect on air quality through reduced emissions, then it may be possible to target environmental budgets as well as transportation budgets.

By developing the ITS Future Big Picture and considering funding options before detailed design, it is possible to revise and amend the proposed solution in light of funding alternatives. This also enables consideration of the terms, conditions, and constraints associated with gaining access to the funds, allowing approaches to deployment to be modified to ensure that requirements for funding are fully met. This is especially important in the funding of ITS, as many existing budget categories may have been established before the advent of ITS, making it necessary to review the requirements for funding in detail and confirm that the proposed ITS is eligible. In some cases, minor modifications or enhancements to the proposed technical solution can open up significant funding alternatives.

## 9.16 Who Pays for the Maintenance and Upkeep of the System?

In addition to the capital investment required to develop and establish the proposed ITS, it is important to identify and confirm the source(s) for operations and maintenance funding. The system may be self-sustaining, utilizing revenue generated from the operation of the system to cover the cost associated with operations, maintenance, and renewal. Alternatively, it may be necessary to provide additional funding input for operations and maintenance. Many government budget allocations are specifically designed to support capital improvements only, so careful review is important.

## 9.17 Who Provides the Subsidy or Takes the Profit?

It is also necessary to identify the need for subsidy or additional funding input to the system and consider who will provide the additional funding. Of course, it is also likely to be the case that profit is generated, and it is equally important to identify who takes the profit. Ideally, this will be directly related to the question of "who pays for the system?" Both questions need to be considered and answers agreed upon and developed. The commercial layer of the ITS Future Big Picture would be used as a tool to support this work and identify the answers which were agreed upon.

## 9.18 Is Private-Sector Funding a Possibility?

This is a very complicated issue, as there are many possible variants for organizational structures. Possibilities include the creation of a new centralized organizational structure to administer and direct the whole ITS, the establishment of a peer-to-peer operation in which a network of equally empowered organizations operate specific parts of the system, or some hybrid organizational arrangement based around existing institutional/organizational structures.

## 9.19 What Organizational Structure Best Supports the New System?

With regard to private-sector funding, there is a great deal of interest in the notion of public-private partnerships in ITS. It typically boils down to one simple notion:

> Can we get someone else to invest resources to satisfy our needs, objectives, and issues?

This is usually conditional on it being possible to define a set of conditions for operating the system under which a private-sector entity can develop a reasonable business opportunity while simultaneously satisfying or addressing our objectives. One of the primary decision factors is the willingness of ITS users to pay through direct-subscription mechanisms for services and information provided. If such mechanisms exist and the users have expressed some willingness to pay, then it is potentially a private-sector opportunity. The private sector may also consider a range of additional factors before deciding to invest the resources required to exploit the opportunity. These include the following:

- *Risk identification, assessment and mitigation:* Can the risk associated with exploiting the opportunity be adequately identified, assessed and managed?
- *Fair sharing of risk and balance of degree of risk against potential reward:* The private sector needs to understand where the risk will lie and the size of the risk compared to the size of the likely return.
- *Public-sector preoccupation with competition and politics:* Will there be some constraint imposed by public-sector regulations on

competition, or political dimensions that would make the opportunity less attractive?

We will address many of these issues in more detail in Chapter 17, which discusses the commercial view of the ITS Future Big Picture.

## 9.20 Who Controls the System and Determines the Effects on the Transportation Network?

This could also be framed as "who controls what part of the system?" as it may well be the case that the overall system is composed of multiple interacting subsystems. Moving beyond the planning, development, and implementation of the ITS, we enter into the realm of operations. Since the way in which the system is operated will have a significant effect on the results obtained, this is a crucial issue. It is also linked to the policy of the operating organization. Therefore, the question of who operates really comes down to the question of who controls the system and whose policies are realized through the operation of the system. As an example, consider the operation of a UTC subsystem, utilizing coordinated traffic signals to control and manage the flow of traffic through an urban area. Using the same subsystem, it would be possible to implement a range of management strategies providing widely varying effects on the traffic. One strategy might minimize traffic emissions, while another might maximize vehicle speeds through the urban network.

What policies will be applied in the operation of the system or subsystem? If there are multiple organizations involved, it may be necessary to synchronize or agree on policies that will be applied. For example, a transit agency may hold a set of policies designed to maximize the use of public transit, while a city or county agency operating traffic signals may have a set of policies designed to maximize the throughput of cars on the urban road network. It may be necessary to carry out major policy reviews, or just make minor adjustments.

An interesting aspect here is that the normal policy definition process, under which policy is defined first then used to drive applications and implementations, may not always work for ITS. The fact that ITS technologies are new and changing rapidly provides the possibility that the policies may be influenced or changed by the policy makers' changing awareness of the possibilities. This accounts for the fact that most policies are not developed in a vacuum but are developed within the pragmatic context of what can be done.

This gets back to the whole "what?/how?" cycle. Policies specify "what" has to be done, but they take good account of "how" it might be achieved; otherwise, they would not be sound, practical policies.

This issue of operation and control becomes particularly relevant in ITS where the interaction between subsystems and the consequential effects on one subsystem operation caused by the operation of another make it virtually impossible to operate your part of the system in "splendid isolation."

## 9.21 Who Will Be the Implementers?

This might sound similar to the question about who pays and owns, but it is really a different question. When we talk about implementers we mean the organization that actually initiates, guides, and controls the planning, development, design, and implementation of the ITS.

## 9.22 What Are the Roles of the Implementers?

This also might seem like a trivial issue, but in a multiagency, multi-organizational context it becomes important. It is likely that many organizations will work in parallel, each implementing a particular piece, or working within a collaborative structure to implement a whole system. Either way, it is important that respective roles and responsibilities in terms of executive actions, funding, operations, and other aspects be clearly defined and understood by all parties.

## 9.23 What Is the Role of the Market?

One of the significant differences between ITS and other transportation initiatives and projects is the potential role of the market. There are many facets of ITS, especially relating to traveler information and consumer electronics, that have significant potential for market dynamics.

## 9.24 Does the ITS Fit With Regional and National Transportation Planning Processes?

One of the critical issues for central and local government agencies is how best to "mainstream" or coordinate ITS planning and development activities with ongoing and existing planning processes for "conventional transportation." Many ITS activities do not fit within existing planning frameworks making it harder to identify budget headings and funding sources for the proposed implementations. This also makes it more difficult to compare ITS projects against conventional transportation projects when it comes to allocation of funding priorities.

## 9.25 Whose Responsibility Is Information Security?

With so much data and information being collected, stored, and communicated, it is likely that some of it could be sensitive or confidential. The issue of responsibility for looking after the security of the data then surfaces as an issue. Even if the data or information is not confidential or sensitive, it may have to be protected to ensure its reliability and accuracy.

## 9.26 Who Owns the Intellectual Property Rights?

This has become a very difficult and complicated area, especially with respect to electronically stored information and software. For example, if a software vendor sells an off-the-shelf software package to a client and a third party enhances and extends the usefulness or functionality of the software, then who owns the new package? Contracts, procurement agreements, and licenses have to be developed to provide satisfactory answers to these questions. It is also important to identify and address these questions before the final decisions on designs and implementations are made. Vendor flexibility with intellectual property rights could turn out to be a significant consideration in design choice.

## 9.27 ITS Application Scenarios

In this section, we try to bring the whole question of institutional/organizational issues to life by moving from the general and abstract to the specific. Through the use of a few ITS application scenarios, we illustrate the kinds of people and organizational interactions required to support real ITS applications. We have chosen two ITS applications or subsystems as examples or cases for this section—traffic management (incident or non-recurring congestion management in particular), which is discussed in Section 9.27.1, and electronic payment services, described in Section 9.27.2. These were selected on the basis of ease of explanation and to highlight significant points. They were also selected because they happen to be among the most popular applications of ITS around the world at present.

### 9.27.1 Incident Management

Figure 9.4 illustrates the general sequence of events that occur in the course of incident management.

These are examined one step at a time as follows:

- *Detection:* An initial indication that traffic conditions are not normal;
- *Verification:* Confirmation that there is actually an incident and that it requires some form of response;
- *Response:* The actions required to respond to the incident and begin the process of incident scene management and dispatching of resources;
- *Clearance:* The actions required to clear the incident from the roadway and return the roadway to normal operation. This would typically involve fire, police, ambulance, tow trucks, and hazardous materials teams as the nature of the incident dictates;
- *Traffic management:* The control and management of traffic flows in the vicinity of the incident and on the surrounding road network both during the incident and during the network recovery period after the incident—including traffic diversions, lane or road closures, and the issuance of appropriate driver and traveler information.

Figure 9.5 takes another look at this incident management process in terms of the people and organizations involved and the interactions required

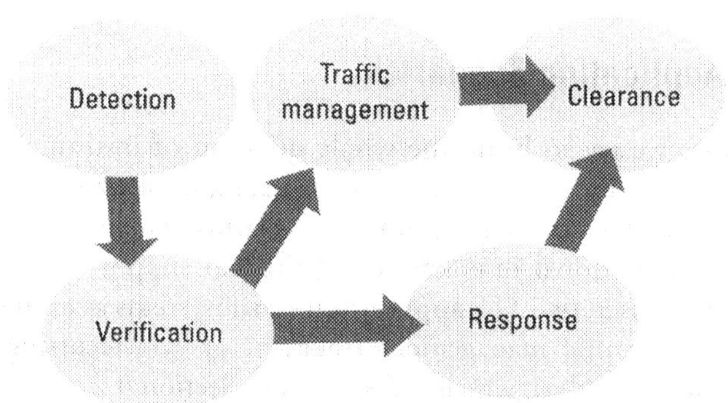


**Figure 9.4** The steps involved in the incident management process.

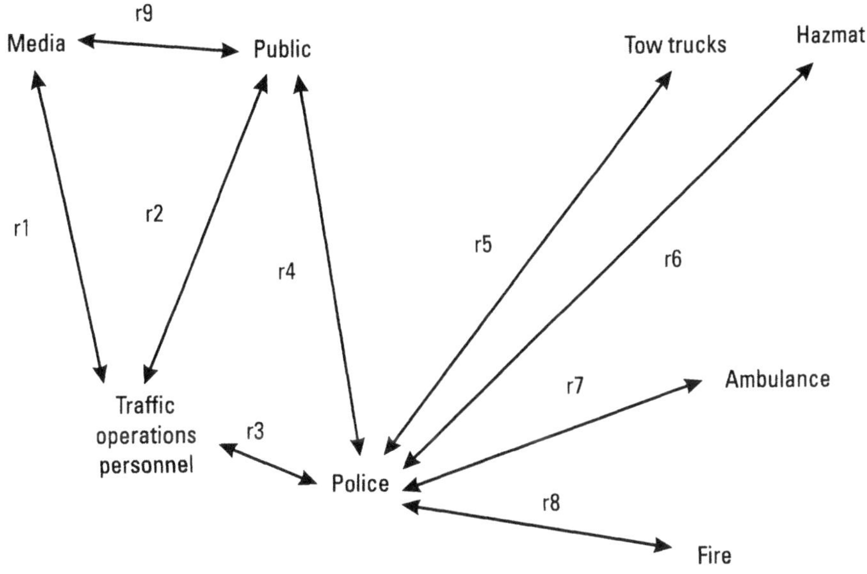


**Figure 9.5** Incident management: "People" relationships.

between them. The figure shows the same process as in Figure 9.4, listing the people who could potentially be involved in each step and the interaction required between them. Note that Figure 9.5 assumes a particular model for interorganizational cooperation during incident management and does not represent a general model for all incident management. Processes have been replaced by people or organizations and data flows by interactions or communications between people or organizations. The logical view of the ITS Future Big Picture would represent these as data flows between processes, suggesting that computer systems do the communications. However, it is equally likely that the processing is carried out by people and that the communication between processes takes the form of fax, e-mail, radio, and telephone messages.

There are many different types of relationships required to support the overall incident management process. Some require that the people and organizations involved be located in close proximity; others require that procedures be planned, predefined, and agreed upon in advance of the incident occurring. For example, the implementation of diversion routes during the incident would require prior planning and agreement among all the road agencies and jurisdictions involved.

Examining the eight relationships shown in Figure 9.5, one-by-one, reveals the following information:

- *Relationship r1:* Between the media and traffic operations personnel. This relationship would be developed to support the traffic management step in the incident management process. Media would be TV, radio stations, or other independent service providers that may provide information to drivers. This would be a predefined relationship and tend to be more formal than informal. The relationship with the information service provider may even take the form of a contract for the supply of incident data.
- *Relationship r2:* Between traffic operations personnel and the public. This relationship would be supported by VMS at the roadside or information dissemination systems such as kiosks. The relationship would be formal and predefined in terms of the data to be exchanged. Data would also come from the public to the traffic operations personnel in the form of reported incidents. The relationship would be utilized during the detection, verification, and traffic management steps.
- *Relationship r3:* Between traffic operations personnel and the police. This is probably the most critical relationship in the process as it would involve frequent exchange of vital information to and from both organizations. The relationship would be formal and predefined so that individual staff know the contacts to be made and the procedures to be followed under different scenarios. The relationship would be utilized during the detection, verification, response, and traffic management steps
- *Relationship r4:* Between the public and the police. This would be an ad hoc relationship to support specific incident reports and inquiries. The channels of communication would be well defined. The relationship would support incident detection and verification.
- *Relationship r5:* Between police and tow truck operators. This would be a formal predefined relationship, perhaps defined in a formal contract for the provision of towing services for specific sections of highway. This would be utilized during the clearance step shown in Figure 9.5.
- *Relationship r6:* Between police and hazardous materials handling team.
- *Relationship r7:* Between police and ambulance.
- *Relationship r8:* Between police and fire.

Relationships r6–r9 would be predefined formal agreements with defined communications procedures and protocols. The relationships would be utilized in the "clearance" step shown in Figure 9.5.

- *Relationship r9:* Between media and public. This would have predefined communication channels but the relationship would be ad hoc, as it would be specific to each incident. The relationship would be utilized during the traffic management step.

Each of the above relationships could be analyzed and characterized using additional parameters as necessary. The objective would be to shed enough light on the nature of the relationship required to support the proposed technical solution. This would then enable the written agreements and procedures to be developed and agreed upon.

### 9.27.2 Electronic Payment Systems

In this context, we are considering electronic payment systems to be region-wide, multiple-application systems capable of servicing the needs of toll collection; transportation payment, including transit fares and commercial vehicle operations fees; and general purpose payment for nontransportation goods and services. Figure 9.6 illustrates this context.

The following three relationships would take the form of a standard customer-service provider relationship. (The driver or passenger pays for parking, tolls, and transit services within the terms of a standard contract.)

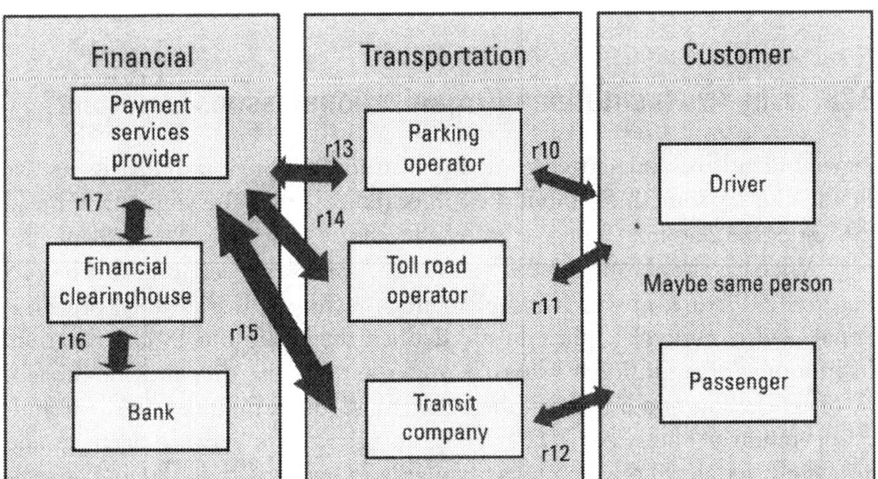


**Figure 9.6** Electronic payment systems: "People" relationships.

- *Relationship r10:* Between the driver and the parking operator;
- *Relationship r11:* Between the driver and the toll road operator;
- *Relationship r12:* Between the passenger and the transit company.

The following three relationships would be configured to the specific circumstances but would be formal predefined relationships similar to the customer/service provider relationship.

- *Relationship r13:* Between the parking operator and the payment services provider;
- *Relationship r14:* Between the toll operator and the payment services provider;
- *Relationship r15:* Between the transit company and the payment services provider.

Financial instruments and agreements would govern the following two relationships, which would be formal, predefined arrangements.

- *Relationship r16:* Between the bank and the financial clearinghouse;
- *Relationship r17:* Between the financial clearinghouse and the payment services provider.

## 9.28 Why Are Institutional/Organizational Issues Important?

Having identified and started to define institutional/organizational issues, we should also say something about their importance; otherwise you may not read the rest of the chapter!

We have already introduced you to the "three-layer" concept for the ITS Future Big Picture, in which complementary technical, institutional/organizational, and commercial solutions are defined separately but in coordination. Institutional/organizational issues are important because they provide the fabric of agreements and understandings that are required to underpin the technical operation of the system. They relate to agreements between agencies and ultimately people who have understood and appreciated the value of cooperation and have decided to identify and define formal mechanisms for such cooperation. For successful implementation of ITS this is crucial.

You may believe that this should be a fairly straightforward job, as agencies responsible for delivering transportation within a specific region probably have such dialogs and agreements already. Well, this is usually true, but the nature of the way we organize ourselves to develop and deliver transportation services makes it difficult or impossible to make use of these existing arrangements. Over the years, transportation providers have, by necessity, become specialized and focused into very polarized interest groups. This has resulted in very effective and efficient mode- and area-specific transportation, but it is not always conducive to the adoption of a holistic view of transportation across a region.

There are a number of factors that describe the importance of institutional/organizational issues. They are detailed in Sections 9.28.1–9.28.5.

### 9.28.1 Systems Are Hardware, Software, Data, and People—We Cannot Do it Without the People

When you consider how ITS operates, it becomes obvious that it would not work without the involvement of people. Typically, systems work as a matrix of people and relationships supported by technology. In many cases, the hardware, software, and data that make up the "conventional" view of systems cannot achieve their objectives without the active involvement, positive support, and cooperation of a range of key people. It is worth making the distinction at this point that the people participation must be valid and worthwhile and not simply "being there." The question of "ownership" in the system and the key people believing that the system works and is effective in helping them to do their jobs, taking them toward objectives and goals that are considered worthy.

### 9.28.2 Efficient Operation, Cost Saving, Maximizing Available Synergy, and Avoiding Costly Duplication

These goals can all be achieved by the technical solution or layer, but only if the required institutional/organizational arrangements are also present and fully functional.

### 9.28.3 Creating New Opportunities for Customer Service

In addressing institutional/organizational issues, we can often identify and exploit new opportunities for enhanced customer service. Consider the implementation of a regional electronic payment system, servicing toll collection, transportation payment, and general-purpose payment. If institutional/organizational issues have been identified and addressed then the toll road operators will be able to offer their customers the convenience of a single form of payment for all modes of transportation. They may also be able to offer toll road users discounts for the purchase of nontransportation services.

Similarly, if the relationships have been developed between the traffic managers and the public transit operators in a region, then they can collectively serve their customers better. This would be achieved through the balanced operation of both traffic and transit management systems and through information and data sharing between the respective organizations.

### 9.28.4 Providing Seamless Service to Customers

Many of the interactions between the ITS and the customer revolve around people and organizations. The customer probably will not see or even be aware of the information and communication technologies being utilized to produce the end results. However, if a customer receives conflicting information from different people in the same organization or in separate organizations then the view is negative.

### 9.28.5 Ensuring That the Needs of the Customer Come Before the Needs of the Organization

Identifying, defining, and exploring institutional/organizational issues in ITS inevitably leads to the critical examination of both existing and future roles and directions of organizations. This, in turn, ensures that the organization maintains a focus on the needs of the customer or user group served by the organization. In fact, one of the ways we have approached the development of consensus on some projects is to focus on the needs of the customer as a convergence point.

## 9.29 How Can You Manage Institutional/Organizational Issues Effectively?

Having identified, defined, and explored all these issues, the question is now how can we successfully deal with them within the context of the development and planning of our ITS.

### 9.29.1 Identify the Key Stakeholders and Participants

This has already been addressed at the requirements definition stage described in Chapter 5, but as discussed there, it is wise to review the cast of characters to ensure that there are no significant omissions. At this point, the review should be focused on making sure that all the participants required to operate and maintain the various subsystems have been identified and included in the development process.

### 9.29.2 Listen to What the Important Stakeholders Are Saying, Understand Them and Their Organizations, and Determine What Drives Them

Most stakeholders will have well-developed views and opinions driven by the objectives and culture of their organizations. These need to be identified, defined, and explored as a step toward identifying common goals and objectives. Sometimes at this stage, you encounter something we refer to us *turfism.*

This is a term we have heard coined to describe organizational intransigence. For their own internal reasons, some organizations decide to be protective rather than outward-looking and decide to protect their "turf" at all costs and with no regard to real objectives and benefits. This is actually a very common phenomenon. We think it is so common because of the polarized way in which we currently organize the planning and operation of our transportation systems. As you may well encounter this in the course of addressing institutional/organizational issues for ITS, here are three general approaches to managing turfism.

1. *Common threat:* The identification and definition of a common threat to bring all parties to a position where they are prepared to work together. This could take the form of a change in technology or a change in the regional transportation context that poses a threat to the current way of doing things and forces consideration of change.
2. *Common objectives:* A more positive approach is to facilitate the identification and agreement of common objectives around which all parties can rally.
3. *Common funding:* A more direct approach is to simply make it a condition of funding that all parties must cooperate. This can be a viable option if local agencies and organizations are seeking central government funding. The funding can be made conditional on regional cooperation.

### 9.29.3 Develop a Detailed Understanding of the Existing Institutional and Organizational Arrangements

This is another one of those data collection and inventory tasks. Much of the information should have been obtained during the requirements definition activities, but it is worthwhile presenting, reviewing, and revising the picture of current institutional/organizational arrangements.

### 9.29.4 Understanding the Stakeholders and Organizations Involved and What Drives Them

One of the objectives in defining and addressing institutional/organizational issues is to find ways to coordinate the cultures and philosophies of separate and sometimes widely differing organizations. These will reside in both the public and the private sectors.

Organizations tend to evolve structures and activities that support their primary focus. However, there can be many variants all capable of supporting that focus. The particular variant chosen is a function of the "culture" of the organization. Here are a couple of cultural variations as examples:

- *Action-oriented:* Organizations focused on immediate actions and results oriented approaches; typically involve fast moving dynamic changes in context with reactive behavior from the organization.
- *Planning-oriented:* Organizations with a focus on longer term effects involving slowly changing parameters. The organization tries to interpret market trends and future customer needs, developing services or products slightly ahead of market demand and invests resources in long-range and strategic planning.
- *Single decision point:* Either one person has sole and ultimate responsibility for taking decisions, or a small team has this responsibility.
- *Matrix management:* A network of peers shares the responsibility for making decisions and providing direction.
- *Service-oriented:* An organization that has evolved to support the delivery of a service to a defined group of service users.
- *Product-oriented:* An organization that has evolved to support the development, manufacture, and delivery of products to defined consumers within identified market segments.
- *Project-oriented:* Organized around the effective execution of defined projects requiring short-term teams to be created to address the needs of specific projects.

This is by no means a comprehensive list of possible organization cultures. We are not experts in organizational management and do not want to give you the impression that this represents the limits of this subject area. As in most of the book, we are trying to provide you with some thoughts and ideas and a few examples so that you can develop you own more detailed approach. As we have stated before, ITS is a multidisciplinary subject with multiple interfaces between professional disciplines. This topic represents the interface between transportation and management science.

### 9.29.5 Share Your Knowledge of What Institutional/Organizational Options Are Available to Them

Based on experiences with other clients on other projects, it should be possible to provide input and suggestions on what institutional/organizational options are appropriate for the regional context to support the potential technical solution. This might include model organizational structures or proposals for addressing identified institutional/organizational issues.

### 9.29.6 Develop a Strawman Institutional/Organizational Framework Based on What You Have Heard and Your Knowledge of the Options Available

This is an extension of the previous step. Structuring suggestions and input on institutional/organizational options in the form of a strawman or model for institutional/organizational arrangements has proved to be a valuable activity for us on other projects. This can then be shared with all parties and used as a tool for communication, understanding and mutual learning. The strawman can either be abandoned or revised as part of the discussion process.

### 9.29.7 Draw Up a List of Organizational and Institutional Issues That Have Been Identified

During the course of discussions on proposed institutional/organizational arrangements, it should be possible to identify and define a series of specific local or regional issues. These should be recorded and agreed upon by all the participants.

### 9.29.8 Evaluate all Issues and Ensure That They Have Been Addressed by the Identification and Definition of Relationships in the Institutional/Organizational Framework

Here again the advantages of developing a Future Big Picture, before launching into detailed design, are manifest. The Future Big Picture provides the framework within which the key stakeholders can be identified and their respective roles and responsibilities defined. We have found it useful to evaluate organizational relationships by considering the following factors:

- *Nature of the relationship:* Ad hoc or systematic; does the relationship have to be supported by a predefined structured agreement, or will a number of ad hoc independent agreements do the job?
- *Formality:* Formal or informal?

Informal relationships are typically supported though ad hoc meetings and telephone calls; these informal arrangements can be very effective as they are typically based on solid interpersonal relationships. They can be effectively focused on specific subject areas of mutual interest.

However, they can be difficult or impossible to coordinate and direct toward a common goal. On the other hand, formal relationships are typically supported through the establishment of a formal predefined mechanism or structure to support the relationship. This would be supported by regular, prescheduled meetings with formal agendas and minutes. This may be defined in a memorandum of understanding or some other form of written agreement.

In a typical organizational structure, a combination of formal and informal relationships would be utilized to support effective operation of the organization. There are a number of additional parameters that may be used to characterize a relationship as follows:

- *Frequency of use:* Frequent or infrequent; the frequency at which the relationship will be utilized to support interaction between subsystem components can be a factor influencing the nature of the relationship. Frequently utilized relationships are also likely to be more critical to overall system performance.
- *Criticality:* Subjective measure of how important the effectiveness of the relationship is to the operation of the whole system. If the relationship under consideration is deemed to be critical to the success or failure of the proposed ITS, then it would be handled with more care and attention than others. Identifying and understanding the critical relationships is key to ensuring that resources deployed in the development of the institutional/organizational framework are invested efficiently. This does not mean that other less critical relationships should be ignored, just that the critical ones should be given priority if resources are limited.

- *Level:* What staff levels in the respective organizations participate in the relationship? For example, board- or director-level involvement could be categorized as high-level contact or interaction between respective boards or directors. Typically, this involvement would address questions of policy or overall direction. Senior management involvement could be viewed as middle-level contact between peers in respective organizations. Typically, this involvement addresses the attainment of higher level common goals and objectives, but it could also be involved in conflict resolution on conflicts that cannot be resolved at the operational level. Operational staff involvement could be viewed as day-to-day issue identification and agreement on issues, conflicts, and operational goals and objectives.

### 9.29.9 Draw Up Drafts of the Agreements That Are Required to Support the Defined Framework

We have found that this works well as a cooperative effort between the system developer and the stakeholder organization staff. Organizations are better able to understand the form and type of agreements that are workable while the ITS developer can bring specialist knowledge of the needs of the technical solution and also model agreements from other projects.

### 9.29.10 Draft Recommendations for Policy Definition Activities and Institutional Changes That Should be Considered

In addition to the agreements, policy and institutional changes can be defined as recommendations at this point.

### 9.29.11 Make the Necessary Revisions to the Technology and Commercial Layers to Synchronize all Three

As discussed in Section 9.2, it may not always be possible to change the existing institutional/organizational arrangements, so it may be necessary to revise the proposed technical solution instead. Even when it is possible to change the existing institutional/organizational arrangements, it is highly likely that you will have learned something more about requirements in the course of developing the institutional/organizational framework. Therefore, it is highly likely that the technical solution will be revised or amended at this stage.

One aspect that is important to discuss at this point is the sequence of events. The steps we have described in Section 9.29 assume that you are attempting to develop consensus and agreement on institutional/organizational issues and arrangements before proceeding with detailed design and implementation of the ITS. This is obviously the most logical way to approach it, as the respective roles, responsibilities and relationships can all be clearly defined and understood prior to major investment. We would endorse an approach using this order.

There is an alternative sequence. This involves a single, strong, well-funded stakeholder who has decided to catalyze the required institutional/organizational arrangements by designing and deploying at least the first stage of an ITS that meets the initial requirements of the single stakeholder. The driving organization uses the existence of the deployment as the focus for further consensus on institutional/organizational arrangements.

While this may seem illogical, it has proven to be an effective approach to managing intransigence and lack of interest from other key stakeholder organizations. Faced with a tangible deployment, rather than a planned one, many action-oriented organizations produce a better response.

In explaining many of the potential institutional/organizational issues and problems, we may have left you with the impression that institutional/organizational issues associated with ITS are overwhelming and close to impossible to counter. However, there are many approaches to managing institutional/organizational issues, some of which we have discussed in the preceding sections. In our experience, these can be utilized to successfully address the issues and problems, provided that the need to do so is identified at an early stage in the development process. We have seen previous examples of development efforts where the institutional/organizational issues were ignored until much too late in the development process, leading to great difficulties. Institutional/organizational issues are not insurmountable, but do require careful management and attention.

# 10

# Developing a Logical View of the ITS Future Big Picture

## 10.1 Introduction

This chapter aims to walk you through the development stage of the logical view within the Cooperative Development Methodology and provide you with some practical guidance on how to develop a logical view of the ITS Future Big Picture.

## 10.2 What Is a Logical View of the ITS Future Big Picture?

The logical view of the ITS Future Big Picture is our first foray into the realm of the *how* part of the "what?/how?" cycle. It tells a particular story about the proposed ITS. Starting on the basis of the requirements defined in the needs modeling part of the Cooperative Development Methodology, this represents the starting point for solution definition, or the first move to the *how* part of the "what?/how?" cycle. As its name suggests, the logical view provides the system developer with a picture that can be used as a tool to illustrate and explore the components required to address all the requirements. As it is the first step into the solution part of the system development process, it does not leap straight into the definition of technologies or the identification of products. Instead, it stays close to the *what* part of the cycle by focusing on what processing should be done and what data flows should be supported if the system is to meet all requirements and work cohesively. This enables the system developer to remain as objective as possible about the proposed solution by enabling the solution to be defined and explained without reference to particular technologies. This is important as the migration to specific solutions narrows down the developer's field of vision. Do this too early and some very good potential solutions may be ignored and too much investment in a particular design path or approach may be committed too early. Here again we are utilizing a tool to support the continued effective use of the "what?/how?" cycle.

You may have noticed that we talked about the logical view in terms of it being a tool for the system developer but have not mentioned the stakeholder. This is because the logical view actually does very little for the stakeholders in terms of direct delivery of results. You could consider the logical view as merely a stepping stone toward a deliverable that provides real benefits to the stakeholder group. However, it is a very important stepping stone that provides some real benefits to both the developer and the stakeholder—eventually.

The logical view builds an initial picture of the emerging ITS by identifying and describing the processes that have to be carried out, where they should be grouped together, and what data should flow between them. As described in Chapter 11, the picture produced is a purely logical representation of the proposed solution in that the solution described is "technology-independent" or in other words, you could follow the logical view using good old-fashioned donkey work with no fancy computers and communications technologies. The logical view sets out a procedural way of getting the job done in the most logical manner. This, in turn, means that the approach is both efficient and comprehensive in that the processing activities are grouped together to make life as easy as possible, and the whole system is depicted as interconnected work activities. Each work activity is optimized to obtain maximum productivity from the data supply and linked to neighboring activities to ensure that the whole work activity is coordinated like a beehive. It is a model showing you how to satisfy the needs of the stakeholder group without selecting technologies or methods.

## 10.3 Why Do You Need a Logical View?

We have had a lot of difficulty in explaining the need for a logical view of the ITS Future Big Picture to stakeholder groups. After many sessions in which we could almost see the shutters coming down over stakeholders eyes, minds, and hearts, we came to a simple realization. Stakeholders do not see any value in this step because it does nothing for them. It does not pass the "What's in it for me?" (WIIFM) test and is consequently a real tough sell to them. In fact, we have come to the conclusion that it is better to not even try to sell it to them. Instead, we prefer to use it as a developer's tool and make it work for us as we move toward meaningful output for the stakeholders to consider. This does not mean that we are recommending that this step be skipped. Quite the contrary, this step is a vital part of the overall Cooperative Development Methodology. It provides some incredibly useful output, including that discussed in Sections 10.3.1–10.3.7.

### 10.3.1 Initial Identification and Description of the Key System Interfaces That Must Be Supported

In developing the logical view, all processing required will be defined and grouped logically, and all data required to support the processing will be defined. This provides our first look at the key system interfaces that must be supported. It does not provide a complete description of the interfaces, but it does describe the data that must flow across them and the processes that will make use of the interfaces.

### 10.3.2 A Good Start on the Way to the Definition of the ITS Future Big Picture or Layered Framework

Having defined the processes and the data flows required, it becomes much easier to envision the required physical devices and communications media to support effective operation of the proposed system.

### 10.3.3 A Structured View of the System Providing the Basis for Modular and Robust Software Development

The availability of a clear picture of the processes that need to be carried out and data that should flow between them provides the software developer with a great platform for the development of modular, robust software to support the operation of the system. While there is obviously a great amount of further effort required to produce the desired code or software, the software development process is made more effective through the provision of this initial platform. This should also help to make the software less error-prone and simpler to develop, document, and maintain as the logical view can be used as a reference framework to guide the development of the software. This should also facilitate the continued user focus right through the software development process.

### 10.3.4 A Clear Picture of the Processes and Data Flows to be Supported by the Institutional/Organizational Arrangements to Come Later

As discussed in Chapter 4, a good ITS Future Big Picture has to provide solutions on three levels—technical, institutional, and organizational and commercial. You can make a really good start on the definition of the required institutional arrangements by studying the logical view and identifying what stakeholders will have to support the data flows with procedures and agreements. For example, if the logical view shows a need for data to flow from a traffic management process to a transit management process, then you can make a start on the definition of institutional and organizational arrangements by identifying which agencies within the stakeholder group would be the logical owners and operators of these processes. This may, in turn, indicate the need for a memorandum of understanding or other agreement between the agencies responsible for carrying out each process.

### 10.3.5 A Clear Check That all Identified Requirements are Being Addressed

The logical view has its roots in user needs and requirements. A good and complete logical view will be able to be traced back to the original needs defined in the needs modeling activities.

### 10.3.6 A Tool to Ensure the Most Efficient use of Data and Optimized System Operation

You will see when we start to describe how to put a logical view together that it provides an excellent approach for ensuring that data is input into the system once only, then used multiple times. It also helps you to identify the most logical way of grouping the processing together, taking account of where the data is coming from and going.

### 10.3.7 A Platform for the Development of Standards

Although the entire ITS Future Big Picture will not provide the information required for the identification, definition, or development of ITS standards, it provides a great starting point. The identification, numbering, and structuring of the data flows and processes required facilitate the further work required to define and develop standards by providing a full view of the entire system and a complete, if superficial, view of the required interfaces.

## 10.4 So What do you do With the Logical View Once you Have Created it?

The first job is to use it as stepping stone to the development of the ITS Future Big Picture. Having developed at least the initial version of the logical view, you can proceed to add technology dependence (but not technology specificity) to the picture to get to the technical layer of the ITS Future Big Picture. It also provides you, as the system developer, with an excellent tool to structure your thoughts and devise questions that need to be asked of the stakeholder group. As we said earlier in the chapter, we would not recommend that you make use of the logical view as a presentation or discussion tool with stakeholder groups. However, it should enable you, or the requirements analyst, to develop the required support materials to illustrate the thinking on system development so far.

## 10.5 What Is a Bubble Chart?

A bubble chart, or data flow diagram, represents "what" the system has to do in order to meet the stakeholders' requirements. The following are the main characteristics:

- Decomposition of the system and its functions;
- Purely abstract model of "what" the system has to do, not "how;"
- Does not constitute a physical model of the system;
- Provides a process specification;
- Is diagrammatic;
- Hides unnecessary detail on any given level, facilitating explanation and understanding and removing ambiguity;
- Has integrated consistency checks;
- Is self-indexing.

## 10.6 Developing a Logical View of the ITS Future Big Picture

In order to develop a logical view of the ITS Future Big Picture, a number of concepts should be utilized. These are described in Sections 10.6.1–10.6.5.

### 10.6.1 General Concepts

One of the key concepts in the development of the logical view is what we call the *bubble chart.* You might hear system engineers refer to these by the more formal and correct term *architecture flow diagram* (AFD) or *architecture interconnect diagram* (AID). They are a form of graphical notation used to illustrate the flow of data and the processing to be carried out within systems and subsystems. As they usually look like a lot of bubbles connected together, we call them bubble charts. The general concept of the bubble chart is illustrated in Figure 10.1; it represents what the system has to do in terms of data flows and processing. There is no consideration of processing time; each process is assumed to be carried out infinitely fast, with data output flows immediately passed on to the next bubble as data input. There are four basic components of the bubble chart, listed as follows:

- Processes;
- Data flows;
- Data stores;
- Terminators.

The first two components are fundamental and must always be utilized. The last two are designed for specific circumstances, which are explained in Sections 10.6.4 and 10.6.5, respectively.

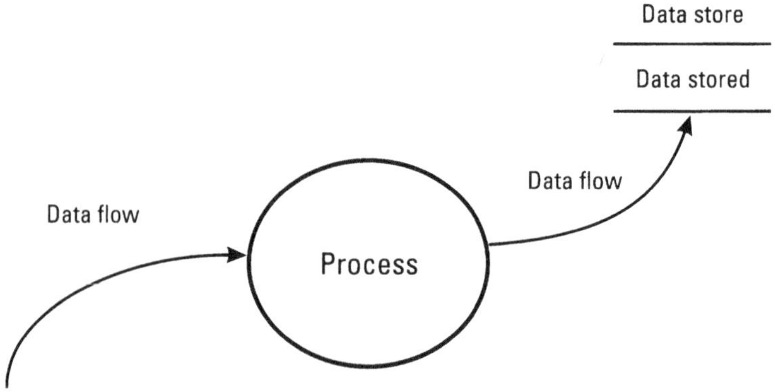


**Figure 10.1** Bubble chart.

### 10.6.2 Processes

These are the bubbles which represent the actions required to convert data input to the bubble, into data output from the bubble. Each bubble should be numbered sequentially from 1, in increments of 1. Each bubble should also have a process label written inside. This should only be as long as necessary to give an indication of the process taking place and differentiate this from any other process. A process label should always consist of a verb followed by an object.

### 10.6.3 Data Flows

These are the arrows, which represent the flow of data around the various components of the system. Data flows can be either data inputs to processes or data outputs from processes. Each data flow should be assigned a short, but informative, label. This should be unique in order to distinguish it from other data flows. Data flow labels never contain verbs.

They can be one-way, or two-way, with one arrow per data group, or item. There are more sophisticated rules governing data flow splitting and merging conventions, but the basic rules given here should be sufficient for most purposes. If you have a particularly complex situation to model then seek further information from *Strategies for Real-Time System Specification* [1].

### 10.6.4 Data Stores

In some cases, the output from a bubble will not immediately be utilized as input to another bubble. In these circumstances, the data output is directed to a data store. A data item placed in a data store is assumed to be available for use by any process, any number of times. The data stays constant until overwritten by new data input from the supplying process.

### 10.6.5 Terminators

These are only used on the overview data flow, or data context diagram. External entities with which the system must interact are represented as rectangles called *terminators.* See Figure 10.2.

## 10.7 Approach

The concepts described in Section 10.6 are utilized to capture the requirements that the system must address in Sections 10.7.1–10.7.7.

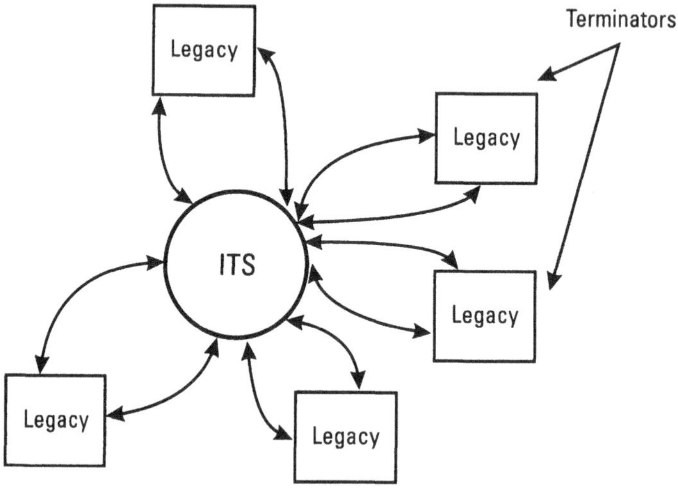


**Figure 10.2** Bubble chart terminators.

### 10.7.1 Construct Overview Data Flow Diagram

This is the first step in the bubble charting process. As illustrated in Figure 10.2, this diagram sets the system to be developed in context with the outside world. External entities such as other departments, outside organizations, and other systems are identified and described by name and by data flow to and from the system. It is assumed that these external entities cannot be influenced or affected by the system development process but that it is essential to interface to them in order to meet the requirements of the system. This diagram is a crucial part of the development process as it shows the major purpose of the system in terms of data input, processing, and data output.

### 10.7.2 Leveling

Once the overview data flow diagram has been prepared and critically reviewed, bubble charts are created in hierarchies, or levels, starting from level 0, which is the most abstract or general level. Once this first level has been completed for the whole system, each bubble can then be further subdivided into more detailed sub-bubbles, or child processes. At each subdivision, or explosion, the newly created bubbles are pushed down to the next level. Hence, the new bubbles created from the explosion of level 0 would be placed on level 1. The new bubbles created from the explosion of level 1 would be placed on level 2 and so on.

The objective is to hide detail in lower levels so that each level can be used to clearly explain the functionality of the system. During each explosion, new bubbles will be created and data flow groups will be decomposed into new data flows. Process and data flow labeling should be carried out as in level 0.

Bubble explosions should be carried out through several levels until the following conditions are met:

- Data flows reach the primitive level where attributes can be associated with each data flow. An attribute is a data characteristic such as units of measurement and acceptable range of values.
- It becomes possible to describe the actions inside the bubble (process specification) in a few lines—and certainly no more than half a page of text.

The resulting output will be a "leveled" set of bubble charts ranging from the data context diagram, through level 0 to level *n*. Note that it is possible to show all the lowest level bubbles or "primitives" on a single chart, but this would be hard to read and understand.

The basic rules that should be borne in mind when creating the bubble charts are summarized as follows:

- Minimize data flows to and from bubbles on any one level;
- No more than five to nine flows per bubble, achieved by two methods:
  - Starving the bubbles: Grouping flows together into single flows;
  - Bubbling up: Splitting the bubble into smaller bubbles.
- Try to concentrate input processing in the same area of the chart to avoid half-processed data flowing around the system.

### 10.7.3 Balancing

One of the most important features of the method is that it enables consistency checking to be carried out very easily. Higher level bubbles must have the same input and output data flows as the next level down or child processes. If a data flow appears from nowhere, or goes to nowhere, then there is an inconsistency, which must be rectified. This graphical method makes it easy to spot these mistakes. If we had relied on text-based methods only, we would have little or no chance of spotting these frequently occurring contradictions, and our credibility with the client would suffer accordingly.

### 10.7.4 Numbering

Each bubble on the level 0 chart will be assigned an integer identification number. This follows classical legal paragraph numbering conventions. See Figure 10.3.

### 10.7.5 Process Specifications

Each bubble created so far has been assigned a unique but rather abstract process label. We need to be more specific and define exactly what we mean by these labels. This definition usually takes the form of a process specification, or Pspec. The process specification text must concisely describe each detail of the functional requirements within each bubble.

### 10.7.6 Data Flow Definitions

We have a similar problem with the data flows. Each data flow on each arrow of the bubble chart has been assigned a data flow label. We need to be more specific and define exactly what we mean by these labels. This definition will take the form of a data flow definition. This should be completed for every data flow created on each level of the bubble chart. The data flow definition text must

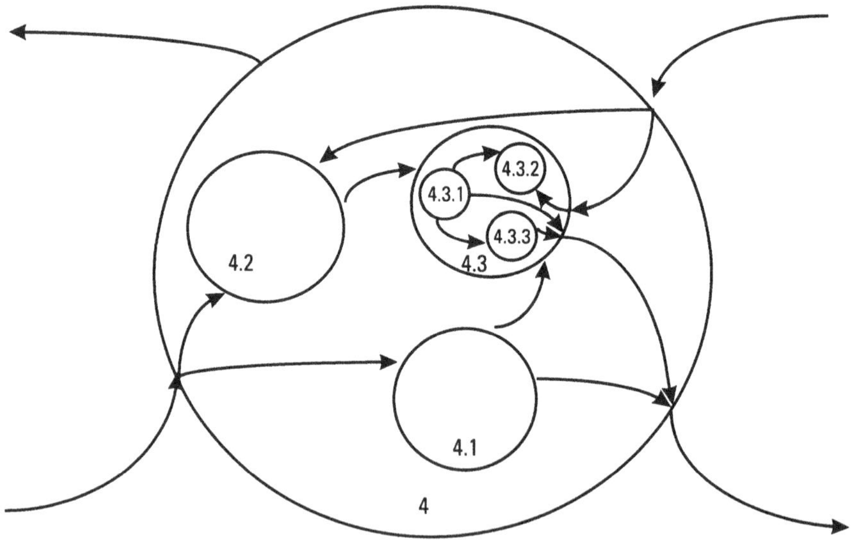


**Figure 10.3** Bubble chart numbering.

concisely describe the data element or data flow group flowing between each bubble on the chart.

### 10.7.7 Response Time Specifications

This is the final component in the needs model. These specifications define acceptable lag times between data input such as requests for information, to data output, or results generation. We typically only consider the time lag between external data inputs and external data outputs (from data input terminals to data output terminals). We usually ignore any time lags introduced between bubbles, assuming that this is a software engineering problem to be dealt with separately.

So we have provided you with a rudimentary introduction to the art of bubble charting by telling you something about the mechanics of the process. You are probably still wondering about the practical aspects of applying these tools. We have found that an effective method of getting started is to have a small team of requirements analysts or experts in ITS get together for a couple of days and put together a very rough first draft of the logical view. The output from this may well be nothing more than a collection of flip chart sheets containing roughly drawn bubble diagrams with half the flows undefined and many of the processes yet to be identified. However, this does provide the raw material for subsequent review by other members of the development team and partial input to a *computer-aided system engineering* (CASE) tool by an operator or technician. As we discussed when introducing the "what?/how?" cycle, this also has the advantage of minimum investment in resources at this point, so the owners of the draft are more inclined to listen and react to suggestions and input. This also provides a mechanism for the utilization of multiple parallel resources to be deployed on the effort as a number of people are empowered to carry out subsequent tasks once the initial draft has been produced. This also has the nice effect of freeing the requirements analysts or ITS experts from the need to be neat and tidy and produce polished output at this stage.

Taking the draft as the starting point, discussion sessions can be held with many review groups and the draft revised until it begins to be credible as a comprehensive logical view of the ITS Future Big Picture.

## 10.8 CASE Tools

Most system developers have their own documentation methods and support tools to facilitate the creation and ongoing revision of the logical view. Therefore, we will not spend a lot of time explaining how we would go about it, other than to say that it is a lot easier to make use of a CASE tool for the job, instead of a white board or a pile of yellow stickies and a big sheet of paper.

Once you get started on the development of the logical view, there will be multiple bubbles on many levels that require significant numbers of consequential changes to be made, should revisions be necessary. We should of course have said "when revisions are necessary" as it will be inevitable. Remember the cyclical nature of the development process! If you are developing even a medium-sized system, it may well save your sanity if you utilize a CASE tool to support the maintenance and configuration management of your ITS. Consider the seemingly small problem where you have five processes required to describe the whole system. Then assume that you will have to decompose each of the five processes into four levels of bubbles until you get to the most primitive level. You then have the job of handling, maintaining, and updating 20 process bubbles and perhaps 60 data flows. With even one small change on one level in one process requiring consequential changes in many other processes and data flows, you can see that even this seemingly small system is difficult to manage from a logical view using manual methods. We would recommend the use of a CASE tool for most ITS planning and development projects for these reasons.

### 10.8.1 CASE Tools That Support Hatley-Pirbhai

To our knowledge, there are currently five CASE tools that support the Hatley-Pirbhai methodology program used in the National Architecture for ITS development program. They all provide some degree of assistance in developing and maintaining the logical view of the ITS Future Big Picture by automating the bubble-drawing process and enabling changes and revisions to be made easily through automatic generation of consequential changes.

Some may be better than others, or only run on specific workstation platforms. The prices for the tools also vary considerably. We did not want to get into a long discourse about the relative merits of different CASE tools as this would distract from the main purpose of this book. We are not system engineers anyway, so our opinion or review of these products would be suspect. Accordingly, where possible, we have provided the current World Wide Web addresses or contact details for each of the suppliers to help you conduct your own review.

Many of the suppliers of these tools provide training on the use of structured system development techniques, including the Hatley-Pirbhai methodology. One training provider of particular note is STG, which provides the services of Derek Hatley, the surviving member of the team that developed the Hatley-Pirbhai methodology.

#### 10.8.1.1 Cadre Technologies

This company has a product called Teamwork that supports the Hatley-Pirbhai methodology. The Teamwork CASE tool from Cadre Technologies was the one utilized by the architecture development teams for the National Architecture for ITS. Details are available at:

Http://www.cadre.com/

#### 10.8.1.2 Iconix

This company's offering is called FreeFlow. Details are available at:

http://www.iconixsw.com/Spec_Sheets/FreeFlow.html

#### 10.8.1.3 STG

The appropriate product from this company is called AxiomSys. Details are available at:

http://www.stgcase.com/casetools/axiomsys.html

#### 10.8.1.4 StructSoft

This company has a product called TurboCASE/Sys. Information on the product is available at:

www.turbocase.com

#### 10.8.1.5 Evergreen Case Tools

This firm's product is known as EasyCASE. We could not find a web site for Evergreen, but contact details are listed as follows:

Evergreen Case Tools
8522 154th Avenue NE
Redmond, WA 98052
(206) 881-5149
Fax (206) 883-7676

## 10.9 Hatley-Pirbhai

The underlying design method adopted by the developers of the National Architecture for ITS was originally developed by Tom DeMarco and subsequently modified for real-time systems by Derek Hatley and Imthiaz Pirbhai. The Hatley-Pirbhai method, as it has come to be known, was used for developing and documenting the National Architecture for ITS and forms the inspiration for the Cooperative Development Methodology described in this book. The method requires the use of specific naming conventions and notations for the processes, data flows, and diagrams and is supported by only a few CASE tools.

### 10.9.1 Data Dictionary

This is another component of the logical view, consisting of a list of data flows defined in the course of developing the logical view. The data dictionary lists all data flows identified in the development of the logical view and parameters associated with each data flow. The exact parameters vary from one CASE tool to another but typically specify the content, format, structure, and label for each data flow in the system. This enables the system developer to quickly check to see if a required data flow has been defined already. It also allows the system developer to quickly check on the format and structure for the data so that it can be used in other processes, if necessary.

## 10.10 Use of the National Architecture for ITS

If you are planning and developing an ITS for implementation and deployment in the United States and plan to attract federal funding to assist at some stage in the project, then you may well want to consider the use of the National Architecture for ITS as a starting point for definition and description of your logical view of the planned ITS. Federal funding may well be contingent on conformance or compatibility with the National Architecture for ITS. A significant component of any guidelines for conformance with the National Architecture for ITS will include support for the data flows and processes already defined as part of this ITS Future Big Picture. Even if your planned deployment will not use U.S. federal funding, it would be worthwhile to review the National Architecture for ITS documents as a starting point for your own work.

The U.S. Department of Transportation has developed a number of tools to help you to gain access to the National Architecture for ITS documents and navigate your way around the information. These include a CD-ROM containing all the documents produced as deliverables for the National Architecture for ITS development program and a hyper-linked version of the logical and physical views of the National Architecture for ITS. There is also a Web site on the World Wide Web, referenced by the CD-ROM but accessible and usable independently. We would recommend that you pay a visit to the Web site and maybe obtain a copy of the CD-ROM. The Department of Transportation has also commissioned the architecture development teams to develop and deliver a training course on how to apply the National Architecture for ITS to your region.

## Reference

[1] Hatley, D.J., and I. Pirbhai, *Strategies for Real-Time System Specification,* NY: Dorset House Publishing, 1987.

# 11

# The ITS Future Big Picture

## 11.1 Introduction

This chapter addresses the development of the central, or keystone, component of the ITS Cooperative Development Methodology: the ITS Future Big Picture. This is developed in step 230 of the ITS Cooperative Development Methodology. It could be argued that the development of the ITS Future Big Picture started much earlier in the process since just about every other step in the process so far has provided information on, and input to, this one. The major activities associated with the development of the ITS Future Big Picture revolve around the synthesis of the output from these other steps in the process to create a conceptual or high-level framework within which our proposed solutions can exist. This is often referred to as a *physical architecture* as it describes the physical or real-life attributes of the proposed solution, in general terms.

Figure 11.1 illustrates the relationship between this step, the development of the ITS Future Big Picture, and the other steps in the process. Note that steps 200, 210, 220, and 250 provide direct input to the development of the ITS Future Big Picture, while 240 and 260 are initially outputs from the ITS Future Big Picture. However, due to the iterative nature of the development process, there is likely to be feedback from steps 240 and 260 to the ITS Future Big Picture, hence the use of two-way arrows between these steps in Figure 11.1.

This chapter aims to provide readers with a description of the ITS Future Big Picture to help in understanding this important part of ITS development and planning. We also hope to explain how and why the ITS Future Big Picture is an extremely useful tool for seeking agreement on key features of the

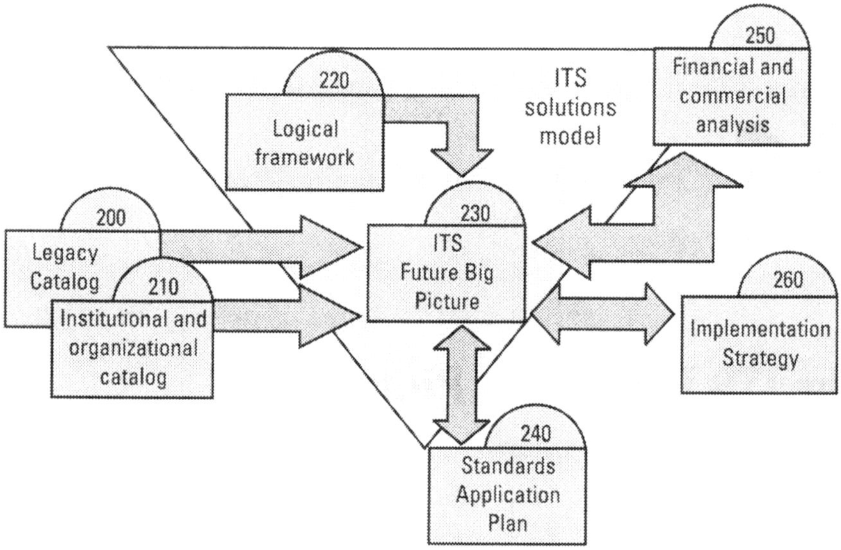


**Figure 11.1** ITS Cooperative Development Methodology steps synthesized to create the ITS Future Big Picture.

proposed solution and sharing information about the planned ITS with others. We have also attempted to provide some practical insight into how to develop one. This is approached from a transportation professional's perspective rather than a system engineering one, so we must caution you that we have not tried to provide all the guidance required to actually develop a robust system engineering architecture. We have avoided detailed technical discussion on how to carry out the system engineering functions required to translate the logical view and knowledge of requirements into a physical architecture. Instead, we have tried to provide an overview of the significant activities. The aim is to develop an understanding of what is happening, rather than equip the reader to actually do the work.

Finally, in this chapter, we explore how to make effective use of the ITS Future Big Picture as a tool during the overall ITS development and planning process.

## 11.2 What Is an ITS Future Big Picture?

The ITS Future Big Picture is a definition and description of all the major subsystems that need to be part of the overall ITS if the requirements identified in the needs model are to be fully satisfied. It also defines and describes the interfaces between each of the major subsystems so that the interaction required can be defined and understood. This includes the identification and definition of appropriate communications media to support the interaction required between subsystems.

This is all developed at a higher level of abstraction than design, meaning that there is a degree of definition on the proposed solutions but that precise definition is not yet possible. Further detailed design work is required to make tradeoff decisions between specific technological approaches. Sometimes engineers and system developers have trouble relating to this fact. Many do not see the need for the development of this type of high-level ITS Future Big Picture and would rather press on to detailed design where the parameters have been established to enable specific solutions to be presented and discussed. The problem with this is that the investment and effort required to move to detailed design can be significant. As discussed in Section 3.2, the main objective of the process is to support the effective use of the "what? /how?" cycle to pin down and stabilize requirements, based on knowledge of technology capabilities and options available.

The ITS Future Big Picture also structures the work activities and processing according to the output from the legacy catalog and institutional/organizational steps in the process. Going beyond the purely logical grouping of the logical view, the ITS Future Big Picture adapts subsystems and communications interfaces to the real way in which organizations and institutions will be configured to develop, implement, and operate the ITS. It also adjusts the logical view to take account of the findings from the Legacy Catalog step in the process, enabling the developer to maximize the incorporation of legacy and the protection of sunk investment.

As discussed in Section 4.3.4, we make use of three separate layers to define and describe the ITS Future Big Picture. We have adopted the three layers indicated in Figure 11.2 for most of our project work. You should note that this differs slightly from the three layers used in the National Architecture for ITS deliverables from the U.S. Department of Transportation. We have chosen a different approach for the following reasons. First, we felt the need to treat the commercial aspects of ITS as a separate layer rather than combine them into the institutional/organizational layer. Second, we have found it easier to explain the development of the ITS Future Big Picture if we combine the "transportation" and "communications" layers into a single "technical" layer. Since the use of the layers is intended to be an aid to communicating and projecting the ITS Future Big Picture, it does not really matter how you slice it up, as long as the approach selected does the job for you. A tool is only as good as the results you

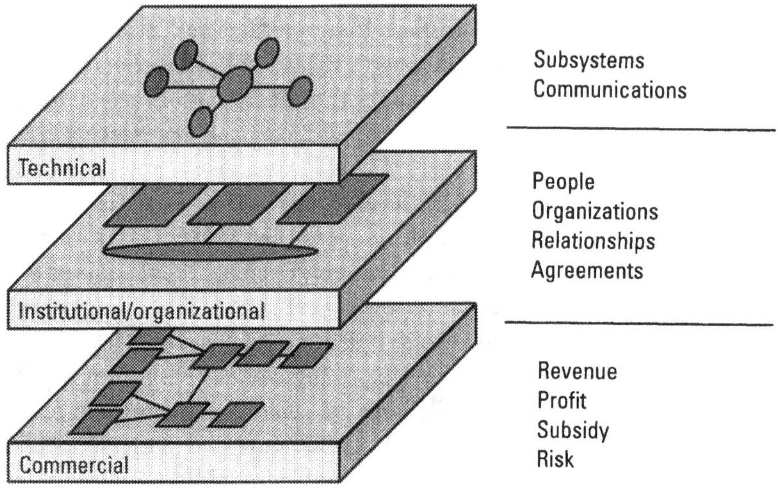


**Figure 11.2** The three layers of the ITS Future Big Picture.

obtain in utilizing it, so we think it is important to evolve and adapt rather than take a "recipe" approach.

We have found this to be particularly important as we set out to apply the principles and methodologies of the U.S. National Architecture for ITS. The deliverables from the development program provide a wonderful source of reference information, concepts, and methodologies for facilitating the development of a regional ITS Future Big Picture or ITS Future Big Picture. However, some of the system engineering principles are new and strange to nonsystem engineers. Taking the original approaches and adapting them to be more relevant and understandable to the group of stakeholders or clients with which we are working has been important in supporting useful dialog and input.

### 11.2.1 Technical Layer

The technical layer defines the subsystems and communication links required for the entire system to work in harmony and address all the stated requirements. The technical layer can be broken down into subsystems and then into smaller units known as ITS Equipment Packages. You may recall from Chapter 7 that an ITS Market Package was defined, from a planning perspective, as the "fundamental building block of ITS": a grouping of technologies aimed at satisfying one or more ITS User Services, the "how" equivalent of the ITS User Service "what." Having moved forward and started to define subsystem boundaries and interfaces, it is necessary to make sure that the building blocks can be procured in a manner that conforms to the newly defined subsystem boundaries. Up to this point, the building blocks (ITS Market Packages) have taken no account of these boundaries. They are configured around the provision of services and the satisfaction of defined needs. Now that we have at least an outline idea of how the physical subsystems will be configured, it is highly unlikely that ITS Market Packages will conform to subsystem boundaries. You may be asking at this point "so what?"

Well, if we try to procure and deploy building blocks or "chunks" of ITS that do not take account of subsystem boundaries then we end up with irrational deployment or implementation sequences. We will cover the development of implementation strategies in detail in Chapter 14, but useful deployment sequences result in interim systems or subsystems that can function and provide benefits. For example, it is highly unlikely that the U.S. interstate highway system would ever have been completed if none of it could have been used until the last section of freeway was completed. Incremental deployment requires that each increment provides some tangible benefit and shows some signs of successful attainment of benefits to encourage continuation to the next phase.

Accordingly, we have to address the issues of ITS Market Package conformance to physical subsystem boundaries. In addition, we have to manage the transition from a user-driven conceptual framework to one that takes account of physical system configuration and deployable "chunks." One way of looking at this is that we are crossing some imaginary line separating planning and engineering. We are moving from the general to the specific.

Unfortunately, this means that we have to introduce yet another strange concept and more jargon. We have to start talking about the ITS Equipment Packages. Consider Figures 11.3 and 11.4.

The first figure shows three ITS Market Packages set within an ITS Future Big Picture. Note how the three ITS Market Packages, A, B, and C, all straddle the two subsystem interfaces. When an ITS Market Package is translated to an equivalent set of ITS Equipment Packages, they are reconfigured to obey subsystem boundaries. This is illustrated in Figure 11.4, which shows ITS Market Package "A" split into ITS Equipment Packages "A1" and "A2" in order to conform to the subsystem interface and become procurable, deployable units.

Accordingly, if an ITS Market Package straddles two subsystems, then it is split into Equipment Packages so that the entire Equipment Package is contained within the boundaries of a subsystem. As subsystems by definition are configured to take account of operating and procurement responsibilities and to take account of functionality, this then provides the deployable and procurable units we are seeking. Equipment Packages, either individually or in groups,

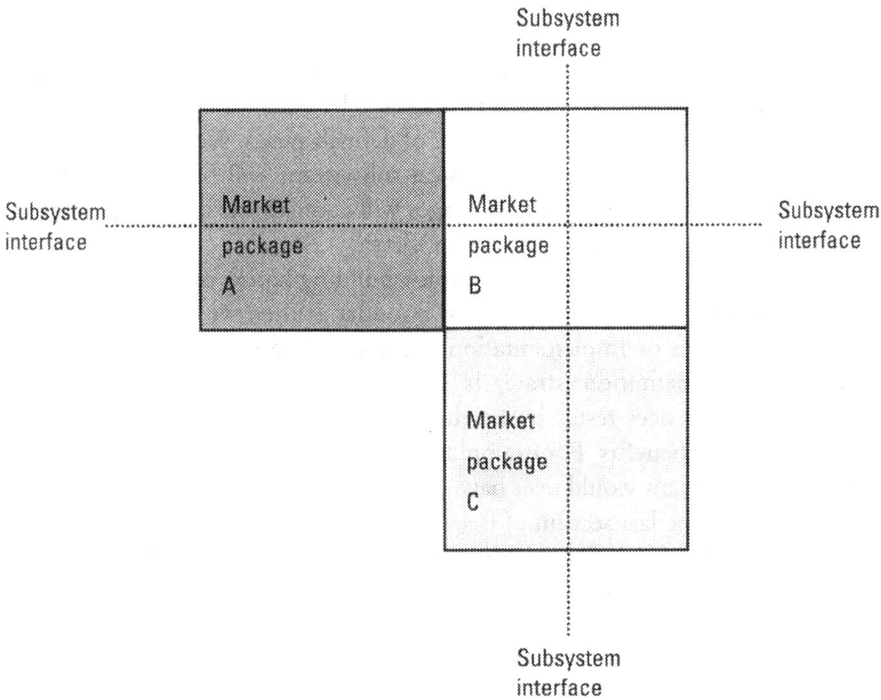


**Figure 11.3** ITS Market Packages within an ITS Future Big Picture.

form the basis for the sequence of ITS projects required to be carried out to result in the ITS Future Big Picture.

To summarize, we have two ways to split the technical layer into smaller units. If we want to look at the technical layer in terms of long-term planning, we can use the ITS Market Packages as the building blocks and use them to define long-term deployment sequences. If, on the other hand, we are considering the needs for short- or near-term deployments, when procurement is imminent, then the added detail provided by the ITS Equipment Packages and their conformance to subsystem boundaries becomes more relevant.

### 11.2.2 Institutional/Organizational Layer

As described in Chapter 9, this layer describes the proposed interactions between people and organizations required to support the technical solution. This defines the roles and responsibilities that each participant would have to adopt to enable effective operation of the entire system on a holistic basis.

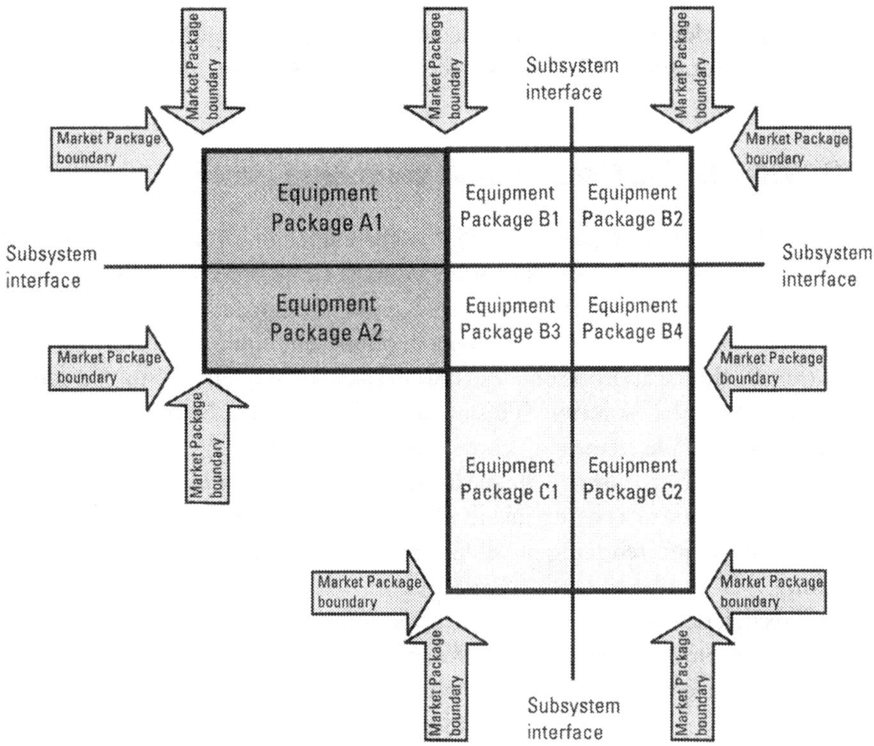


**Figure 11.4** ITS Market Packages split into ITS Equipment Packages.

### 11.2.3 Commercial Layer

Commercial aspects of ITS are explored in detail in Chapter 17. This layer of the ITS Future Big Picture would depict the flow of revenue associated with each subsystem, the capital and operational costs associated with each subsystem, and communications link and the likely profit or subsidy associated with each subsystem. In addition to the identification of public- and private-sector roles in the institutional/organizational layer, the commercial layer would provide additional definition on the public/private relationships and the specific role of the private sector in the overall ITS.

The ITS Future Big Picture is characterized as being "technology-dependent" but not "technology-specific." This means that some progress has been made toward identifying the technologies that need to be deployed to provide the desired results. For example, if we consider communications links between subsystems, then the ITS Future Big Picture would indicate whether wireline or wireless communication technologies should be utilized but would not, at this point, specify the precise technology application.

## 11.3 Why Do You Need an ITS Future Big Picture?

At this point in the process, we are still in the course of "zooming in" toward exactly what is required, so it is enough to make some broad choices regarding categories of technologies to be deployed. System engineers often refer to the process of "zooming in" as the "system development spiral" since the gradual reduction of choice and options leads, in ever-decreasing circles, to the definition of a single "best" solution. This could be thought of as "coarse filtering" of the options available, removing those choices that definitely will not support the objectives, while pushing detail decisions back until later. If you think of the proposed solution as being inside a box, then we are trying to make the box smaller at this point, reducing possible choices but not yet identifying the single solution.

This deferral of decisions is an important feature of the overall ITS Cooperative Development Methodology. We are still learning more and more about the real objectives to be satisfied by the ITS as we move through phases of "mutual learning" involving stakeholder and system developers. As we incrementally provide more information regarding the potential solution and maintain a dialog between all the key participants, more information appears regarding the requirements to be satisfied and the most appropriate solutions.

Maintaining a degree of choice as to which technologies to utilize also provides us with flexibility for the future. As ITS technology development is moving at a fast pace with many new information and communications technologies becoming available, it is very important to maintain some level of future flexibility. If the time horizon for the full system implementation is on the order of 5–15 years, as is typical, then it is almost certain that new technology options will be available before the complete system is implemented.

Another way of looking at the need for the ITS Future Big Picture is to consider what might happen if we did not have one. Experience on ITS development projects that adopted the "bottom-up" approach to design has shown that some of the problems encountered may include the following.

- The design of communications systems for short-term needs and narrow requirements, resulting in significant reinvestments as new needs emerge. Assessing probable data loading to be handled by the communication network, based on a restricted view of the future, can result in premature overload for the communications network.

- Infrastructure that become redundant before its intended life due to unexpected technology development and an inflexible ITS Future Big Picture.
- Many opportunities for sharing costs and benefits with other agencies and the private sector remain unidentified and unexploited. Overlap develops between plans in adjacent agencies and with the private sector.
- Agencies and organizations focusing on short-term needs and a narrow view of their own requirements make overlapping investments.
- The lessons learned and the experience of other implementing organizations are often lost or not taken into account as a narrower range of participants are involved in the development and implementation.
- At a late stage, a key participant is identified yet excluded from the planning and development activities. Having no ownership in the current plans, it may be difficult or impossible to bring the participant to the table, resulting in a suboptimum solution for the future.
- Potential funding sources are ignored or not identified.
- Members of the general public are not made fully aware of all the likely impacts on them and on the transportation facilities within the region.

## 11.4 Developing the ITS Future Big Picture

In developing the ITS Future Big Picture we are trying to add a little bit more detail and a little bit less choice to our proposed solution. In doing so, we can turn the proposed solution into a more meaningful picture, containing tangible real-world components and communications media.

As described in Chapter 10, the picture we have developed up to this point, the logical view, is an abstract description of the proposed solution. It is an idealized picture of the proposed ITS based on the most logical approach to data collection, storage, processing, and communications between processes. The physical view takes us one step closer to a real picture of the proposed solution by adding additional perspectives. First, the physical view as the name implies starts to turn the purely logical "paper process" defined by the logical view into an outline framework of physical devices and communication links. Processes in the logical view are transformed into subsystems in the physical view. Individual processes may be combined to reflect common ownership or operation.

Second, the physical view takes account of the way transportation in the region is currently organized; grouping activities and processing them together into appropriate configurations. Third, the physical view takes account of the legacy that has to be incorporated or protected as part of the new ITS development. One way to think of the transformation from the logical view to the ITS Future Big Picture is to consider the logical view as the perfect, idealized solution and the ITS Future Big Picture as the real-world compromise. Many times during the course of the U.S. National Architecture for ITS development program, we heard the same phrase used as guidance to the team: "The best architecture is the one that gets implemented." The ITS Future Big Picture adds practical perspective to the idealized solution, maximizing the chances that our solution will be implemented.

From a data flow and subsystem interconnection viewpoint, the physical view progresses to a definition of the physical connections that will be required between subsystems by collecting data flows defined in the logical view and assigning them to physical connections. The physical view has interconnections defined between "real" subsystems, rather than the logical data flows between processes. The subsystems have real tangible substance as they are described in terms of physical entities. They can now be described in terms of the "functions" they will perform and support. These "functions" describe real-life activities to be performed by humans, or by hardware/software, and represent the first step toward defining formal procedures for humans to follow or the early basis for software development. The interconnections also take on real-life attributes, being defined in terms of wireline or wireless communications channels. Each communication channel may be shared by multiple data flows.

The development of the ITS Future Big Picture also takes us beyond the purely technical considerations of the logical view to the institutional/organizational, commercial, and technical considerations of a real-world, practical view. While the logical view has a single technical layer, the ITS Future Big Picture has three layers.

Note that on larger, more complicated projects the development and maintenance of the logical view and the ITS Future Big Picture would be supported by a software application such as a CASE tool. The U.S. National Architecture for ITS was developed with the assistance of the "Teamwork" case tool described in Section 10.8.1.1.

To summarize, the development of the ITS Future Big Picture combines the developer's knowledge of technologies, communications, and how systems work with an understanding of the way in which the key participant organizations operate. The resulting framework provides a structure within which more detailed designs can be developed.

## 11.5 Using the ITS Future Big Picture

Once we have at least the initial version of the ITS Future Big Picture, what do we do with it? What activities can we support through the use of the ITS Future Big Picture, and how do we utilize it to the best effect?

There are a number of activities that can be launched and sustained once the ITS Future Big Picture has been put together. These represent a combination of "people" and "technical" activities as described in Section 11.5.1.

### 11.5.1 Stakeholder Outreach and Consensus Formation

After the outreach has been performed for this step in the ITS Cooperative Development Methodology, you should check that any new objectives uncovered in the course of discussions are addressed in the proposed ITS Future Big Picture on all three layers. The ITS Future Big Picture supports good communications between all parties involved in the development and planning process by providing a focal point for information about the proposed implementation and the likely effects on the region. The ITS Future Big Picture acts as the "common language" or reference point, enabling multiple professions and organizations to gain access to information required to assess their potential roles and responsibilities and the effects of the proposals.

### 11.5.2 Confirmation That all Requirements Have Been Addressed

Development of the ITS Future Big Picture provides us with a special opportunity to take an initial overall view of our plans for ITS for the region and ensures that all of our original objectives have been fully satisfied. Each subsystem defined in the ITS Future Big Picture should be capable of being mapped, or traced back to one or more objectives. This is referred to as *traceability* and is normally summarized in the form of a table or matrix showing the objective summarized as ITS User Services mapped to corresponding subsystems in the ITS Future Big Picture. There is an interesting asymmetry associated with this part of the process that is worth discussing as it sheds some light on the usefulness of some of the steps in the process. To get to the point where we have defined an ITS Future Big Picture, we started by identifying, defining, exploring, and confirming requirements in terms of needs, problems, issues, and objectives. We then transformed these into ITS User Services and developed ITS Market Packages to address all of the defined ITS User Services. Subsequently, the ITS Market Packages were used as the basis of the ITS Future Big Picture. Figure 11.5 illustrates this sequence.

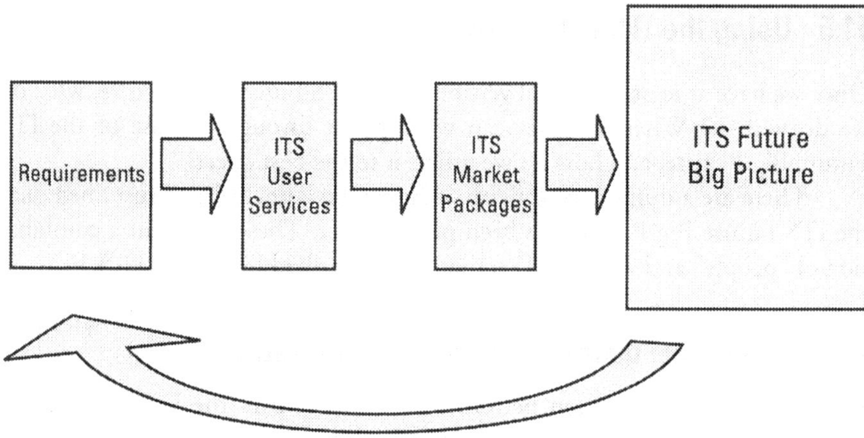


**Figure 11.5** From requirements to ITS Future Big Picture and back.

When we go back the other way from the ITS Future Big Picture toward objectives, we can map the ITS Market Packages (and the ITS Equipment Packages) directly to objectives, omitting the ITS User Services step. This may seem strange, but is does highlight the utility of ITS User Services. They were developed as a tool to help us to structure requirements and facilitate the development of the solution through a user-driven process. Having arrived at the proposed solution, at least in outline form, we do not need to use the ITS User Services as a tool when tracing back to the original requirements.

### 11.5.3 Project Definition

One of the technical activities supported by the ITS Future Big Picture is the identification and definition of a range of projects or individual implementations that will be carried out over a defined time period and at specific geographic locations or areas within the region. Having developed the ITS Future Big Picture, we now have to break it down into manageable "chunks" that combine to form an implementation plan or deployment strategy.

There are multiple parameters to be considered when determining the most appropriate project sequencing for the ITS Future Big Picture, and we will discuss these and the development of an ITS implementation plan in detail in Chapter 14. One of the parameters to be considered relates to the benefits and costs that can be attained by each individual project implementation. While the technically optimum approach may dictate a particular sequence of complementary projects, the need to show standalone benefits from each individual implementation may dictate another sequence. In making the transition from the ITS Future Big Picture to defining projects that can be evaluated with respect to costs and benefits, we cross an important boundary. To provide the level of detail required to develop accurate cost estimates and assess likely benefits, we have to move from "architecture," or ITS Future Big Picture, to design. Technology choices or tradeoffs must be taken in order to pin down the specific nature of the project.

One way to manage this transition is to develop an evaluatory design that encompasses the design choices and detail required for evaluation purposes, yet fits within the overall ITS Future Big Picture. This was the approach adopted in the U.S. National Architecture development program. To develop some specific cost and benefit figures, three design scenarios were defined;—Urbansville, Mountainville, and Throughville. These represented "typical" urban, rural, and through corridor regions in terms of travel demand patterns and demographics. The architecture was then translated into an evaluatory design and applied to each scenario, enabling costs and benefits to be determined. The main point to bear in mind is that the ITS Future Big Picture or architecture can support multiple design approaches, so the evaluatory design represents only one of a number of possible approaches.

## 11.6 Supporting the "What?/How?" Cycle

A large part of the effort so far has been invested to maximize the probability of successful designs and projects. Much of the resulting output from the ITS Cooperative Development Methodology will appear to be uninteresting and unnecessary detail to the transportation community, but it is of vital importance to the information technologists and system developers responsible for producing the end products. As we have stated several times in the course of unfolding the ITS Cooperative Development Methodology, the ITS Future Big Picture will mean different things to different people. In terms of moving forward to designs and project definitions, the information provided in the ITS Future Big Picture gives the developer the information required to define and direct successful hardware and software development and implementation projects. This detailed technical information is best assimilated and utilized by these technology-oriented types. On the other hand, the ITS Future Big Picture provides a fairly concise summary of the proposed solution and a clear trace back from the solution to the original needs that were identified, defined, explored, and confirmed during earlier steps in the process.

As the key participants or stakeholders are the experts in this part of the process, they should be less concerned about the technical information and more focused on the implications and effects of the proposed solution on their organizations. This should not be interpreted as hard and fast demarcation but as helpful advice on moving forward from the ITS Future Big Picture toward implementable projects and strategies. In fact, one of the benefits of having the ITS Future Big Picture is that multiple participants with varying viewpoints can share the same cohesive set of information, deriving their own particular subset from the whole.

This touches on the "boundary effect" of ITS. As ITS requires multiple professions and disciplines to work together to achieve agreed on results, communication across professional- and experience-related boundaries becomes crucial. The ITS Future Big Picture straddles most of the major boundaries between people involved in the process. The key to successful use of the ITS Future Big Picture lies in identifying the parts that are relevant and appropriate to an organization's perspective and needs.

In very broad terms, the ITS Future Big Picture provides a solid platform for further system development, project definition, and the creation of an implementation strategy for the proposed ITS. It also provides the focal point where all the relevant information required to assess the organizational and commercial effects of the proposed technical solution can be accessed.

### 11.6.1 Identification and Discussion of Roles and Responsibilities

One of the "people" activities that can be supported by the ITS Future Big Picture is the identification and discussion of the relative roles and responsibilities to be taken by each organization and stakeholder in the ITS Future Big Picture. Having defined the proposed solution in technical, institutional/organizational, and commercial terms, it is possible to use the picture to discuss roles and confirm acceptability and feasibility. This is the point at which potential changes to prevailing institutional/organizational arrangements can be identified and considered. Alternatively, it may be necessary to amend the technical solution to take account of institutional/organizational constraints.

### 11.6.2 Coordination With Conventional Transportation

The ITS Future Big Picture also provides an opportunity to consider and assess the relationship between the proposed ITS implementation and parallel development and implementation of conventional transportation plans and initiatives. For example, major infrastructure developments such as new highways or transit facilities can have a significant impact on travel patterns and the demand for travel in a region. Have these effects been taken into account in the development of our proposed ITS solution? Similarly, the implementation of new infrastructure may afford synergistic opportunities for the design and installation of ITS hardware and devices. New communications facilities such as fiber optic cables may be implemented as part of the transport infrastructure initiatives, providing substantial cost savings.

### 11.6.3 Alignment With Private-Sector Objectives

Having developed at least an initial idea of what you want and how you plan to go about it, you are armed with some good information that can help you to communicate and negotiate with the private sector. This could initially take the form of an information exchange under which you share your plans with the private sector, and private-sector entities indicate what their own plans are. Having refined your own plans in the course of developing the ITS Future Big Picture, it becomes easier to share these with the private sector in a meaningful manner.

### 11.6.4 Platform for Standards Development and Participation

Depending on the timing for the development of the ITS Future Big Picture, it can be used to direct attention to where standards need to be developed and adopted or to point out the need for participants to be actively involved in the standards development and application process. Currently in the United States, more than 100 ITS-related standards are currently under development. Many standards activities are built on the requirements identified in the U.S. National Architecture for ITS. It may well be the case that your architecture development program comes at a later stage in the maturity of the ITS field. In this case, the ITS Future Big Picture may point less to the need for standards development and more to the need for the application of certain specified standards. Existing ITS standards may even form part of the input to the development of the ITS Future Big Picture.

### 11.6.5 Checking Out the Consistency Between Layers

This is the first time in the process that the whole ITS Future Big Picture has been assembled. Therefore, it is the first opportunity to review the synchronization between each of the three layers. Taking each of the three layers in turn, it is important to check that they fully support each other and mesh together. For example, does the institutional/organizational layer support the operation of the technical layer? Do the arrangements detailed in the commercial layer enable the technical layer to be financed and operated? Will the proposed institutional/organizational arrangement depicted in the institutional/organizational layer fully support the needs of the technical layer? Will the proposed commercial arrangements work with the proposed technical and institutional/organizational solution?

### 11.6.6 Identifying Roles and Responsibilities

The physical view provides the clearest pictures so far of the roles and responsibilities that will be assigned to each participant in the ITS. Therefore, one of the important uses of the physical view is the support of consensus on these roles and responsibilities. Building on the institutional/organizational layer development activities described in Chapter 9, the institutional/organizational layer is assembled with the technical and commercial layers and all three are subjected to a preliminary check for consistency and synchronization. The complete ITS Future Big Picture can then be utilized as a communications tool for exploring roles and responsibilities at a more detailed level. If necessary, the results form the consensus formation activities that can be applied to create revisions and amendments to the institutional/organizational or the other layers in light of new information or clearer direction.

### 11.6.7 Scanning for New Technologies, Products, and Services

The ITS Future Big Picture enables us to move ahead with short-term development and implementation, while maintaining a monitoring and review function for future technologies. As we have a framework defining how all subsystems inter-relate, it is possible to feed in new technologies and standards as they are developed. We can use the ITS Future Big Picture to see how these new developments might fit into our overall plan and measure likely impacts on what we have already deployed.

# 12

# Applying ITS Standards

## 12.1 Introduction

This chapter addresses the activities required to complete step 240, "Standards Application Plan," in the ITS Cooperative Development Methodology.

This chapter attempts to provide readers with some guidance and insight on the application of ITS standards within the ITS Cooperative Development Methodology. This includes a discussion on the relationship between the ITS Future Big Picture, standards, and design. Standards, or the lack of them, have long been recognized as one of the major constraints to widespread and effective deployment of ITS. Another advantage of the ITS Cooperative Development Methodology approach to the development and planning of ITS projects is the ability to harness existing and emerging standards in a structured manner. The opportunity is also provided to play an active role in the development process for standards that do not yet exist.

Standards are really a design and procurement issue as the ITS Future Big Picture should not have sufficient detail to enable you to specify required equipment and interfaces. However, we have found the earlier the intending implementer begins to assess the need for standards application, the better. Particularly in the ITS field, where standards are still under development, an early look at the requirements can afford an opportunity to influence and affect the emerging standards.

The chapter begins with a brief discussion of the relationship between the ITS Future Big Picture, standards, designs, and implementation. This is followed by an exploration of the various types of standards that might be developed, adopted, and applied to ITS. The current status of ITS standards development activities is then addressed, followed by a discussion of the importance of standards. The chapter is rounded off with a brief section on the relationship between standards and interoperability, followed by some advice on Standards Application Plan development.

## 12.2 The Relationship Between the ITS Future Big Picture, Standards, Designs, and Implementation

The ITS Future Big Picture provides us with a conceptual framework within which all the essential pieces required to address requirements for now and the future have been identified and defined at a low resolution of detail. This gives us the ability to identify pieces of the overall system and interfaces between those pieces that would give benefits, or allow more efficient operation, if standards were applied. As we leave the high level of the ITS Future Big Picture and add more detail by taking more design decisions and exercising choice between technology options, we migrate into design and away from the ITS Future Big Picture.

It is in the realms of design, procurement, and implementation that standards have their true place and provide most value. Remember that the ITS Future Big Picture is a high-level framework that can support many detailed design variations. Therefore, it is not possible to specify, procure, implement, or be "compliant with" the ITS Future Big Picture.

## 12.3 Types of Standards for ITS

There is a range of standard types according to the scope of coverage of the standard and the nature of the item to be standardized. The range includes those described in Sections 12.3.1–12.3.7.

### 12.3.1 Ad Hoc Standards

These are developed by an implementing organization when no current standard exists or is under development. They may be specific to a single project, or to all projects carried out by a single organization.

An example of an ad hoc standard covering all projects by an agency is the Title 21 standard for dedicated short-range communications developed by the California Department of Transportation. This document defines the basic compatibility requirements for automatic vehicle identification (AVI) equipment for electronic toll collection applications throughout the state of

California. It has provided the basis for manufacturers to resolve compatibility problems uncovered during development.

Note that there is a fine line between ad hoc standards and no standards at all. Ad hoc standards that have no defined migration path to widely agreed standards or that are based on proprietary approaches may not be standards at all.

There can also be a situation where one ad hoc standard becomes so pervasive that it becomes the de facto standard. The approach and definition have not been the subject of a consensus formation process such as that applied by the various standards development organizations. The approach has been developed and defined by a single, industry-leading organization, yet has evolved to become a national or international standard. A classic example of this is the Windows operating system developed by Microsoft.

### 12.3.2 Interface Standards

These define the protocols, transmission methods, data, and information to be transferred across a boundary between two independent subsystems within a larger system. These specify data content and structure and the method for coordinating transmission and receipt of data from one subsystem to another. They may also specify the physical details of the connection, such as number of pins and the shape of the connector.

Typically, these standards relate to communications techniques using wireline or wireless communications technologies. A commonly encountered example of an interface standard is the IEEE RS232 standard for the serial interface connecting computer equipment and peripherals.

### 12.3.3 Product Standards

These define the parameters to be satisfied when providing a product to an agreed upon standard. Typically, these specify the features and functions to be provided if the product or service is to comply with the standard. These standards act as a kind of shorthand for procurement documentation as the specification of the standard ensures that the product or service complies with a previously agreed upon description of functionality, obviating the need to detail this in the procurement documentation.

### 12.3.4 Service Standards

These define the parameters to be satisfied when providing a service to an agreed upon standard. The features of the service would be defined and the content of the standard would be similar to that for product standards as described in Section 12.3.3.

### 12.3.5 Regional Standards

These go beyond organizational boundaries and set common standards for a geographical region. The region would be defined on the basis of common needs and interoperability requirements. Another way to look at the geographic region is to consider it as a defined market area, encompassing customers with common needs and perspectives. An example of a regional standard would be an electronic fare payment or ticketing system for regional transit use. It would be important that the system provide interoperability across the geographic area defined as the region, but there may be no need for interoperability between adjacent regions, as little or no travel takes place between the two.

### 12.3.6 National Standards

As the name implies, these standards cover entire countries. This level of standardization would be applicable to products and services that require wide national interoperability in order to work efficiently and generate the desired benefits. A good example of this is the DSRC application used to support commercial vehicle operations such as electronic preclearance at state boundaries and borders, or streamlining trucking administration. Many trucking companies operate national freight services, requiring the truck and driver to cross the entire country. The DSRC application is most efficient if it is interoperable on a national scale. Therefore, a national standard would be appropriate.

National standards may also be of benefit in situations where there is no need for national interoperability. The existence of a national standard may have the effect of reducing product and service costs by enabling providers to focus on a narrower range of products, achieve economies of scale in manufacturing, and amortize research and development costs over a larger customer base. Users also benefit through reduced prices and the ability to procure products from multiple, competing sources.

National standards are developed and supported through a large number of national standards development organizations around the world. In the United States alone, there are more than 350 standards development organizations.

### 12.3.7 International Standards

This level of standardization is typically considered when there are potential economies of scale attainable through the development and production of common products or services designed to serve an international community. International standardization efforts are supported by the ISO. An example of an appropriate application for international standards is the specification of freight container identification approaches. Freight containers are transported around the world; therefore, an identification system would be required to support international interoperability. There is, in fact, an existing ISO standard addressing this topic.

There are also needs for international standards based on safety and human factor considerations. The development and agreement of common "look and feel" interface standards on an international basis has the potential to make a major impact in transportation safety. For example, the internationally standardized octagonal stop sign enables drivers to recognize this instruction almost universally.

## 12.4 Current Status of ITS Standards Development

Standards development for ITS is very dynamic around the world. Efforts in Europe, through CEN and the national standards bodies; in Japan through HIDO; and in the United States through the Federal Highway Administration and the standards development organizations (SAE, ASTM, IEEE, ITE, and ANSI for ITS); and internationally through the ISO a stream of new standards are being produced. Some of the significant standards topics under development at the time of writing include the following:

- DSRC for vehicle-to-roadside communications (toll collection, commercial vehicle operations, access control, and driver information systems);
- Wireline communications between field devices and control centers and control center-to-control center communications;
- In-vehicle data bus for supporting the attachment of multiple ITS devices to factory installed in-vehicle wiring and electronics.

There are many ITS standards initiatives under way, including almost 80 individual activities in the United States alone.

In summary, the development of ITS standards is partially complete with intense international activity producing the required standards at an accelerated pace. As discussed in Section 12.9, we have not attempted to provide a detailed account of current standards development and status, as this would be futile in a book of this nature.

## 12.5 The Importance of Standards

When considering the application of ITS standards and developing a plan for doing so, it is important to consider the desired outcome or objectives. Sections 12.5.1–12.5.6 describe a few objectives you may wish to consider as the basis for your Standards Application Plan objectives.

### 12.5.1 Protecting Legacy

Most ITS implementers have some kind of legacy system or implementation that they wish to preserve and enhance as part of the new implementation. This is detailed in Chapter 8. The use of and adherence to standards maximizes the probability that both current and future legacy can be integrated into new deployments. We consider the system you are currently planning, developing or deploying to be "future legacy."

### 12.5.2 Reducing Product and Service Unit Costs Through Volume

The adoption of widely accepted standards enables providers to increase volumes of similar products and services, providing them access to economies of scale. These economies consist of manufacturing and administrative cost saving through higher sales volumes and the ability to amortize research and development costs over a larger customer base.

### 12.5.3 Reducing Total Costs Through Sharing and Minimizing Risk

The chances are that available standards were developed at the cost of practical experience from previous implementers and providers. Adoption of standards enables you to avoid this cost of "pioneering" and leverage these previous efforts. It may also reduce total costs by reducing the risk of implementing products and services with short service lives. The fast pace of technological development could leave you out on a limb when technology moves ahead. Through the adoption of standards, you can avoid this and gain access to the advantages of "collective possession," producing significant influence on the development of products and services through size of customer base.

### 12.5.4 Avoiding Provider "Lock-In"

If you look at life purely from a technical perspective, you would think that most providers would be developing their products and service offerings to be so-called open systems. This would enable them to give their customers the best of all worlds, the short-term ability to buy systems and components to attain the desired benefits and the long-term guarantee of flexibility and choice—conforming to available standards and designed to fully support interoperability and interfaces to other products and services. However, there is a paradox associated with new and emerging markets for goods and services.

Put yourself in the shoes of a product or service supplier addressing a new or early market such as ITS with new products, services and other offerings that have been developed especially for the market using significant resources and investment. The riskiest time for you as a vendor is when you have just launched your product or service, because typically this will be your point of maximum exposure where your sunk costs are very large and your return on investment is small or zero.

You may be risking your own money, or more typically, you will be making use of funds from investors. Either way you will be strongly motivated to take all possible steps to maximize the probability that you get the return on investment anticipated when you started. In fact, your investors may well have insisted that you define your plans for this before they committed to providing the necessary research, development, and marketing resources in the first place.

So this initial market entry is focused on establishing a share of the market and ensuring as far as possible that you get a return within the forecasted period. You are faced with customers who are new to the concepts and technologies on which your offerings are based, and it is difficult to establish key success factors such as requirements and features because the market has not yet made up its mind about what is needed.

The last thing on your mind at this point is making it easier for another vendor to come along and enter the same market with rival products and services or making life easier for existing competition to keep up. An obvious strategy for increasing the chances of a return on investment is to exclude competition if possible or at least make it difficult. Your primary focus is on making sure that the market develops and continues to exist.

Your attitude may well change over time. As you get established and start to attain the desired return on investment, others will have entered the same market and increasingly you will be competing for market share. At some point you have two choices in terms of business direction:

- Continue to fight with the competition, competing for larger shares of the same fixed market;
- Look for ways to increase the size of the market so that everyone can have a larger slice of the pie.

One way to increase the size of the market is to promote the development of standards, reducing the risk to the buyer and increasing the potential sales volumes available to you. Sale volumes can also increase at this point in the market due to the new ability of the customer to avoid "lock in" to proprietary systems, They can gain technical and price advantages through the use of standards.

### 12.5.5 Streamlining Procurement

The adoption of recognized and well-known standards streamline the process of procurement by reducing the need to prepare customized documentation. A good standard that addresses elements of your procurement will enable you to specify the description of the products and services by referring to the standard for a description of required features, functionality and testing regimes.

### 12.5.6 Fostering a National ITS Industry With a Competitive Edge

This may not be of immediate concern at a regional and local level, but is worth mentioning as an economic factor. Through the adoption and application of ITS standards, it may be possible to support the development of regional or national ITS capabilities with respect to manufacturing and service provision. As providers supply standardized products and services to their local, or "home," markets, they also develop a competitive edge by having the ability to offer them on a wider basis.

## 12.6 Standards and Interoperability

Interoperability in ITS has been defined by the ISO in ISO TC204 Document N271 as:

> The ability of systems to provide services to and accept services from other systems and to use the services so exchanged to enable them to operate effectively together.

Interoperability should also be considered on the same three levels we have been advocating for the ITS Future Big Picture:

- *Technical:* The technical ability of interoperable equipment to interface and communicate with each other;

- *Institutional/organizational:* Formal establishment of procedures and mechanisms for the exchange of data and information;
- *Commercial:* Commercial agreements and formal legal relationships between subsystem operators and between operators and users.

Unfortunately, this means that the existence and application of an ITS standard does not guarantee that independent subsystems complying to standards will be interoperable. Standards often emerge as a result of a compromise among vendors of existing products and providers of current services. The documentation of the standards can sometimes be ambiguous and is often produced before thorough testing for compatibility.

Given the level of detail present in most standards, it is possible for products, services, and systems to fully comply with a given standard and still not be interoperable. Most ITS standards will have been developed in isolation from other ITS standards development activities by a small group of focused technical people. There is rarely a strong coordinating mechanism to ensure that individual standards take account of each other. It is necessary to define testing and certification procedures to ensure that procured systems are truly interoperable.

## 12.7 Some Advice on the Development of a Standards Application Plan

Obviously, the specific Standards Application Plan you require will vary considerably depending on the application you intend to incorporate and the circumstances or context within which you will implement. Sections 12.7.1–12.7.8 describe things you may wish to consider when putting your plan together.

### 12.7.1 Develop a Plan; Do not Let it Happen by Default

The single most important piece of advice we can give you with regard to standards application for ITS is to plan ahead and develop your strategy in advance. We have found that this provides much better support for your objectives and enables you to address standards issues more efficiently and effectively. As standards for ITS evolve and emerge, it is very useful to have a game plan already thought out in terms of objectives and possible actions. This not only enables you to take a more proactive and effective role in the standards development process but also allows you to provide a more informed reaction to new standards and standards development initiatives.

### 12.7.2 Review the Current Status of ITS Standards Development

Take the time to conduct a thorough review of standards development activity at regional, national, and international levels. This should include consideration of what standards are already available, what standards are under development, and the current state of ITS standards development. Most central governments involved in the development and application of ITS have ongoing initiatives in standards development and can provide a summary of current activities on request.

### 12.7.3 Use the ITS Future Big Picture to Identify the Need for Standards and Define the Type of Standards Required

Referring to the ITS Future Big Picture, carry out a careful review of standards requirements for your planned deployment. This should consider the criticality of standards application for the different parts of your ITS Future Big Picture. For example, if a particular interface is fundamental to the successful operation of the entire system, then it should be high on the list of interfaces for which widely agreed standards will be adopted or applied. It should also be possible to identify the type of standard that would be appropriate by considering the three levels of the ITS Future Big Picture. Whether the standard needs to be regional, national, or international can be deduced from the information provided in the ITS Future Big Picture.

### 12.7.4 Consider the Implications of Applying Existing Standards

Based on an understanding of your needs and the attributes of existing standards, decide if the current standards are appropriate to your context. It would be desirable to ask a couple of questions when making this decision:

- Are providers in your market able to offer products and services in compliance with the standard?
- Will the application and adoption of the standard achieve your previously defined objectives?
- Is the market at a state of maturity where standards will be beneficial in stimulating the market, rather than stifling innovation?

### 12.7.5 Define an Action Plan for Involvement in the Development of Other Standards That Are Important to You

Where standards are still under development in areas that you have defined as important to your implementation, establish specific strategies and actions for proactive involvement in the development of those standards. Typically, standards development work is dominated by a small number of vendors or service providers with little or no input from potential implementers or end users of the standardized products or services. Even minimal involvement at the development stage can yield significant benefit through the development of standards that are more practical and address your needs as the final implementer.

### 12.7.6 If no Standards Exist, Then Review Experiences of Other Implementers and Develop De Facto Standards

Due to the evolving nature of ITS and the embryonic state of many application areas, you may well encounter the situation in which no standards exist and no standards development work is under way. In these circumstances we would advise a thorough review of experiences from other implementing agencies and organizations. It is very rare to find that you are trying to do something so unique that there is no previous experience around the world.

### 12.7.7 Set Up Mechanism for Evolving De Facto Standards to Regional, National, or International as Required

If it is necessary to establish de facto or ad hoc standards due to the pressure to implement and the lack of standards or standards development activities, then it is important to consider the future evolution of these interim measures. It would be prudent to follow the establishment of the interim standard by also establishing a mechanism for cooperation with other implementers, on developing and agreeing on a more widely accepted standard.

### 12.7.8 Define and Establish Certification and Testing Procedures

Many standards emerging from the development process do not have certification and testing guidelines incorporated. If standards are to be specified and adopted as part of the implementation of your ITS, it is important to develop such guidelines. You may wish to incorporate formal testing and certification procedures into the procurement documentation to ensure that the providers understand how the deliverables will be tested for compliance with the standard.

This is a topic that is the subject of much development work around the world at the current time. In some cases, the development of testing and certification procedures is an integral part of standards development activities, while in others the standard is first developed, followed by the testing and certification procedures.

## 12.8 Specific Advice on DSRC Standards

At the time of writing, many developing countries, the so-called economies in transition, are addressing the question of DSRC standardization. We believe that it might be helpful to readers from those countries and informative to others if we provided some thoughts on how to approach the DSRC standardization issue. Three different standards are emerging for DSRC around the world. The United States, Europe, and Japan all have their own specific approaches using different frequencies for transmitting the data, different approaches to wireless communications, and different communications protocols. This makes it a difficult and complex task to decide whether to adopt one of these emerging standards for your country or to try to develop your own. Our approach to this difficult issue revolves around a requirements-based process. First, we would ensure that an ITS Future Big Picture had been developed at a national level, providing the information about major subsystem and interface requirements as described in Section 3.6. We would use this information to develop a data-loading analysis indicating the volume of data to be transmitted and the required data transfer rates. We would also establish the message content format and structure necessary to fully support all ITS applications within the ITS Future Big Picture, both at the start of the implementation strategy and through to the end.

With this information in hand we would then carry out a thorough analysis of existing and emerging DSRC standards from around the world. This would be used to determine approaches taken by others and advantages and disadvantages of each approach.

Finally, we would identify, define, and analyze the important parameters pertaining to DSRC standardization, relate these to our needs, and carry out a multivariate or framework analysis on these parameters. The output from this analysis would characterize our needs with regard to DSRC so we would compare this with current and emerging standards to find a close fit. If a close enough fit was not available, we would develop our own standard. To give you some idea, the kind of important parameters would be those discussed in Sections 12.8.1–12.8.3.

### 12.8.1 Technical

There are a number of technical parameters that are relevant to DSRC standardization. These include the following:

- *Transmission frequency:* What frequency is used to transmit the data?
- *Communications approach:* Active two-way communications or passive reflective?
- *Footprint:* What distance around the beacon does the communications zone stretch to?
- *Transaction speed:* How many bits per second or baud can be transmitted?
- *Message length:* The maximum message length that will be required to support your applications;
- *Expandability:* Does the technology have an upward growth path in terms of capabilities and data carrying capacity?

Many of these parameters would be used to establish the maximum amount of data, or message length that could be supported during one beacon to transponder communication session.

### 12.8.2 Institutional/Organizational

The following institutional/organizational issues need to be considered:

- What frequencies are available? Does the standard make use of frequency bands that will be available in your country both now and in the future?
- Beacon spacing: How far apart will the beacons be? In other words how many beacons can you afford?

### 12.8.3 Commercial

The following commercial issues need to be considered:

- *Beacon cost:* How much does the beacon or reader cost?
- *Transponder cost:* How much does the tag, transponder, or in-vehicle unit cost?

- *Suppliers:* How many suppliers are manufacturing or have committed to manufacture products in compliance with the standard?

While the analysis would have to be specifically tailored to the specific needs of the country in question, these are the general principles of our suggested approach. Note that the analysis process would involve multiple decision makers as well as multiple decision factors, making it a suitable case for the use of social decision analysis techniques.

## 12.9 Sources of Further Information on ITS Standardization Activities

We were presented with a dilemma when writing this chapter. On one hand, we wanted to be as helpful as possible in giving you information to help you develop your ITS Standards Application Plan. On the other hand, the international situation with respect to ITS standards development and evolution is so dynamic that any attempt at documentation will be out-of-date before this book is published.

As a solution, we decided to provide some Internet information sources rather than attempting to document the current situation in full detail. For detailed information on the current status of ITS standards development activities around the world, visit the Web site of the ISO, which supports international ITS standardization efforts. The address is

http://www.iso.ch/

Additional details of national ITS standards development activities may be obtained at the following Web sites:

- United States: ITS America at:

  http://www.itsa.org/home.nsf

  or the Federal Highway Administration at:

  http://www.its.dot.gov/standards/

- Europe: The *Comité Européen de Normalisation* (CEN) or European Committee for Standardization at:

  http://www.cenorm.be/

- Japan: The Highway Industry Development Organization at:

  http://www.nihon.net/hido/ITS/index_e.html

# 13

# Effective ITS Outreach

## 13.1 Introduction

This chapter addresses step 800 on the ITS Cooperative Development Methodology. It describes a generic approach to ITS outreach activities, incorporating guidance and advice on developing and taking part in your own specific outreach program in association with the planning and development of an ITS project. Note that step 800 runs throughout the length of the ITS Cooperative Development Methodology and has a few firmly defined links to the other activities shown in Figure 4.2. This is intended to indicate a continuum of links between the activities and step 800.

## 13.2 Objectives

ITS developments are typically directed or steered by a small core group of knowledgeable people who work for the client organization, are close to the development process, and have a good understanding of the development activities. From past experience, we know that the ultimate success or failure of the ITS proposals depends on how well this core group can identify, define, and address target audiences whose reaction to the proposals will have a significant influence on the success or failure of the proposed ITS implementation.

Outreach within the context of ITS development is the process through which we identify and confirm our objectives, define the messages to be projected, identify and define the target audience groups, develop tools and materials to support the desired messages, and inform and convince the wider audience or community. It is vital that the wider audience is brought along in terms of information about the ITS project, appropriate participation in the project, and the development of ownership in the final proposals. This group consists of people who will be affected by the ITS proposals and who have important inputs to make to the project. The various audience groups need specially tailored information and answers concerning many aspects of the proposals, including specially configured answers to the questions listed in Sections 13.2.1–13.2.6.

### 13.2.1 What?

This category includes the following questions:

- What is ITS?
- What are we planning?
- What are the details of the proposals?
- What likely effects will the proposals have on me?
- What possible solutions are available to address our requirements?
- What are the benefits?
- What are we trying to achieve?

### 13.2.2 Why?

This category includes the following question:

- Why are we doing all this work?

### 13.2.3 How?

The following questions belong in this category:

- How are we going about the work?
- How will we achieve our objectives?
- How will we solve the problems?
- How should we adapt our original requirements in light of new information about technological possibilities?

### 13.2.4 When?

The following questions belong in the "when?" category:

- When will the proposal be implemented?
- When can we expect to see results?
- Over what time period will the ITS be implemented?
- What will come first and why?

### 13.2.5 Where?

It is necessary to answer the following "where?" questions:

- What corridors and geographic regions are being studied?
- Where will the ITS begin?
- What area will it eventually cover?

### 13.2.6 Who?

In this category, it is necessary to answer the following question: Who will be responsible for the development, design, building, and operation of the system?

The exact way in which these needs are satisfied through the development of a workshop and meeting schedule will vary considerably from one project to the next.

## 13.3 ITS Outreach Plan Objectives

The primary objectives of the ITS outreach effort are usually the following:

- To support the successful implementation of the ITS through management of the people aspects of the planning and development phase;
- To identify the essential relationships that must be established, developed, and supported during the course of ITS planning, development, and subsequent implementation;
- To identify the major influences and threats to the success of the project and develop appropriate strategies for avoiding or mitigating the threats and taking full advantage of the opportunities;
- To identify and define the audience groups to be addressed;

- To identify and define the messages to be conveyed to each audience group;
- To develop a coherent and integrated approach to the development and delivery of the information required to each audience group.

## 13.4 Approach

We typically adopt a coordinated, integrated ITS outreach plan approach designed to achieve the objectives listed in Section 13.3 and consisting of the following steps:

- Identification and definition of primary audience groups;
- Identification and definition of essential relationships between each audience group;
- Definition and description of proposed actions to facilitate the establishment, support, and development of each relationship.

The overall approach involves the identification and definition of the important groups in the development project and the relationships that must be supported between them. Sections 13.5 and 13.6 detail these steps.

## 13.5 Identification and Preliminary Definition of Primary Audience Groups

This is an important activity. On previous projects we have encountered three generic audience groups, as illustrated in Figure 13.1.

Each of these audience groups requires a different type of ITS outreach. The content and format of the message projected to the group and the way in which they are engaged are all dependent upon the nature of the group, the needs of the group, and the desired role of the group. Sections 13.5.1–13.5.3 look at each group in turn.

### 13.5.1 The Technical Steering Committee

Comprised of representatives from the primary or lead stakeholder agency, the technical steering committee is usually the one that is paying for the planning and development activities or has responsibility for managing this phase of the

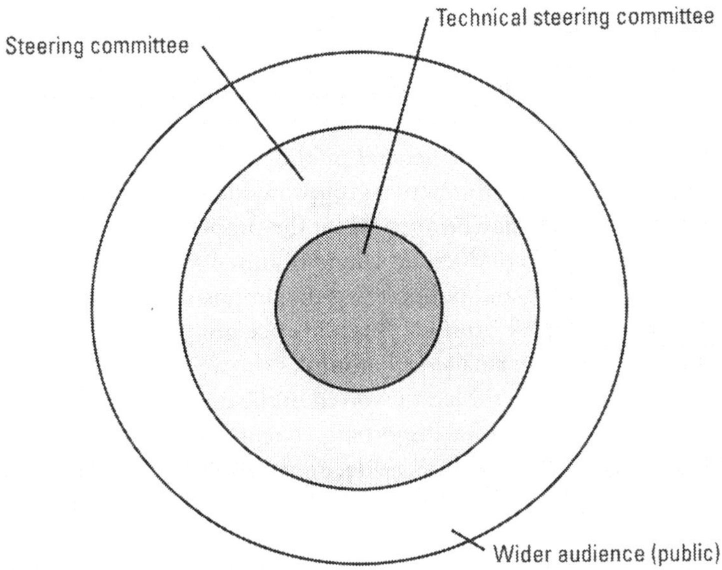


igure 13.1 Primary audience groups.

ITS project. Members of the committee are responsible for the daily decisions regarding the project. Their expertise can vary widely and range from engineering to economics. Most have had limited exposure to ITS already. They would receive the most detailed information necessary for day-to-day interaction and effective decision making regarding the development of the ITS and the overall direction of the project.

### 13.5.2 The Steering Committee

The steering committee consists of a wide variety of individuals representing collaborating transportation agencies and other stakeholders—for example, the police, local private road operators, turnpike authorities, and the federal government. They will be involved with the project at key decision points. This group is very important, because it represents the stakeholders that may be impacted by this project. It is critical that the group members possess enough knowledge to make informed decisions enabling us to jointly develop a system that is beneficial to all stakeholders and has minimum negative impacts. Accordingly, they would receive detailed information to enable them to provide input, guidance, and suggestions at key decision points in the course of the project.

### 13.5.3 The Wider Audience

The wider audience consists of various audience subgroups external to both the technical steering committee and the steering committee, such as other transportation agencies and the general public. It could also include the local chamber of commerce, environmental groups, residents associations, and other pre-existing groups that may be affected by the proposed development. It will probably be necessary to conduct some kind of initial survey activity to identify these groups. Advertising and publicity for the proposed development may also draw responses from these groups. This audience group will typically need to be broken down into even smaller subgroups as the work progresses. The wider audience, even though it is the least involved in the development of our proposals, will ultimately be the most important. It represents the everyday users of the transportation system. Most of the decisions made in the project will have a direct impact upon the audience members' everyday use of the transportation system and ultimately their satisfaction with the system.

Our experience from previous ITS developments indicates that ignoring this group can lead to significant problems related to resistance to the proposals and misunderstandings or misconceptions regarding the likely effects of the proposals.

The members of this group would receive more general information about the project and would be consulted to obtain a general understanding of the external perception of our proposals, problems to be addressed, objectives to be achieved, and suitability of proposed solutions. It is also helpful if both steering committees have a significant role in supporting ITS outreach to the wider audience.

## 13.6 Identification and Definition of Essential Relationships Between Each Audience Group and Proposed Actions to Establish, Support, and Develop Them

The relationship between the three audience groups and the ITS development team is illustrated in Figure 13.2.

As illustrated by Figure 13.2, there are three primary relationships that have been identified as follows:

- *Relationship 10* between the technical steering committee and the ITS development team;
- *Relationship 20* between the technical steering committee/ITS development team and the steering committee;

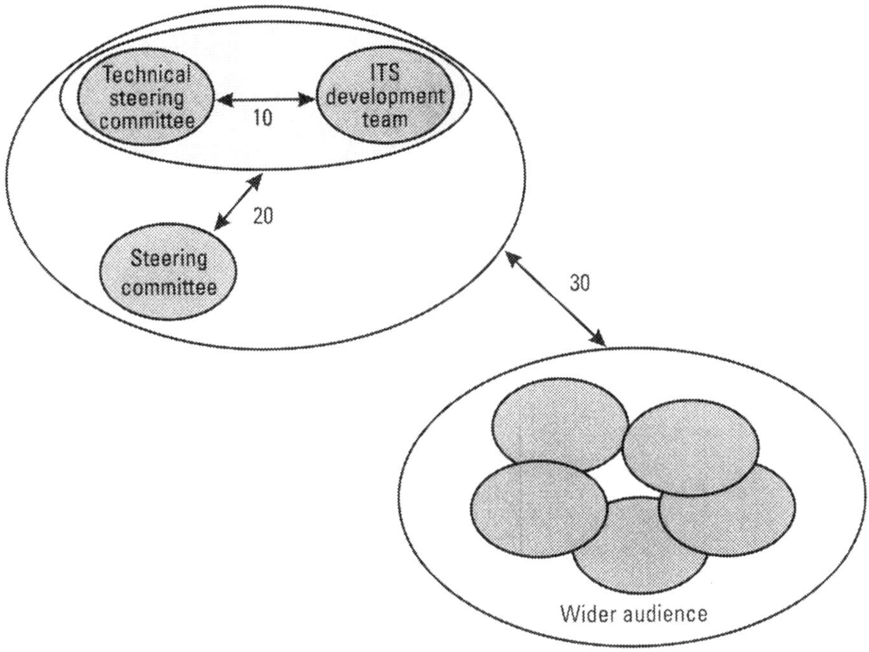


**Figure 13.2** Primary relationships.

- *Relationship 30* between both steering committees/ITS development team and the wider audience.

Each of these primary relationships is defined and described, and proposed actions to establish, support, and develop each relationship are described in Sections 13.6.1–13.6.3.

### 13.6.1 Relationship 10: Between the Technical Steering Committee and the ITS Development Team

Figure 13.3 illustrates the relationship between the technical steering committee and the ITS development team.

#### 13.6.1.1 Relationship Description

Of necessity, this needs to be a close relationship. On one hand, the ITS development team needs to understand the goals and objectives of the technical steering committee and its perception of the problems to be solved and the policy objectives to be addressed. On the other hand, the technical steering

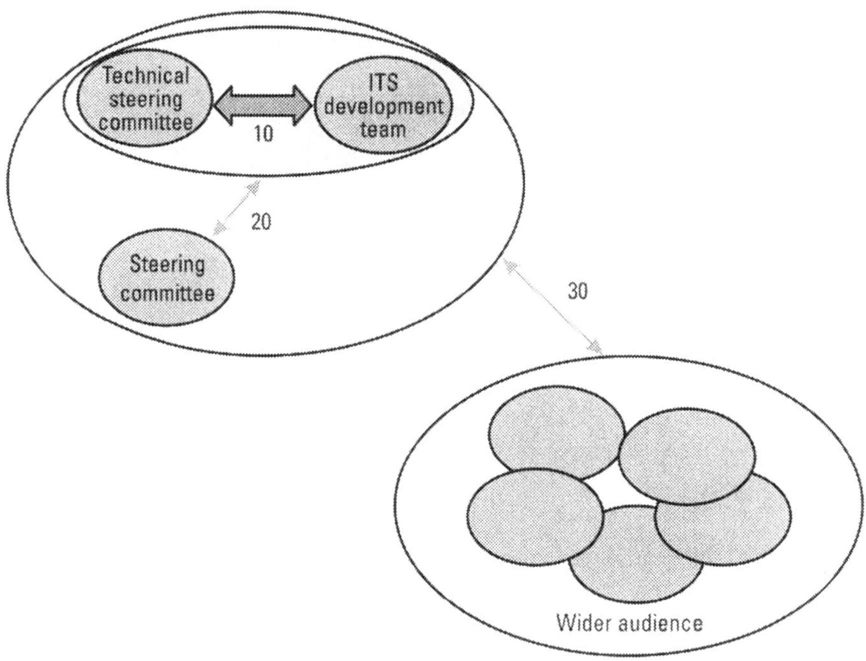


**Figure 13.3** Relationship 10—between the technical steering committee and the ITS development team.

committee needs to have a clear understanding of the capabilities and constraints associated with technological solutions and the likely impact of proposals. We establish and develop this relationship using the tools described in Section 13.6.1.2.

#### 13.6.1.2 Proposed Actions To Establish, Support, and Develop the Relationship

There are a number of actions that can be used to support the establishment and development of the relationship. These include the following:

*Regular Technical Steering Committee Progress Meetings*

These have usually been scheduled and resourced in the current work plan and represent an important support action.

*Workshop: SWOT, SWIIFT, and Problem Management*

This workshop is carried out as a brainstorming session, with support materials such as flip charts and "getting started" graphics prepared prior to the workshop. The outcome of the session is recorded and produced as concise notes. The workshop addresses three topic areas, listed as follows:

- Strengths, weaknesses, opportunities, and threats (SWOT) analysis;
- "See what's in it for them" (SWIIFT) analysis;
- Definition of problem management strategies.

### *SWOT Analysis*

It is very important to identify the strengths, weaknesses, opportunities, and threats related to our proposed approach and the work we will be carrying out. The SWOT analysis would be conducted as an interactive session with the steering committee. A SWOT would consist of the following elements:

- Inventory of organizations and agencies that could potentially pose threats or provide opportunities to us;
- Identification of individuals: List of individuals that represent either threat or opportunity;
- Identification and definition of project strengths;
- Identification and definition of project weaknesses;
- Identification and definition of opportunities to promote the project;
- Identification and definition of threats to the project.

### *SWIIFT Analysis (Identification of Major Motivating Factors)*

This is complementary to the SWOT analysis and is designed to determine why a person or organization should want to help our project and determine what benefits may accrue to them if we are all successful. A SWIIFT analysis would consist of the following elements:

- List of organizations and persons that may be candidates to take part in or support the project;
- Identification and definition of major motivational factors such as:
  - Prestige;
  - Promotion;
  - Community benefits.

### *Definition of Problem Management Strategies*

In the course of the SWIIFT and SWOT activities, many issues and concerns may be raised by the participants. A simple generic strategy can be employed to address and resolve issues and concerns. It involves the six steps outlined below.

1. *Listen and Identify Problems.* Listening to and identifying the problems is the first step. Doing this well provides a strong foundation for decision-making and project development.
2. *Inform.* Once the problems are finalized and agreed upon, solutions to address these problems may be explored.
3. *Debate.* Engage in discussion and try to persuade to our point of view or have our point of view modified.
4. *Convince.* Convince them that our way is the right way to do it or accept that we need to make changes.
5. *Mitigate.* Accept that our approach is correct but we cannot convince them. Develop strategies to mitigate the effect.
6. *Engage.* If we have convinced them, then engage them in active participation. This generic approach must be tailored for the local context, through the use of some workshop time to discuss the SWOT and SWIIFT analysis results and agree upon the local problem management strategy details.

There are other actions that can be taken to establish, support and develop this relationship. These actions also support relationship 20 between the technical steering committee/ITS development team and the steering committee decsribed in Section 13.6.2, which addresses this relationship.

## 13.6.2 Relationship 20—Between the Technical Steering Committee/ITS Development Team and the Steering Committee

Figure 13.4 illustrates relationship 20.

### 13.6.2.1 Relationship Description

This must also be a reasonably close relationship, as the steering committee must develop a consensus of agreement with proposals as they develop. Although we do not usually look to the steering committee to make executive decisions on the direction of the project or provide influence at an operational level, members of the committee will be closely involved at major decision points in the project. Consequently, they will require the same type of

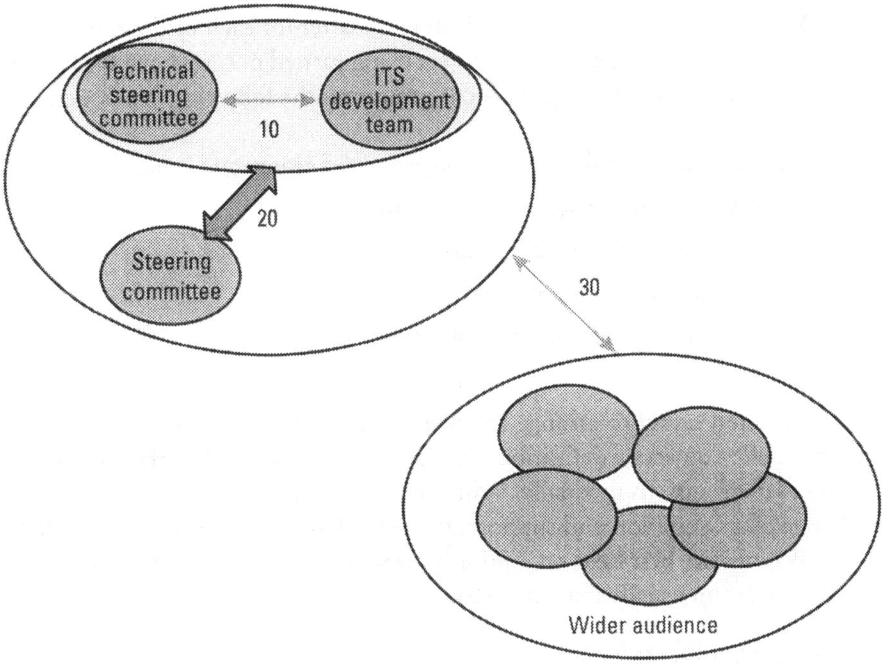


**Figure 13.4** Relationship 20: Between the technical steering committee/ITS development team and the steering committee.

information about ITS technology capabilities as the technical steering committee and the background information required, enabling them to make effective assessments and evaluations of our proposals.

### 13.6.2.2 Proposed Actions To Establish, Support, and Develop the Relationship

There are a number of actions that can be used to support the establishment and development of the relationship. These include the following list.

*Steering Committee Focus Groups*

These would be held to coincide with the major decision points in the project. These can be steering committee meetings at which a presentation would be given by the ITS development team, and the committee would then offer input. Alternatively, these meetings can be held in the form of a focus group at which a structured discussion among the committee members is facilitated, based on input from the ITS development team.

While the number, timing, and subject matter for each of these meetings must be customized to the specific project being carried out, we have previously supported at least five meetings or focus groups on the following key issues:

- Requirements analysis, policy objective definition, and legacy inventory;
- ITS technology review and selection;
- ITS Future Big Picture selection;
- ITS Implementation Strategy;
- Phase 1 ITS design review and summary.

It is often useful to arrange the first of the focus groups on requirements analysis, policy objective definition, and legacy inventory with each of the main stakeholder groups individually. This provides the opportunity for candid exchange of information and supports the clear definition of needs. The rest of the workshops are best held on a joint basis with the whole stakeholder group. Each focus group usually lasts for a day.

### *ITS Discussion Workshops*

These discussion workshops are designed to provide both the technical steering committee and the steering committee with the opportunity for detailed discussion on a range of ITS topics. They should include short, illustrated presentations covering the information to be discussed and interactive group discussion sessions. Each workshop lasts for approximately half a day involving both the technical steering committee and the steering committee as the discussion group.

Each participant should be provided with a prediscussion information package containing paper copies of all slides used in the workshops and other relevant background materials.

The following sections describe workshops we have defined on previous projects.

#### Workshop 1

The objectives of workshop 1 would be as follows:

- To provide an overview of the development plan and activities proposed;
- To explore the nature and definition of ITS and discuss the relevance of ITS to the region's transportation policy objectives, transportation problems and issues;

- To identify current trends in ITS from around the world and make a preliminary assessment of technologies and applications that may be appropriate for the region.

Proposed topics to be addressed at workshop 1 are listed as follows:

- *An overview of the development process:* A brief introduction to the process being used with schedule and milestone information. This should provide participants with a clear indication of what input is required from them, when it will be required, and for what it will be used.
- *The nature and history of ITS:* ITS are substantially different from conventional transportation solutions, having a high information and communications technology content. This means that ITS has to be handled in a special way compared to other more familiar approaches such as designing roads, bridges, or other transportation infrastructure. This part of the workshop would include a short presentation of discussion material describing ITS and summarizing the evolution of the whole subject area, followed by a short group discussion session on the relevance to the region context.
- *Current ITS trends around the world:* ITS is currently being developed and implemented in many countries around the world. In this part of the workshop, a brief presentation of discussion materials describing current trends in ITS around the world will be followed by a group discussion on what trends may be of interest in the region.
- *ITS project examples from around the world:* Building on the previous part of the workshop, this part will enable us to explore international ITS developments and implementations in more detail. A number of examples will be described and discussed.

### Workshop 2

The objectives of workshop 2 are listed as follows:

- To develop a common understanding of the primary objectives to be addressed in implementing an ITS in the region and outline a future vision for an ITS in the region.
- To discuss and agree upon the best approach for applying ITS technologies in an integrated coordinated manner.

Proposed topics for workshop 2 are listed as follows:

- *ITS Objectives Statement and Vision:* A common understanding of the overall objective of the proposed ITS development and implementation and a shared vision of the transportation future is essential to the success of the project. This part of the workshop will support the development and agreement of both items. A short, illustrated presentation describing the proposed ITS Objectives Statement and ITS Vision will be followed by a group discussion on the suitability of both for the local context.
- *From transportation policy to ITS User Services:* This workshop will provide an opportunity for discussion on the proposed approach to identification and confirmation of transportation policy objectives, problems, and issues. This also includes the subsequent capture and translation of requirements into ITS User Services. A short illustrated presentation will lead off the discussion. This will describe the process of identifying and confirming requirements, then developing and applying ITS User Services. Also, examples will be presented and discussed. The methods described in Chapter 5 would be applied in this workshop to support the identification, definition, and confirmation of ITS User Services.

### Workshop 3

Workshop 3 objectives are listed as follows:

- To explore the range of ITS technologies that are currently available and being deployed around the world as well as to discuss the potential suitability of such technologies for the local context.
- To identify ITS Market Packages, based on the range of technologies identified and discuss their application in the region.
- To discuss the proposed approach to applying ITS technologies within a coherent integrated framework for the region.

The presentations can be followed by breakout discussion groups to be facilitated by the system developer and the lead stakeholder agency.

Workshop 3 topics would include the following:

- *ITS Market Packages (the concept):* A brief presentation on the ITS Market Package concept and its application to developing a structured view of the wide range of ITS technologies will be followed by a group discussion on the advantages of adopting the concept for this project.

- *ITS enabling technologies:* A range of ITS technologies, including sensors, display devices, communications, and vehicle location technologies will be presented, illustrated, and described. This will be followed by a group discussion on the potential application of such technologies in the region.
- *ITS Market Packages:* Building on the earlier discussion of ITS Market Packages, details of a range of ITS Market Packages will be presented and illustrated to support a group discussion on their suitability for application in the region.

### Workshop 4

Workshop 4 objectives are listed as follows:

- To discuss the proposed approach to applying ITS technologies within a coherent integrated framework for the region;
- To discuss ways in which the implementation of ITS in the region could be phased in increments, taking account of technical, economic, and political issues.
- To describe and discuss the proposed approach to be taken in determining the benefits and effects of ITS on the regional transportation system.

Workshop 4 topics are listed as follows:

- *Building an ITS Future Big Picture:* The steps required and the issues to be resolved in building a comprehensive framework, or ITS Future Big Picture will be presented, illustrated, and discussed. This will include discussion of technical, economic, commercial, and institutional/organizational aspects of the ITS Future Big Picture for the region;
- *ITS Implementation Strategy:* Ways in which the implementation of ITS in the region could be phased in increments, taking account of technical, economic, and political issues, will be discussed. ITS Market Packages will be used as "building blocks" for the Implementation Strategy, and the appropriate sequencing of these over time and geographical location will be discussed.
- *ITS benefits evaluation and assessment:* This part of the workshop will support discussion and information exchange on potential approaches

to the determination of benefits and costs for proposed ITS implementations in the region. A short presentation will be made on methods and models, and this will be followed by a group discussion.

#### 13.6.2.3 Workshop Types and Group Dynamics

As stated earlier, this chapter is not intended as a recipe for ITS outreach. It is intended to illustrate the principles and enable you to form your own approach tailored to the needs of your stakeholder group. There is another set of advice we can offer to help which concerns selecting meeting types and dealing with group dynamics. We are not psychologists, so this will not be a technical discussion on the theories of meetings or group dynamics. We will simply tell you about some of the techniques that have worked for us and other people we know. With all of these techniques there is one important piece of advice we would provide. It may seem simple, but it is vital that you establish clear objectives for your meeting before deciding on the type of meeting it should be. You should use the objectives to guide your choice of meeting type and the development of your materials. You should also share your objectives with the participants right at the start of the meeting, so you all know at the end of the meeting if you have been successful or not.

Also, we have found it very valuable to remind the meeting participants of the overall process we are going through and indicate where this meeting fits into the overall picture. We also like to show the participants how their input will be used. If the workshop or meeting is a follow-up from a previous activity, we would also like to show how the input from previous activities has been used, so that the participants understand we are not just following a process but that what they say and do count and will be acted upon.

#### 13.6.2.4 Workshop Types

In the same way that you select the most appropriate ITS products, technologies and services to meet your requirements and objectives, you choose the type of meeting you are hoping to have, to maximize the chances of hitting your targets.

##### *Discussion Workshops*

Roundtable discussion workshops in which the system developer or analyst interacts with the stakeholder groups has been a useful tool on some of our projects. We have found them useful with groups of up to 15 people, although larger groups have also worked provided that breakout group sessions are included in the agenda. Some stakeholders will not become engaged in large groups no matter how hard you try to facilitate it. Also, the larger the group, the fewer the opportunities for an individual stakeholder to play a significant role. We use an overhead projector and a flip chart for these workshops. We keep the lights up to encourage participation and follow the rules of brainstorming as a means of encouraging dialog and interchange. We have also limited these to one-day or even half-day sessions. In terms of staffing, we typically use three people from the analyst developer side—a facilitator whose job it is to get the dialog going and keep it going in the right direction, a reporter whose role is to say nothing but record everything, and a meeting manager who has an oversight role for the facilitator and recorder.

### *Viewpoint Platforms*

Sometimes in the interest of achieving consensus within the stakeholder group, we have given up the floor at a meeting to one of the stakeholders. This enables an individual to state his or her position and their reason for taking it. This may require some high levels of facilitation in order to keep the meeting on track.

### *Information Sessions or Briefings*

We have found these to be most effective when prepared thoroughly in advance and limited to half-day sessions. The objectives of these meetings has been to inform the stakeholder group and encourage them to move towards a decision that may be facilitated at a subsequent discussion workshop. The information may be relative to the project at hand or may be background information from other projects aimed at providing information for better decision making. We have successfully used this type of meeting for briefing stakeholders on what technologies are already being deployed and what the benefits are that have already been attained. We tend to use 35 mm slides or LCD direct projection of the computer screen to support these sessions. We dim the lights and encourage the audience to listen to the briefing rather than be interactive.

### *Exercise Workshops or Cooperative Development Workshops*

These are not aerobics sessions for your stakeholders, but meetings in which you expect them to work with you, hands-on, and create something together. We have conducted exercise workshops in which ITS User Services were initially defined, for example. The key to success in this type of meeting is to ensure that the participants know what to expect before they come to the meeting and have the necessary background material to fully participate.

### *Peer Workshops*

This is a variant on the information briefing but delivered by someone who has been a client or stakeholder on another project. Sometimes it is much more powerful for a peer group or agency to make the points than for the system developer or analyst. In the case of this meeting type, it is very important that the system developers, analysts, and the people delivering the peer-to-peer briefing have a solid common understanding of the subjects to be covered, the points to be made, and the objectives. It does no harm to share this information in advance with other meeting participants.

#### 13.6.2.5 Group Dynamics

It is also important to bear in mind that there are usually complicated group dynamics or interactions that will take place at each meeting. In addition to attaining your primary objective, you should also be aware that each and every meeting is a potential opportunity to progress or retard your stakeholder consensus formation efforts. Managing the people interaction at meetings by providing the opportunity for freeform informal discussion and carefully facilitating more structured sessions can provide big dividends.

It may also be the case that the ancillary activities to the main meeting, such as people having tea or coffee together in the lobby between sessions, may be the source of your greatest strides forward in consensus. Therefore, you should not try to make the agenda too "efficient" by cramming too many sessions in and squeezing the break time.

Exercise type meetings have great potential for consensus formation and the development of an understanding between disparate groups. The process of working together on a common task toward a common goal at the workshop is a great way to get to know each other and to practice real-life cooperation.

### 13.6.3 Relationship 30—Between the Steering Committees/ITS Development Team and the Wider Audience

Figure 13.5 shows the relationship between the steering committees and ITS development team and the wider audience.

#### 13.6.3.1 Relationship Description

This could be a very difficult interface to manage. If the activities are not carefully planned, agreed upon, and managed, it would be very easy to send out conflicting messages to the wider audience and cause confusion and misconceptions regarding our work and our plans. The exact manner in which you organize and manage this is up to you and should be determined by local and project needs. However, we would recommend the type of configuration illustrated in Figure 13.6.

This configuration involves the designation of a single point of contact with the wider audience, as indicated in Figure 13.6. This could be one spokesperson or a small team of spokespersons. This person would then be responsible

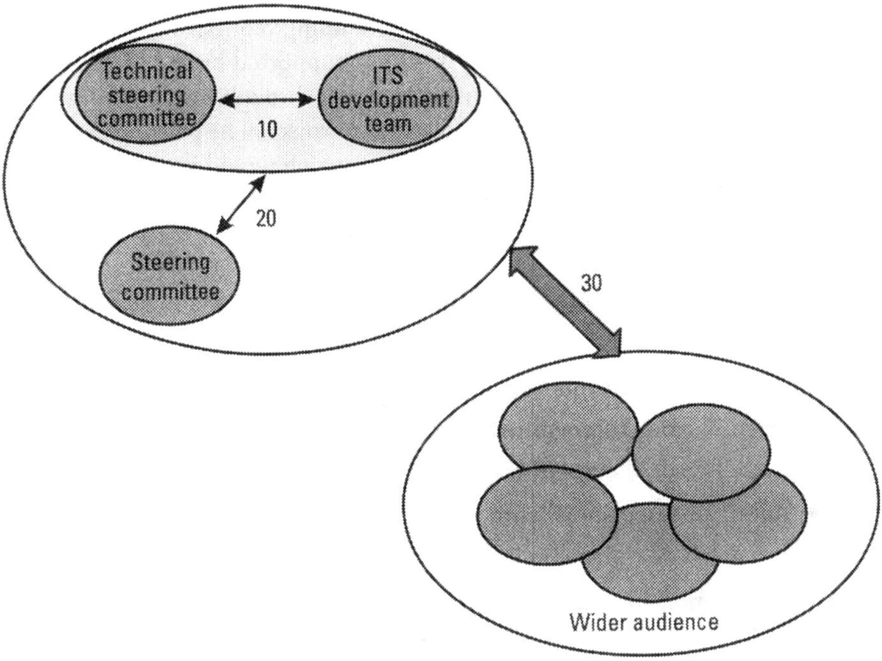


**Figure 13.5** Relationship 30—between the steering committees/ITS development team and the wider audience

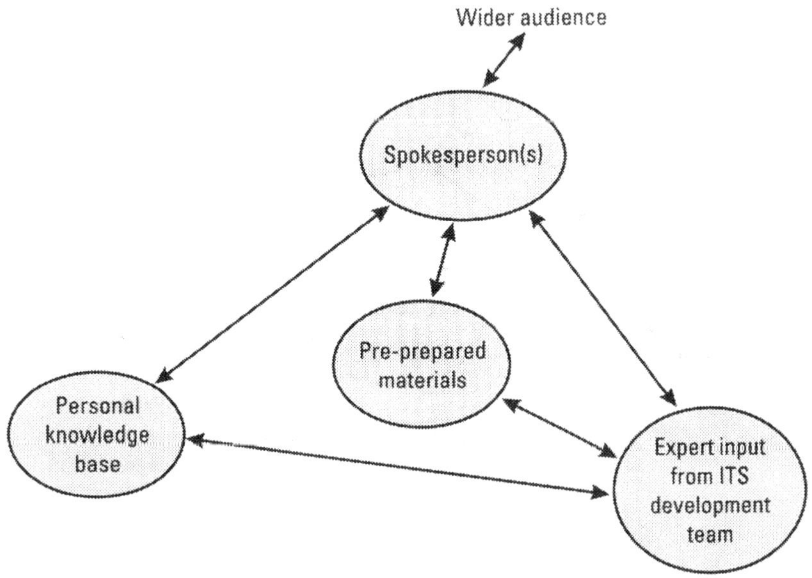


**Figure 13.6** Outreach organizational structure.

for dealing with information provision and handling requests for information from the wider audience. The person would be supported by the ITS development team either directly on a case-by-case basis or through the development of a selection of prepared materials on ITS and the project. The person would also draw on his or her personal knowledge of ITS, which would be supported and enhanced by the ITS development team. The detailed relationships required to support this configuration are shown in Figure 13.7. They are listed as follows:

- 30.1: Spokesperson(s) with ITS development team;
- 30.2: Spokesperson(s) with wider audience;
- 30.3: Spokesperson(s) with pre-prepared materials;
- 30.4: ITS development team with prepared materials;
- 30.5: ITS development team with personal knowledge base;
- 30.6: Spokesperson(s) with personal knowledge base.

These relationships can be supported through the actions described in Section 13.6.3.2.

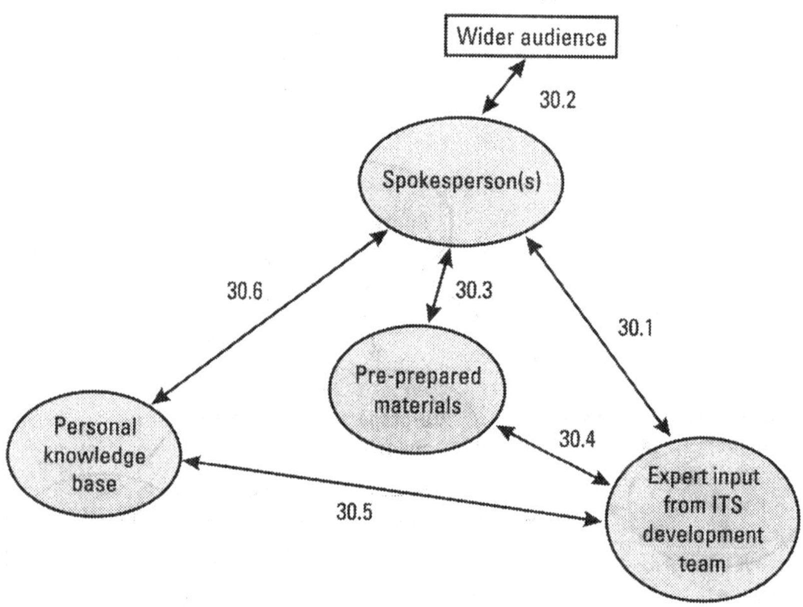


**Figure 13.7** Relationship 30: Detailed relationships between the steering committees/ITS development team and the wider audience.

#### 13.6.3.2 Development of Preprepared ITS Materials

This would provide the wider audience with a range of information on ITS generally and the project specifically. The objectives for these activities are to take the wider audience through several steps leading to either active involvement in the project or positive support for our work and our proposals. These steps are described below.

*Awareness*

Raising awareness about the ITS typically includes the following activities:

- Raise overall awareness;
- Describe the potential benefits of what we are planning;
- Provide details of what we are planning;
- Indicate how plan might affect various groups.

*Interest*

Building on the raised awareness of the ITS, the following activities would be conducted to generate and maintain interest.

- Encourage interest in the project and make requests for more focused detailed information;
- Show how our plans satisfy the SWIIFT analysis findings.

*Desire*

This builds on the "awareness" and "interest" steps through the following activities:

- Bring them around to wanting to actively help and support the planned ITS;
- Help them to understand the benefits and effects on their daily lives.

*Action*

The cycle is complete by conducting the following activities designed to elicit action from the group.

- Participation in system requirements analysis;

- Active participation in institutional/organizational arrangements;
- Positive support for our work and proposals.

This sequence of steps will be supported through actions aimed at developing prepared ITS materials, described below.

### Newspaper and Magazine Articles

A bimonthly newsletter could be developed and produced containing articles on the ITS project and other ITS topics of interest. The articles in the newsletter could also be developed into suitable material for press releases and newspaper/magazine articles.

### Discussion Papers

A number of discussion papers could be developed and circulated to a selected audience including agencies and individuals identified in consultation with the steering committee. Topics could include the following:

- Introduction to ITS;
- Regional or national transportation objectives and problems;
- ITS technologies;
- ITS applications around the world;
- The proposed approach for this project;
- An ITS vision for this project.

### Development of Speaker Scripts

Prepared speaker scripts and presentation materials could be developed to assist in describing the ITS project goals, objectives, and current status. It is anticipated that two separate scripts or presentations would be required:

1. *Politician:* Enabling a nontechnical decision maker to describe the project to a nontechnical audience;
2. *Technical:* Enabling technical management of the project to report on the work being carried out.

### Establishment of ITS Media Resource Library

A centralized media library supported by the client or a contractor with location and staffing determined by agreement. The library would contain prepared slide presentations, videos, support papers, speaker scripts, information packs, and brochures on a wide range of ITS topics, including the proposed implementation and details of the current study's status and results to date.

This facility would be used to develop and support local champions for ITS generally and for the proposed implementation specifically. These champions would be drawn from the existing transportation community and be provided with information to carry the ITS message to others in their community and to the wider general public. It is also anticipated that this facility would support the development and maintenance of multimedia tools such as Internet applications and Web pages.

## 13.7 Summary

This chapter aims to provide you with a model for approaching ITS outreach needs and issues. We recommend a needs-driven approach based on the identification and understanding of the relationships that need to be supported for a successful outcome to the project work. We have shown some examples of such relationships as encountered on previous projects and described some ways to support the key relationships.

# 14

# Developing an ITS Implementation Strategy

## 14.1 Introduction

This chapter addresses the activities to be carried out under step 260, "Implementation Strategy," in the ITS Cooperative Development Methodology. In general terms, this step involves the decomposition of the ITS Future Big Picture into smaller units that can be deployed over a period of time. The chapter begins by describing what an Implementation Strategy is and why it is useful. The various factors affecting the Implementation Strategy or deployment sequence are then described and discussed. The chapter concludes by providing some advice on the development of an ITS Implementation Strategy.

Sections 14.1.1–14.1.6 discuss important concepts associated with the development and implementation of an ITS Implementation Strategy.

### 14.1.1 Supporting an Evolutionary Approach

It is highly likely that lessons learned and experiences gained in the early part of the ITS implementation will be of great value later. Therefore, it is important to support an evolutionary approach to implementation, providing the mechanisms and opportunities to feed such early experiences into subsequent activities.

### 14.1.2 Managing Uncertainty

As the time frame for the implementation of the complete ITS Future Big Picture may be on the order of 15–20 years, there is bound to be some degree of uncertainty. This could be related to changes in the region, population, or in the technology available. It is important to manage this uncertainty through the development of a robust Implementation Strategy that provides a stable implementation approach.

### 14.1.3 Opportunistic Leveraging of Federal Funding

Having a clear idea of where you are today, where you want to be tomorrow, and how you plan to get there are important tools to use for funding purposes. For example, having these tools enables you to take an opportunistic approach to the use of federal funds, through the identification of such funding sources and the use of local funding leverage to gain access to them.

### 14.1.4 Public/Private Partnership

In a similar manner to the leveraging of federal or central government funds, private-sector funds can also be leveraged for ITS implementation. The ITS Implementation Strategy supports this through the ability to identify what will happen where and when and synchronization with private-sector goals and objectives.

### 14.1.5 Developing Synergy Through Integration

The ITS Implementation Strategy also provides support for cost saving and resource sharing through integration of development and implementation activities. Having an Implementation Strategy developed, agreed on, and understood makes it easier to identify the potential synergies and exploit them effectively.

### 14.1.6 Consensus Building Among the Major Stakeholders

An ITS Implementation Strategy can also be an important consensus-building tool. In the first place, the development of the Implementation Strategy requires agreement between all the major stakeholders. This provides one of those focus opportunities where you have something tangible that requires discussion and agreement. In the second place, the output from the process, the Implementation Strategy itself, is a very valuable communications tool that can be used in combination with the ITS vision described in Chapter 6 to communicate plans to other agencies and organizations.

## 14.2 What Is an Intelligent Transportation Systems Implementation Strategy?

In simple terms, an ITS Implementation Strategy is a defined migration path or transition plan, with the objective of getting you from where you are today to where you want to be tomorrow with respect to ITS. Based on the ITS Future Big Picture as the definition of where you want to be tomorrow and the Legacy Catalog as the definition of where you are starting from today, the Implementation Strategy lays out the sequence of deployments or building blocks that you intend to put in place. The sequence will act as a "roadmap," defining the chosen route from today to tomorrow and describing a smooth transition from today to tomorrow in sensible, attainable, and sustainable steps. As such, it has to address two primary dimensions—time and space. Figure 14.1 illustrates this point.

From a time perspective, the Implementation Strategy must define the sequence of building blocks in terms of a series of deployments, showing how each deployment relates to the previous and the next and in what time order they will be deployed.

From a space perspective, the Implementation Strategy has to show the geographical location of the deployments and the proposed evolution of the entire system across the region. This highlights the location of projects and the development of the system over the region. Therefore, the Implementation Strategy takes the ITS Future Big Picture and translates it into a program

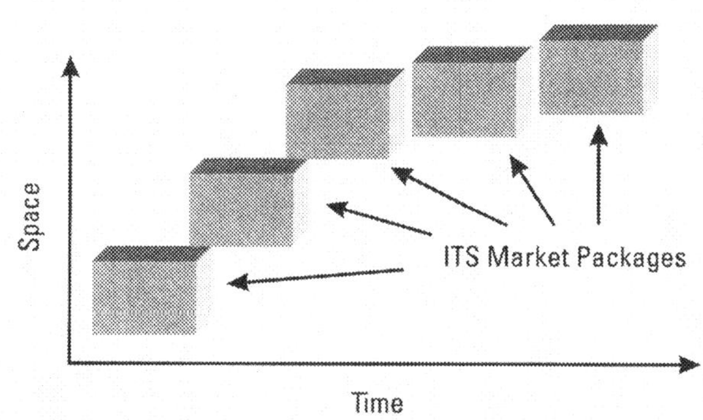


**Figure 14.1** The two dimensions of an Implementation Strategy.

of activities defining what will happen, where it will happen, and when it will happen within a region.

There are also two levels of detail associated with an Implementation Strategy—planning and deployment. The Implementation Strategy has to support two sets of objectives. On one hand, the Implementation Strategy has to show how the planned ITS will fit with the existing regional transportation planning process. On the other hand, the Implementation Strategy has also to indicate the most logical sequence of deployment units to be procured and deployed. This second objective provides the support required to progress from the ITS Future Big Picture to detailed designs, procurements, and project implementation.

It is worth taking a few minutes to talk about these two levels, as it is an important concept running like a thread through the whole ITS Cooperative Development Methodology.

### 14.2.1 Planning

This level involves the use of ITS Market Packages as the fundamental building blocks, developing a sequence of ITS Market Packages over time and space for the evolution of the ITS Future Big Picture. Bearing in mind that this will be from a planning perspective only as ITS Market Packages cannot be deployed without the addition of a further level of detail. However, the ITS Market Package descriptions and the defined sequence can be used as input material for regional transportation plans.

### 14.2.2 Design and Deployment

This level of detail involves the use of ITS Equipment Packages as the building blocks. Adopting this approach, we can support the transition from the ITS Future Big Picture to detailed designs and project definitions. Since the ITS Equipment Package is a procurable unit, it can form the basis for detailed design and subsequent deployment. Detailed design issues and features to look for in good ITS designs are addressed in Chapter 15.

As discussed in Chapter 7, the ITS Market Package is the fundamental building block of ITS from a planning perspective, while the ITS Equipment Package does the same job from a deployment perspective. ITS Market Packages support progress toward integration and mainstreaming of your ITS planning and development efforts, into the transportation planning process by providing program headings and planning-level deployment descriptions. These are supported by Measures of Effectiveness and cost-benefit figures developed in steps 740 and 750 of the ITS Cooperative Development Methodology.

ITS Equipment Packages support the transition from ITS Future Big Picture to detailed designs by enabling the definition of procurable ITS deployment units that obey ITS Market Packages and subsystem boundaries. This enables standalone Equipment Packages to be procured and attain measurable benefits.

In conclusion, we would advocate the use of ITS Market Packages as the fundamental building blocks for the development of an Implementation Strategy. This does not provide the level of detail required to procure or deploy as the ITS Market Packages do not conform to subsystem boundaries as defined in the ITS Future Big Picture and do not have the design tradeoff incorporated yet. However, they do provide the information required to make the basic decisions about what should happen, where it should happen, and when it should happen within the region. Once this is defined, it becomes easier to progress to detailed design, project definition, and procurement.

## 14.3 The Utility of an ITS Implementation Strategy

This is the point at which the full benefit of the ITS Future Big Picture can be realized through the staged development and deployment of the whole carefully planned ITS. The Implementation Strategy draws on many of the previous activities and focuses these into an action plan for getting the job done in a systematic and coherent manner. An Implementation Strategy is useful for a number of reasons, including those described in Sections 14.3.1–14.3.3.

### 14.3.1 Programming the Use of Resources

The development of an Implementation Strategy enables the whole deployment sequence to be viewed and resources to be allocated accordingly. The ITS Future Big Picture allows you to see how it will all fit together and what potential roles and responsibilities may be assigned. The Implementation Strategy provides the tool to program the resources needed to make it all happen.

### 14.3.2 Synchronizing With Other Programs

The need for integration has a temporal aspect in addition to the technical, institutional/organizational, and commercial aspects discussed previously. Planning deployment sequences within an Implementation Strategy enables consideration of coordination issues with other programs and initiatives.

### 14.3.3 Matching the Pace of Deployment to the Capability To Operate and Sustain

The development of an Implementation Strategy supports the adoption of a sustainable deployment pace. The whole sequence of planned activities can be reviewed and coordinated with the capabilities of the deploying organizations.

## 14.4 Factors That Influence the Choice of What, Where, and When

Starting from the ITS Future Big Picture as previously developed, it is necessary to take multiple factors into account when developing the most appropriate Implementation Strategy for a specific region. A combination of factors unique to the region and the point of departure from which the region begins structured ITS deployment affect the choice of Implementation Strategy. For this reason, it is not possible to define or specify a single path that is applicable to all regions and circumstances. Each region should be studied individually and a specific Implementation Strategy defined, based on consideration of a range of factors. Sections 14.4.1–14.4.24 list the factors we have encountered in the course of the various projects with which we have been involved.

### 14.4.1 Technical Feasibility

Some ITS Market Packages are well understood, tested, and proven; others are still emerging and require further research development and pilot testing, so these would come later in the deployment sequence. For example, most traffic signal control applications have been implemented in many locations and a considerable body of knowledge has been accumulated. On the other hand, applications such as automated highway systems or advanced vehicle control may be emerging from research and development programs but still be light on deployment experience.

### 14.4.2 Protecting Legacy and Sunk Investment

As discussed in Chapter 8, an important consideration in the development of the ITS Future Big Picture is the protection of previous transportation investments, whether they be in ITS or conventional transportation infrastructure. What you already have on the ground will be a major influence in the sequencing of deployment. For example, if your ITS Future Big Picture includes an electronic payment system for transit fare payment and you have a fairly new transit system, then it may be appropriate to move the electronic payment systems deployment to the earlier phases of the ITS Future Big Picture deployment. This would support the attainment of additional benefits from the previous investment and serve as a clear indication that legacy is being managed and preserved as part of your current efforts.

### 14.4.3 Early Winners

More often than not, ITS deploying agencies are under considerable pressure to show initial success in their ventures. ITS especially draws this type of pressure since many of the technologies, products, and services are unfamiliar and have not yet developed a comfort or trust level with nontechnical decision makers or the voting public. Therefore, it may be appropriate to include the more proven ITS products and services in the early stages of deployment, thus maximizing the probability of early success.

### 14.4.4 Return on Investment

This is sometimes referred to as "biggest bang for the buck." It is closely related to the "early winners" factor. It may be important to consider early implementations that are capable of providing clear, tangible evidence that a great deal of benefit was achieved through the expenditure of a relatively small amount of resources in terms of time and money.

### 14.4.5 Risk Management

Building on the previous two factors, this involves consideration of the various risk factors associated with ITS deployments. Risks should be evaluated according to two primary factors:

- The probability of the risk becoming a reality;
- The severity of the consequences.

A robust risk management strategy can then be devised encompassing risk avoidance and risk mitigation actions. These considerations may move some ITS deployments back along the timing sequence due to immature technologies, or lack of deployment experience, for example.

### 14.4.6 Evolutionary Deployment Factors

Due to the dispersion pattern for legacy systems, or other local factors, it may make sense to adopt a particular evolutionary deployment or roll-out strategy. Two major variants that we have encountered in the course of previous project work are discussed in Sections 14.4.7 and 14.4.8.

### 14.4.7 Geographic Deployment Strategy

This entails the identification of problem areas where there are one or more transportation problems that might be addressed successfully by ITS deployments. The ITS Future Big Picture is then translated into a series of designs and projects to be deployed in clusters around the problem areas. Later stages of the deployment strategy increase the geographic coverage of the previous deployments until the whole region is covered.

### 14.4.8 Functional Evolution

Instead of adopting a "spatially related evolution of the ITS Future Big Picture," the choice here is to deploy across the entire region as quickly as possible, but with limited functionality. Not all the ITS User Services identified during the requirements analysis activities are provided, with a limited subset of ITS Market Packages and Equipment Packages deployed across the region on an area-wide basis. Subsequent deployment phases then extend ITS User Services and functionality until the entire region is addressed and provided with all the required ITS User Services.

These are, of course, extremely over-simplified evolutionary deployment options. In reality, it is highly likely that a hybrid combination of both approaches would work best. This is due to the fact that some ITS applications such as traveler information lend themselves to functional evolution, while others such as advanced traffic management lend themselves to geographic evolution. In either event, the selection of a roll-out approach will have a significant influence on the Implementation Strategy.

### 14.4.9 Cost Considerations

These include the basic consideration of what you can afford. It also covers the ability to split the proposed deployment unit down so that each deployment cost is contained within a "single spend" figure. ITS Equipment Packages that consist of a large raft of hardware, software, or communications equipment may have to be phased to take "maximum spend" constraints into account. These spend constraints may be a regulatory or legislative issue or may simply be related to the comfort level of the deploying organization in procuring and deploying on a large scale.

### 14.4.10 Staff Resources

This is closely related to the cost issue discussed above. Staff can be considered as part of the resource set required to fuel ITS planning, development, and deployment. Consideration of both the quantity and quality of the staff available to support various deployment sizes and types may have an influence on the selected Implementation Strategy. For example, if available staff have little or no experience in ITS procurement and deployment, then it may be appropriate to select early deployments based on similarity to previous projects. Another option would be to select smaller, yet unfamiliar deployment types to provide the opportunity for staff to be trained and gain experience before the larger, more demanding deployments.

As will be discussed in Chapter 16, the choice of a procurement approach is also influenced by the availability of appropriate staff resources. In turn, the availability of certain procurement options may influence the Implementation Strategy.

### 14.4.11 Leveraging Efforts of Others

As we have stated many times in previous chapters, one of the big advantages of developing an ITS Future Big Picture is the ability to identify and exploit synergy with other public and private organizations. The availability of this synergy may well be time dependent as the other organizations implement their plans. In order to synchronize your efforts, it may be necessary to adjust your ideal schedule or Implementation Strategy to fit with others, maximizing the leverage effect. For example, if the local cable TV company is about to deploy a large-scale fiber optic communications network in your region, you may want to bring your wireline communications Equipment Packages up to the front of the Implementation Strategy time sequence and match with their roll-out approach.

### 14.4.12 Survivability

This is a terrible term, but we could not think of anything else to call it. This one is based on pure common sense and goes something like this: There is no point in being technically, institutional/organizationally, or commercially sophisticated with early deployments, if you do not survive until the whole ITS Future Big Picture is complete. This consideration may well make you decide to do the "no brainers" first, increasing your chances of survival to do the smart stuff later.

### 14.4.13 Technology Maturity

The ITS Future Big Picture is, by definition, a complete future ITS addressing identified as ITS User Services. Inevitably, it will incorporate ITS Market Packages that are not "technologically mature" when the deployment work starts. This means that either the technology does not yet exist or it is still too new to have a good chance of successful deployment. You may well be wondering why such ITS Market Packages would be included in the ITS Future Big Picture in the first place. They are there because the idea is to focus on user needs over the long term and be as comprehensive as possible. The existence of these immature technologies means that we have identified some need for them and have left the flexibility in our ITS Future Big Picture to include them when they are ready. We have not constrained our view of the future by available technology (well partially—as it is nearly impossible to avoid completely).

### 14.4.14 Technical Logic

This is the first thing that many engineers will consider as an important factor in selecting the sequence for the Implementation Strategy. In an ideal world, it would probably be the only factor that needs to be considered. This looks at the whole ITS Future Big Picture and breaks it up into deployment units that make sense from a purely technical perspective. Consideration is given to the relationship between the various ITS Market Packages and the interdependencies between them. For example, from a purely technical perspective you would deploy the traffic control center and communications Equipment Packages before deploying the VMS to display the messages from the control center via the communications.

### 14.4.15 Results

Here we encounter the difference between perception and reality. It may be necessary to bring Equipment Packages forward in the deployment sequence if they have the ability to produce high-impact, perceived benefits, rather than actual ones. Related to the survivability and early winner factors, this factor may influence the choice of deployment sequence away from the technically logical.

### 14.4.16 Modularity

Particularly in situations where consensus formation has been difficult or even impossible, it may be necessary to consider what we call modularity. Modularity refers to the ability to deploy and support the Equipment Package by a single organization, not relying on institutional/organizational arrangements from other external organizations. This may seem strange, but we have encountered situations in which the best way to make progress on consensus formation was for one agency to "bite the bullet" and go ahead with a modular deployment that was completely within their control and jurisdiction.

### 14.4.17 Timing Factors

These can include the overall time it will take to fully deploy an Equipment Package or the timing of your efforts in contrast with other transportation initiatives in the region. They may also include the overall anticipated duration of the deployment phase. It may be appropriate to choose early deployments that are shorter in duration so that early benefits are obtained in a prompt manner, in line with the discussion under "early winners."

### 14.4.18 Ease of Implementation

This is a general purpose factor that acts like an umbrella for several subfactors. In the scheme of things, just how easy will it be to implement one Equipment Package compared to another?

### 14.4.19 Local Politics and Momentum

It could well be the case that you find yourself with the budget to plan, develop, and implement ITS because some nontechnical decision maker has made a public commitment to make something happen and has made the necessary budget available. In such circumstances, the re-election of a public official may be at stake, due to a public promise of a specific action. It would be somewhat less than prudent to delay that action too late in the deployment sequence. This is definitely another "survivability" factor!

It may also be that there is no single official on the line, but that local politics is such that a particular application has prominence and high visibility. This is again a time for realism to prevail over technical elegance!

### 14.4.20 Institutional and Organizational Issues

The big question here is:

> Is institutional/organizational change required to support the deployment phase?

If the answer is yes, then it may be wise to push this deployment phase back down the timeline. This is somewhat related to the "modularity" factor described above. Even if consensus is achieved and agreements have been struck to support the planned ITS, organizational inertia is such that it may take considerable time and effort to make the necessary institutional/organizational changes. It may be appropriate to move such Equipment Packages back in the timing sequence to avoid stalling and delay in implementing the overall ITS Future Big Picture.

### 14.4.21 Choosing a First Step That Provides Leverage for Later

For example, installing a data collection or surveillance system initially for your own use may provide valuable data and information that could be used to bargain with other agencies and the private sector.

### 14.4.22 Commercial Aspects

There are a number of commercial factors that might influence your choice of deployment sequence:

- *Partnership availability:* Is the private sector at the point in their plans where they are willing to enter into leveraged public/private partnership agreements, and is your sequence in line with theirs?
- *Public agency funding availability:* Sometimes the availability of public funding is cyclical. Some themes may be hot at the moment and funding may be available. You may have to grab it while you can!
- *Private-sector investment opportunities:* Perhaps the consumer market, for a number of external factors outside your control, is ready for a

particular product or service. The private sector is likely to have figured this out and will be ready to exploit the market. Will their desires match yours?

### 14.4.23 Standards

We covered this in some detail in Chapter 12. Are the standards required to support a successful implementation available now or are they coming later? This might significantly affect your choice of deployment timing sequence.

### 14.4.24 Training

Staff need to be trained to operate and maintain the new technologies. A systematic training and staff development program should be defined and synchronized with the deployment sequence. The schedule or budget constraints associated with the training program may create factors that influence the Implementation Strategy.

## 14.5 Advice on the Development of an Implementation Strategy

Now that we have explored what an Implementation Strategy is and what factors you should take into account when developing one, we thought you might appreciate some thoughts on how to go about developing one for your region. We have to begin by saying that this topic, more than any other in the book is much more of an art than a science. Taking account of multiple influencing factors and a varying starting point for deployment requires a great deal of skill and judgment. Accordingly, what follows is the best advice we can give you based on our own experiences. As with many other things in this book it definitely is not a "how to" recipe for you to follow blindly or naively. The advice is designed to give you the benefit of our experience on which you can expand and develop your own particular approach.

The starting point for the development of the ITS Implementation Strategy consists of two components that were created in early stages of the ITS Cooperative Development Methodology:

- Legacy Catalog;
- ITS Future Big Picture.

These represent the two endpoints for the migration path to be defined by the Implementation Strategy. As illustrated in Figure 14.2, the Legacy Catalog represents "today" by describing the unique context for the region as the launch point for further ITS deployment. The ITS Future Big Picture represents "tomorrow" as it encapsulates all the technical, institutional/organizational, and commercial initiatives that you wish to carry out in order to satisfy all requirements. There are many possible paths from "today" to "tomorrow," and the purpose of the Implementation Strategy is to define the most appropriate one for your circumstances.

Taking these two elements as the starting point, the next job is to start to identify the single migration path that is best for the region. This involves identifying and considering the many factors described in the previous section and tries to balance the needs derived from consideration of each factor to create an attainable and sustainable deployment sequence.

The deployment sequence will consist of building blocks and activities on the three levels contained in the ITS Future Big Picture:

1. Technical: Procurements, projects, and deployments;
2. Institutional/organizational: Agreements, relationships, and communication between partners;
3. Commercial: Public/private partnerships, leveraged private-sector activities.

To assist in defining the Implementation Strategy, it may be useful to refer to the output from many of the other steps in the ITS Cooperative Development Methodology. For example, the financial and commercial analysis

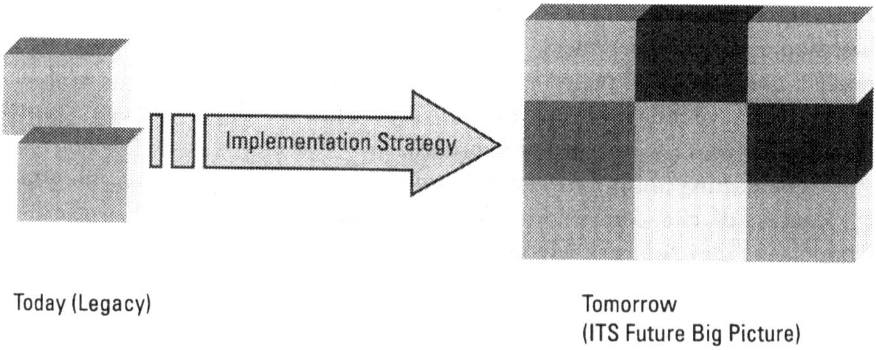


**Figure 14.2** From today to tomorrow.

conducted in step 250 may provide useful information on commercial aspects of the deployment.

Other input for defining the Implementation Strategy comes from consideration of the multiple factors as described in Sections 14.4.1–14.4.24. We have had some success in managing the identification, consideration, and review of these factors through the adoption of a "weighting and ranking" approach. This involves working with the stakeholder group to agree on weights or priorities for each influencing factor, then determining a score for each factor for each deployment unit. A weighted score can then be determined for each deployment unit and used to help position the deployment units in the implementation sequence.

The process we have used on previous projects, involved the following steps:

- *Step 1:* Identify all potential influencing factors. The list we provided earlier in this chapter could be used as a starting point for this. We would not claim that the list is complete, but it may give you some initial material to stimulate thought. This activity is best treated as a 'brainstorming" or "brain dump" type of activity, where the emphasis is placed on the identification of as many relevant factors as possible, postponing structure and consistency until the next step.
- *Step 2:* Review the identified influencing factors carefully, with a view to the later development of weights for each factor. This will involve the combination of factors that are similar or closely related and the redefinition of factors to make them more suitable for deriving agreed weights. For example, in the list we provided earlier the "early winners" and "return on investment" factors are very closely related to each other, so we would probably combine these into a single influencing factor at this point.
- *Step 3:* Inform the stakeholder group about the list of influencing factors and encourage them to participate in a "weighting and ranking" exercise. We have found that this is most effectively carried out in the form of an invited workshop, facilitated by the system development team that put the list of factors together. Stakeholders are sent an agenda and objectives for the workshop and asked to review the materials in advance. This preworkshop information package should also contain a list of influencing factors with definitions.

  During the workshop, the system development team explains the objectives and the format for the workshop and then describes and explains each factor in turn. In the course of this part of the workshop, there may be further rationalization of the factors, including the combination or deletion of factors based on input from the stakeholder group. It is important to welcome this kind of input and clearly incorporate it into your list of influencing factors. There is a temptation for the system development team to be rigid, inflexible, and not open to new input at this stage, especially if substantial time and effort has been invested in the development of the influencing factor list. It is important to avoid this by thinking of the initial list as a "strawman" for discussion. No matter how well you have researched the influencing factors there is a strong possibility that the stakeholders, with their deeper understanding of what they want and the environment they work in, will have additional valuable input.

The stakeholders are then asked to put their own weightings on each factor. This can be a take away and complete later exercise, but we prefer to get it done during the workshop, providing the opportunity for further discussion and clarification from the system development team or between stakeholders. This also significantly increases the chances of getting a completed response from each stakeholder. The weights can then be averaged across the stakeholder group and provided as input to the next step. A variation on this theme would be to apply a weighted average to the stakeholder weights, taking account of differing levels of importance from one stakeholder to another. For example, weights determined by stakeholders who might be expected to play a leading role in financing, implementing, or operating the ITS may be given a higher weighting than those who are playing an advocacy role.

- *Step 4:* Post-workshop analysis. This could be done as part of the workshop, but we have had more success treating it as a separate, expert evaluation exercise. This is where each deployment unit is assigned a score, to be multiplied by the previously defined weights, to derive a *weighted score.* The deployment unit should be ITS Market Packages as discussed earlier.
- *Step 5:* Second, follow-up workshop to communicate results, explain, debate, and seek final agreement from the stakeholders. You should bear in mind that the output from the weighting and ranking process is merely a first approximation to the Implementation Strategy. Applying the weights and scores does not provide a "scientific" or rigorous single solution. The adoption of the process helps to take some of the subjectivity from the consideration of the influencing factors and provides a framework in which all factors are given a reasonably equal

treatment. The follow-up workshop is a very important part of the Implementation Strategy development work as it provides the opportunity for the stakeholders to review the first attempt at the Implementation Strategy and apply more judgment to the final selection.

We have quite deliberately chosen to suggest a process in which the stakeholders provide the weights but do not take part in the initial scoring of each deployment unit. We have had experience in attempting to have the stakeholder participate in both the weighting and scoring process and have had limited success. The stakeholders seemed to lose patience with the longer weighting and scoring process and had difficulty completing the process.

## 14.6 Relationship to the Regional Transportation Planning Process

The Implementation Strategy step is one where there is a clear interface with the regional transportation planning process. Unfortunately, it is not a simple relationship, since the transportation planning process can either take the Implementation Strategy as an input, or it can provide input to the Implementation Strategy.

Planning is a cyclic process; therefore, depending on where in the planning cycle you conduct your ITS Cooperative Development Methodology work, the plan might come before or after the Implementation Strategy. If the plan is reasonably mature and we are aiming to put the results of our ITS Cooperative Development Methodology work into the next generation of the plan, then the Implementation Strategy will come first and everything will seem logical. If the plan is brand new (say within 12–18 months of revision) then the plan will come first and the Implementation Strategy may have to take account of the programs included in the current plan. This is a consequence of the newness of ITS and the use of separate dedicated funding. As the process of "mainstreaming" ITS continues, we will encounter fewer examples of transportation plans that do not already take account of ITS, and this will make it easier to relate the Implementation Strategy to the transportation planning process.

## 14.7 Monitoring and Updating the Implementation Strategy

The ITS Future Big Picture and the Implementation Strategy used in combination provide the tools to enable you to apply another system engineering technique. This is sometimes referred to as fast-track, or evolutionary development. The overall philosophy calls for the rapid development of an ITS Future Big Picture for the system, followed by the deployment of a small module of the overall system. The experience gained from deployment is then used to revise and enhance the ITS Future Big Picture and prepare for further deployment. This technique has been utilized successfully in the past in dynamic situations where it has been very difficult to develop an accurate view of the ITS Future Big Picture, or where technology is changing very rapidly. The ITS Cooperative Development Methodology lends itself to the application of this type of philosophy.

In more detail, the steps you would take would be as follows:

- Develop the ITS Future Big Picture as quickly as possible, aiming to have a coarse view of the future that may be imperfect.
- Review the ITS Future Big Picture carefully and taking account of all three levels, select a single deployment unit that is likely to be required no matter what changes in technology or requirements take place in the future. This is known as "module 1" and is typically deployable over an eight- to ten-month period, providing a manageable opportunity to learn from the deployment and incorporate the lessons into future deployment.
- Deploy module 1, gain experience from the deployment, and use it to revise the ITS Future Big Picture. This would involve the support of a continuous review of the ITS Future Big Picture, aimed at incorporating the following:
    - Lessons learned from earlier deployments;
    - Technology developments;
    - Institutional/organizational developments;
    - Commercial developments.

The revisions made to the ITS Future Big Picture are incorporated into future deployment plans, and the cycle is repeated until the whole ITS is deployed.

# 15

# ITS Design and Implementation

## 15.1 Introduction

We have made the same point many times throughout this book regarding the difference between the ITS Future Big Picture and detailed designs for ITS deployments. This chapter addresses this point again through a discussion on what makes a good ITS design. It is not our purpose to provide a detailed guide on how to design ITS; this would be far beyond the scope of this particular book. The objective is to provide the reader with some insight and information regarding the use of the ITS Cooperative Development Methodology to improve the quality of ITS designs and subsequent procurements. Therefore, the information in this chapter will not empower you to go out and be an ITS designer, but it will facilitate your understanding of what ITS design is and what to look for in a good ITS design and a good ITS designer.

The relevant steps in the ITS Cooperative Development Methodology are step 400, "Design" and step 410, "Implementation." The "Implementation" step is covered in this chapter due to the close relationship between the way you design and the way you implement ITS (and most other things).

## 15.2 What To Look for in a Good ITS Design and a Good ITS Designer

Let's begin by describing some general features that should be present in any good ITS design. Then we will go on to talk about software and hardware designs specifically, as they have their own particular features.

### 15.2.1 User-Focused, Requirements-Driven ITS Design

This is the main thrust of the ITS Cooperative Development Methodology. The information provided by carrying out the steps in the ITS Cooperative Development Methodology acts as the basis for user-focused, requirements-driven ITS design. However, the requirements specified up to this point have been higher level, aimed at establishing the overall framework within which the detailed designs will exist and relate to each other. Detailed design requires detailed requirements to be defined and agreed upon.

Therefore, the first step in the detailed design of an ITS will be the "layering" of additional levels of requirements detail on to the existing higher level ones. The current user requirements are specified in terms of ITS User Services. Using these as a starting point, the ITS designer will add levels of detail to each ITS User Service until it is possible to specify a detailed solution based on the requirements. For example, the ITS User Services do not take account of the specific management strategies to be supported, so the designer would have to add detail to the ITS User Service to take account of this.

A good ITS designer will ensure that the functions and features in the selected design solution are fully traceable back to the detailed requirements established on the basis of this work. Here again we can see a cyclic "what?/how?" situation. Within step 400, a good ITS designer will recognize the need to support a cycle of detailed requirements and proposed solution definition. Those that approach this as a linear sequential activity, where detailed requirements are defined, closely followed by detailed design and implementation, do so at their peril as inevitably detailed requirements will shift in light of increased knowledge of what can be done.

### 15.2.2 Willingness and Capability to Revise Earlier Work in Light of New Information

A good ITS designer will develop the design on the basis of the stated requirements but also be prepared to revisit the requirements and make modifications in light of new information. ITS is such a new subject area, it is highly possible that something new will be learned or a new product/service will emerge, causing the original requirements to be rethought. In close cooperation with the client or stakeholder group, it is possible to revise and enhance requirements at this point.

Many system designs are based on the results of earlier field and operation tests involving smaller scale deployments of similar technologies, products, and services. Experience gained in the course of such tests is often extrapolated into subsequent designs. While this extrapolation may not be fully warranted, it is usually the best information available at the time. When better information becomes available later, a good designer will incorporate this new information into his or her design process. For example, as ITS product and service market penetration increases, the impact and benefits measured in a small field or operational test may not increase proportionately. Information or data from subsequent deployments may confirm this.

### 15.2.3 Use and Incorporation of Relevant Standards

As discussed in Chapter 12, careful review and use of the relevant standards for ITS is an important feature to look for in any good ITS design.

### 15.2.4 Coordination Between Systems and Agencies

The continual reference back to the ITS Future Big Picture indicates how the subject design will fit inside the overall framework. This will include the definition and support of current or future interfaces to other subsystems operated by the same agency or by another agency. The support of such interfaces should be provided in a seamless manner through the integration of the various components and subsystems in the overall system implementation. This means that information displays, methods of operation, and the entire interface between the human system user and the system itself is implemented in a coherent, consistent manner.

### 15.2.5 Design Options Explored and Explained

Within each ITS market package there are multiple design options available. A good ITS design will be derived from consideration of the ITS Market Package and a thorough exploration of the design choices or tradeoffs available. The choice of which tradeoff to make and what detailed design choices to settle on are very dependent on specific circumstances such as the availability of products and services in the marketplace.

### 15.2.6 Clear, Unambiguous Definitions

Taking full advantage of the requirements definition work that has taken place previously, it should be possible to minimize the ambiguity in detailed designs and specifications.

### 15.2.7 Modularity

Good ITS designs will be easy to procure and implement—if they are modular. What do we mean by modular? We mean reasonably self-contained in that the implementation of a particular project has a clear beginning and an endpoint that is not too far over the horizon. We also mean projects that can be isolated from each other and that are not too dependent on each other for success.

## 15.3 Software Designs

The design and implementation of software projects has been the subject of much study and development experience since the invention of the computer. In a coarse attempt to summarize the current state of the art of software development, we would characterize it as an "immature" discipline. This is particularly the case when compared to engineering disciplines such as mechanical, civil, and chemical engineering, where the development procedures are well-documented and matured over a large number of iterations.

In many ways, software development encapsulates the issues associated with ITS in general. An emerging discipline, profession, or subject area has many frontiers where new things are being learned and tools and methodologies are being developed. In such circumstances, it is important to focus on the objectives of the project in hand and not get drawn into the research and development aspects of the emerging topic.

It is also important to avoid falling victim to the "hackers," the "fly-by-the-seat-of-the-pants and just-get-the-job-done brigade." This approach, which we have encountered so often in ITS projects, inevitably leads to software that has errors that are difficult to find and correct and is very difficult to maintain, enhance, or expand. While some software developers believe in the evolutionary, make-it-up-as-you-go-along approach, others have learned that there are no shortcuts to the development of robust, high-quality software. See Figure 15.1.

Figure 15.1 shows the relationship between the stage in the life cycle of the project and the amount of resources required to fix a bug or error due to ambiguity in requirements. It shows that time and effort invested early produces the most leverage or "bang for the buck." The systematic, structured, disciplined approach to software development provides this kind of build-up supporting truly high-quality software.

As you would expect, the system engineering community has been addressing this very issue and has come up with a number of useful approaches. One of these is known as the *capability maturity model for software.* This is a tool that organizations can use to assess the areas in their software development

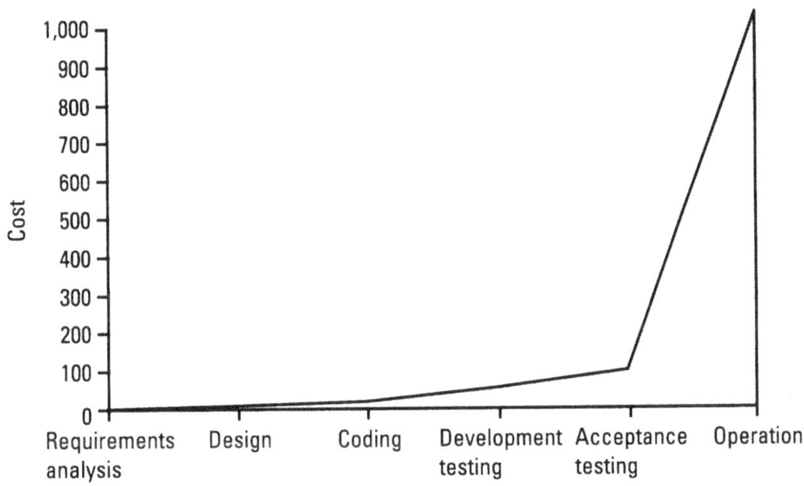


**Figure 15.1** The relative cost of fixing ambiguity at each stage of the development process.

process that could be improved. There is also another interesting methodology called *a method for assessing the software engineering capability of contractors* (CMU/SEI-87-TR-23). Both of these models, or methodologies, provide a set of procedures and yardsticks that could be used to gauge the performance of a particular software developer, against the performance of the industry as a whole.

As has been our philosophy in the rest of this book, we will not get into more detail about these methods; we have done our job if you are now aware of them and interested in making use of them. For further information, the following Web site may be of use:

http://direct.asset.com/wsrd/product.asp?pf_id=ASSET_B_1382

The choice of methodology and the specific yardsticks to be adopted we leave up to you. In fact, it is entirely up to you if you use the methods and yardsticks or not. No matter what you choose to do, we would encourage you to include the factors discussed in Sections 15.3.1–15.3.16 in your assessment of the capability of a software developer:

### 15.3.1 Systematic, Structured Development Approach

Find out if the developer has experience in and plans to use some kind of software engineering or development approach based on a disciplined and structured approach.

### 15.3.2 Testing

Ask the developer about the plans for software testing at each stage in the development process.

### 15.3.3 Maintainability

What steps does the developer plan to take to ensure that the software can be easily modified at a later date by another company, if necessary? This mainly relates to the plans for software documentation and the use of annotated software code.

### 15.3.4 Software Reuse

Does the developer operate a policy of maximum software reuse where commonly used subroutines and algorithms are stored in a centrally maintained library for common use?

### 15.3.5 Careful Objective Setting and Requirements Analysis

It should go without saying that you would only select a developer who employs these principles, especially as you would have invested significant resources in the development requirements and the ITS Future Big Picture.

### 15.3.6 ITS Future Big Picture Reaction

Does the developer relate to and embrace your ITS Future Big Picture with enthusiasm, or is it treated as something to put up with and get on with the job? The reaction should give you some clue about the culture of the software developer's organization.

### 15.3.7 Evaluation of Risks and Required Resources

Check if the developers are aware of the subject of risk management and if they routinely make assessments and evaluations on the risks involved in their software development projects. Also find out if they are in the habit of quantifying the resource required to carry out particular projects. If they do not, this might indicate that they do not really know what they are doing and what they are getting themselves (and, unfortunately, you) into.

One important aspect of the risk associated with the project is the use of new software tools, techniques, and approaches. If the developer plans to use your project to try out or introduce a new approach that has not been used by the developer before, you may want to review this very carefully.

Don't get us wrong; we don't see anything wrong with innovation in software engineering just as long as the client knows that he or she may be a guinea pig. We have been involved in software development projects such as these and in retrospect, we would prefer the use of tried-and-trusted techniques. There are times to stretch out for the future and times to lean back on the safety of the past.

### 15.3.8 Thorough Preparation for the Project

Determine the developer's approach to the beginning of a project. Does he or she just launch right in, or are there indications that a thorough preparatory phase is implemented prior to project execution?

### 15.3.9 Continuous Monitoring and Evaluation of Software Development Progress

Ask developers to share information about their plans for monitoring the incremental progress of the software development project. Determine if they plan to deploy adequate resources to checking and monitoring and have an appropriate organizational mechanism to support effective progress monitoring and course correction.

### 15.3.10 Flexible Development Approach

Building on the above, establish the developers' planned reaction if course corrections are required. Does the methodology they plan to employ properly support such course corrections?

### 15.3.11 Careful Review of Final Deliverables

Find out what the plans are for deliverable review. This should be the developers' internal review plan and be separate from any testing and certification required under the terms of the procurement contract. In other words, does the developer plan to trust that the final deliverable will pass your testing and inspection, or do they plan to carry out their own tests?

We have worked with software developers that have sent out companywide e-mails jubilantly proclaiming that the software has passed client testing and hailing it as a major victory. While it is nice to celebrate the passing of a milestone, such a reaction suggests some degree of surprise that the software passed the client test. In our opinion, the jubilation should have been as a result of internal testing prior to the client tests.

### 15.3.12 Feedback on Lessons Learned

This is a good indicator of the nature of the developer. If the company is in the business for the long term and is serious about providing high-quality software, it will be a "learning" organization. That is, there will be established procedures for feedback on lessons learned and amending the way things are done. It is one thing to make a mistake, or be naïve; it is something else to go on making the same mistakes over and over again.

### 15.3.13 Hardware

Many of the factors discussed under software apply equally to the hardware design components of an ITS project, with a few specific additions.

### 15.3.14 Ability To Specify Concisely and Procure Successfully

Check the hardware designs and make sure that in addition to meeting your requirements, they provide concise specifications with minimum ambiguity. Also ensure that the resulting designs are practical from a procurement perspective.

### 15.3.15 Maximum Use of Off-the-Shelf Components

In most cases, the designer will gravitate naturally toward the use of off-the-shelf products and services wherever possible. This would typically represent the lowest cost and lowest risk approach. However, it is worth checking that this is the case. Occasionally, the system designer may have a desire to develop a new product or service and while there is nothing wrong with that, you should be aware that this approach has been selected.

### 15.3.16 Design With the Whole System Life Cycle in Mind

A good ITS design will not just meet the requirements, it will look beyond implementation and initial commissioning to operations and maintenance. The full system life cycle from initial conception through implementation to operations maintenance and renewal will be taken into account. There will be evidence of tradeoffs between up-front capital investment and long-term operational costs. The design will also accommodate operational considerations such as operator and user training and ease of use issues such as human factors.

## 15.4 Translating ITS Market Packages Into ITS Equipment Packages

A feature of the ITS design should be a smooth transition from the high-level descriptions of potential solutions or system building blocks provided by the ITS Market Packages, through the ITS Equipment Packages, to final detailed designs. This should start with the sequence of building blocks defined in the implementation strategy and move progressively into greater detail to the point where there is enough information to specify, procure, and implement the ITS.

## 15.5 Project Definition

This will ultimately lead to the definition of ITS projects so it is worth taking a few minutes to talk about the difference between a design and a project. In simple terms, a project is a design or several designs that have had the procurement perspectives added. As discussed in Chapters 14 and 16, there are many procurement mechanism choices and many possible deployment or implementation sequences. These are affected by the numerous influencing factors described in Chapter 14. Once design work has reached an advanced stage it is appropriate to look at the these factors and devise projects that can be successfully implemented.

## 15.6 Practices and Procedures From ITS Experiences

We would like to round off this chapter by passing on some general guidance on managing ITS projects, gleaned from the experiences we have had so far. This is by no means a comprehensive set of guidelines as we are still learning new things in the course of our continuing project work. However, it may give you some food for thought and a platform for building your own guidelines.

### 15.6.1 Change Control or Configuration Management

Once the detailed requirements have been established and agreed on through support of the "what?/how?" cycle, we move on to design and implementation.

At this point we start to invest significant resources in hardware and software development so further changes to requirements will start to get very costly in terms of budget and implementation time. There is also no point in kidding ourselves that the requirements are completely set in concrete. There may be legitimate reasons discovered once the implementation is under way for changing requirements. Having recognized that this is the case, it is absolutely vital that some mechanism be established to manage this aspect of the implementation.

In our experience, many of the problems associated with cost and time overruns in ITS implementations have their origin in this issue. While recognizing that a few late changes are inevitable, these need to be managed and subjected to a high degree of discipline. The system engineering profession has identified this as an important issue and developed management techniques to address it. These are referred to as *change control* or *configuration management procedures.*

We'll take some time to give you some general information on these techniques at this point. The most important advice we can give you is to recognize that as we transition from the ITS Future Big Picture to design and implementation, there is a need to maintain and support the final design through implementation and even beyond. Even after the final system has been handed over and is no longer your responsibility, change control should remain in place.

Figure 15.2 illustrates a generic change control process. Note that this is not intended to be a single solution for all implementations but to give you some starting thoughts on developing your own. The resources allocated to supporting the process will vary according to the scale of the project. For example, if the work involves designing and implementing a small-scale traffic signal project, the resources may consist of a single person executing each step in the process. If the work involves a multimillion dollar ITS implementation, there may be a number of people involved with both technical and financial expertise. The resources would be scaled in proportion to the implementation budget and the complexity of the implementation.

#### 15.6.1.1 Formal Change Request

A formal change request mechanism should be defined. This would include the identification and definition of the people involved and the procedures to be followed. It is useful to include a formal dialog between the change requester and the system designer in this mechanism. This provides the opportunity for clarification of the requested change and acts as a "prefiltering" opportunity.

It may be possible to accommodate the "real" need being expressed in the change request without changing the requirements. This is because many users have a perception of the new system that is based on how things are currently

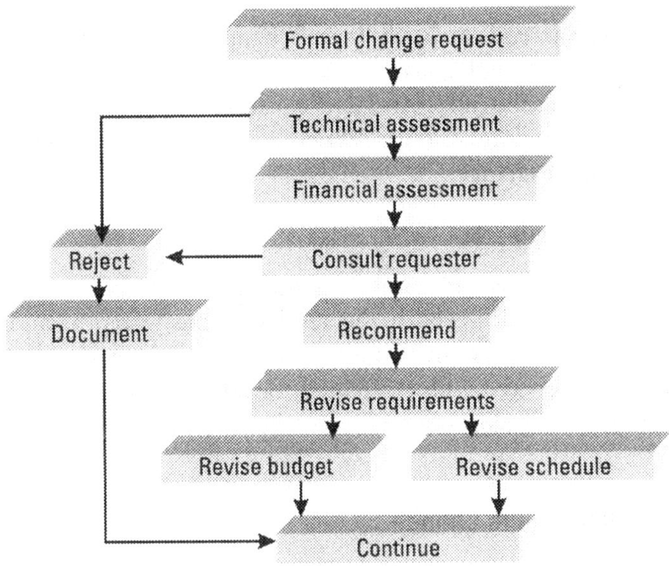


**Figure 15.2** A generic change control process.

carried out. A change in user perception may be what is really required. As usual, there is a thin line to be walked here between supporting a truly user-focused, needs-driven implementation and introducing some discipline into the process.

### 15.6.1.2 Technical Assessment

This involves an assessment of the technical feasibility of the requested change and the implications and consequences. If the requested change is considered to be technically unfeasible, then the request should be rejected at this point and the requester informed of the reasons. If the change is considered to be technically feasible, then the implications and consequences of the change need to be considered. As most ITS developments are composed of closely inter-related components or modules, a change in the design of one component is likely to have a consequential effect elsewhere. The assessment of the consequences may lead to a review of technical feasibility or may provide input to the financial assessment step.

### 15.6.1.3 Financial Assessment

This applies a large dose of reality to the situation by attaching a cost dimension to the requested change. We have encountered many situations in ITS projects where the client or stakeholder has been allowed to separate functionality from cost. This leads to an endless "wish list" of change requests with no consideration of the financial consequences. This step is intended to avoid such situations by ensuring that the full financial implications of the requested change are understood and taken into account. Very often the cost of a change will affect the desirability of the change.

#### 15.6.1.4 Consult Requester

After the financial assessment, it is a good idea to go back to the change requester and communicate both the results of the technical and financial assessments. This provides the requester with the information necessary to decide whether to proceed with the request or not. The requester is also given the opportunity to develop ownership in the eventual reject or recommend decision

#### 15.6.1.5 Recommend the Change

If the requested change passes through the mechanism to this point, then a formal recommendation to adopt the requested change should be made. This should be documented and authorized by the appropriate client or stakeholder representative. Note that from a contractual perspective this may have to be authorized by the client's nominated contract administrator, if the subsequent budget and schedule changes are to be officially recognized as changes to the contract.

#### 15.6.1.6 Reject the Change

This is self explanatory but should be formally acknowledged with appropriate documentation. The requester should also be formally notified of the rejection, with justifications for the decision.

#### 15.6.1.7 Document the Decision

As referenced above, the decision should be documented for future reference. This provides an "audit trail" for the change control process and a reference library for subsequent change request considerations. It is not uncommon, especially on larger projects, for the same, or similar change requests to be made several times in the course of implementation. A reference library documenting previous change requests may save considerable resources in these circumstances.

#### 15.6.1.8 Revise Requirements

When a change is formally adopted, the requirements also need to be formally revised, agreed on, and documented.

#### 15.6.1.9 Revise the Budget

The appropriate changes should be made to the budget, including any consequential contract amendments or variation orders.

#### 15.6.1.10 Revise the Schedule

The appropriate changes should be made to the schedule, including any consequential contract amendments or variation orders. Impacts on the overall implementation schedule may also be communicated to the client at this point.

### 15.6.2 Testing and Certification

This is an important element that has been overlooked on some of the projects with which we have been involved. It should really be part of the requirements definition work, because if you cannot devise a test procedure, then maybe the requirements are not defined well enough. It is also an important factor in the eventual success or failure of a project, since it is hard to be successful if no one has defined success.

### 15.6.3 Clear Functional Specifications

Establishing clear, unambiguous, stable requirements is very difficult and needs constant attention and effort throughout the ITS Cooperative Development Methodology and especially during design and implementation.

### 15.6.4 Good Project Management

Good project management is a product of both the client side and the developer or contractor. Good project management, in our experience, will be indicated by the presence of the following factors:

- *Clear, concise communications:* This covers communications between client and developer and within each party's respective organizations.
- *Cooperation:* Fully operational and practical cooperation between all parties, involving mutual support and problem-solving rather than competition and finger pointing.
- *Responsibility:* Responsibility being assumed by the key players in the ITS development on both an individual and a collective basis.
- *Scheduling, budget tracking, documentation, monitoring, and review:* Good project management will have all of the above activities arranged so that there are no sudden surprises in the course of the project. While

things may go wrong, leading to delays and increased costs, these are identified and anticipated, and appropriate corrective action is defined and implemented. This also means that the project manager will not wait until the end of the project to conduct a retrospective analysis of what happened.

Project monitoring and review will be embedded in the fabric of the project and form part of the project culture. Formal milestones and even "inchstones" (more frequent milestones) will be identified as part of an overall project plan. Typically advanced software tools are used to support project tracking and revisions along the way.

The project planning work does not just define a plan and leave it on the shelf. It also establishes the necessary information flows for project tracking and defines clear roles and responsibilities. Most important of all, good project management will involve getting the project people onboard in terms of ownership in the goals and objectives of the plan and practical ownership of the plan elements.

# 16

# ITS Procurement Strategies

## 16.1 Introduction

This chapter addresses the development of procurement strategies for ITS. We consider this to be part of the work required in step 410, "Implementation," in the ITS Cooperative Development Methodology. In this step we have moved beyond the ITS Future Big Picture, developed an Implementation Strategy for phasing the design and implementation of the ITS Future Big Picture, and carried out detailed design. It could be argued that the development of a procurement strategy would take place during the development of the Implementation Strategy or during the detailed design work required to define projects. The development of a procurement strategy is likely to have an impact on these other activities since design, phasing, and procurement have mutual effects.

We have chosen to discuss the development of the procurement strategy within the context of the implementation step in the ITS Cooperative Development Methodology primarily because this is where procurement is traditionally carried out. If the typical approach of defining requirements for a specific project then going straight to design and procurement is adopted, then procurement is considered during the implementation stage.

Having developed the ITS Future Big Picture, we are now in the position to proactively seek to procure what we need in order to put it together. We are in the position we tried to attain right from the start of the process. We now know what we want and can focus on the most effective way of asking for and obtaining it. This element of knowing in some detail what we want is important. If we do not really know what we want then we are at a disadvantage when it comes to opening a dialog and negotiating with vendors.

Accordingly, we have positioned ourselves to take maximum advantage of our knowledge and experience if we can identify the most appropriate way of asking for what we want. This is an important benefit of using the ITS Cooperative Development Methodology. The ability to review the ITS Future Big Picture and a mature set of requirements facilitates the positive selection of the most appropriate procurement mechanisms for the circumstances, rather than relying on a default or historical procurement mechanism.

This chapter explores ways in which we can ask for what we want with regard to ITS. We begin the chapter by discussing the elements and features of ITS procurements that make them different from conventional asphalt, concrete, and steel procurements. This leads us to the definition of a "buyer/seller" system or process that helps us to understand the dynamics of the ITS procurement process and the respective roles and needs of the participants. We then explore the range of procurement mechanisms that are available, discussing the advantages and disadvantages of each, giving an indication of the choices that are open to the procuring agency. In accordance with the philosophy that runs through this book, we attempt to show you how to select the most appropriate approach based on your needs and objectives. This is addressed Section 16.3 describing a structured analysis approach to selecting procurement mechanisms.

Finally, we discuss how the choice of procurement mechanisms may affect the proposed implementation strategy and consequent ITS designs.

Note that we have referred to the procurement organization as the "buyer" and the vendor as the "vendor" throughout this chapter. This is an attempt to simplify the language used in the chapter and should not be considered as an attempt to constrain the procurement mechanism. The "buyer" could just as easily lease or rent the products and services.

## 16.2 Why Is ITS Procurement Different From Asphalt, Concrete, and Steel?

We have to say that in many ways ITS procurement is exactly the same as procurements for asphalt-, concrete-, and steel-related transportation projects. There are many similarities between the approaches for each type of project and the transportation profession has a great deal of expertise in the procurement approaches.

In fact, many aspects of ITS projects may be comprised of familiar elements such as asphalt-, concrete-, and steel-related procurements. Combined with these, however, will be the advanced technologies such as communications, sensors, software, and other hardware devices.

However, there are significant differences relating to a number of aspects of ITS.

- *ITS employs advanced technologies that are still emerging and evolving.* ITS makes use of a range of information and communications technologies. Some are mature and proven; others are relatively new and still evolving. This makes it difficult to establish processes and procedures that can become standard and familiar.
- *Need for specialized skills and knowledge.* The application of information and communications technologies requires a new set of skills and experiences. These are not traditionally found within the transportation profession.
- *Complexity.* The need to integrate multiple components from different vendors and make all the subsystems talk to each other usually leads to added complexity. While the transportation profession is not unfamiliar with complex projects, such projects seem to occur with more frequency in the ITS context.
- *Standards do not exist or are still emerging.* Many mature standards are available to assist in the specification of asphalt concrete and steel type procurements. Unfortunately, the same cannot be said for ITS. The whole subject area is still too new to have a complete set of standards. Consequently, the buying agency has a more difficult job in developing specifications and less access to lessons learned by others. This also makes it more difficult to be precise when developing initial requirements as a full picture of technology capability and market offerings may be hard to obtain.
- *Difficult to find single vendors with all prerequisite skills and experience.* This is also related to the newness of ITS as a subject. Just as the buyer agency may have problems in not being familiar with the technology, the vendors in the market may not possess the skills and experience required to address the full range of activities required. Many vendors in the market may have been there for a number of years and have offered products and services that the transportation market has historically desired. The shift to ITS may have caused many of the tried and trusted traditional vendors to have the "wrong" skills set. On the other side of the coin, the new information and communications technology vendors may have little or no experience of the specific needs and characteristics of the transportation markets.

Most of the differences relate to the buying agencies' ability to specify exactly what is to be bought and the expertise and experience required to evaluate offers, monitor supply and delivery, and ensure value for money. As ITS is relatively new to the transportation profession, many of the control and performance measures for supply and delivery have either not been defined or are not part of standard practice.

Another difference lies in what we call the *tangibility* of the delivered product or service. Conventional procurements result in the delivery of a tangible, visible deliverable such as a new road, bridge, railway, or light rapid transit system. We can perform measurements and tests on these items to make sure we got what we paid for and that our requirements were satisfied. Many ITS procurements result in the availability of new information or data, the delivery of new software, or new means to provide information to the traveler or influence travel behavior. Now many of these items have their own performance parameters and metrics also, but they may be unfamiliar to the transportation profession.

Intellectual property rights can represent another new set of issues. Information and communication systems and technologies may well have intellectual property aspects. Many developers offer predeveloped systems or system modules on a license basis.

## 16.3 The Buyer/Vendor System

The best way to approach the issues associated with procurement of ITS products and services is to consider both the buyer and the vendor as part of a single system. In this system the buyer and the vendor (or provider in the case of services) are both required to play specific roles that in turn require specific expertise and knowledge to be held by each party. The roles, in an ideal system, would be directly related to the objectives of the buyer. The buyer would have perfect knowledge of the experience and capabilities of the available vendors and the available products and services. The vendor would have perfect knowledge of the needs and nature of the buyer. Figure 16.1 illustrates this ideal system.

In practice, many things get between the buyer and the vendor, causing distortion in the communications between the two. These distorting factors include the following:

- *Procurement regulations:* Many central and local government regulations impose constraints on procurement activities within the public sector. These typically revolve around the requirement for fair and

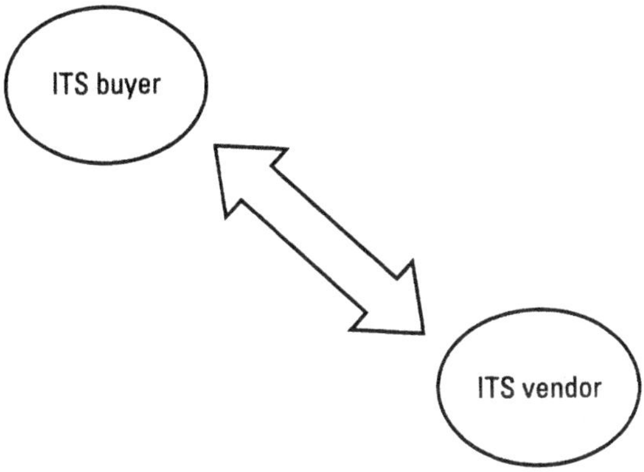


**Figure 16.1** The ideal buyer/vendor system.

open competition. These are very necessary to ensure fairness in the procurement process and foster competition and the ensuing downward pricing pressure; however, they impose barriers to innovation and risk sharing through public/private cooperation.

- *Translation of the initial requirements by the contracts or procurement department of the buying organization:* Somehow in the process of translating our initial requirements into procurement specifications, we often introduce errors and distortion. As illustrated in Figure 16.2, a cloud emerged over the buyer/vendor interface.

The aim is to avoid distortion as we go through the cloud, if possible. Many distortion factors are unavoidable; therefore, a management strategy has

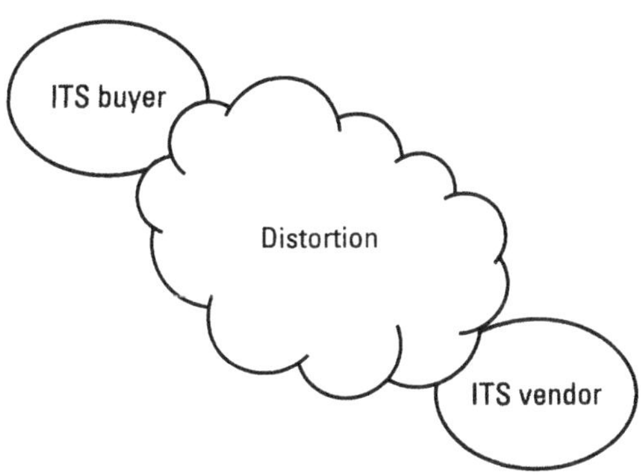


**Figure 16.2** Translation of requirements.

to be developed. It is interesting to note from the "process" view of the buyer/vendor relationship that shared knowledge and understanding of what ITS technologies are actually capable of and are best suited to, is a prerequisite for success. There has been much discussion about raising the awareness of transportation professionals, especially in the procurement area. It is also vitally important that the vendors are aware of the technology capabilities in terms of solid practical knowledge of what will and will not work. The vendors also need to have a detailed understanding of the procurement mechanisms that the buyer will employ and the motivation of the buyer in terms of the objectives the buyer is trying to achieve.

This points to the need for information sharing and perhaps mutual learning between the buyer and the vendor. Perhaps we are creating an unnecessary division between the buyer and the vendor when it comes to ITS.

## 16.4 Exploring Procurement Mechanisms

There is a range of procurement mechanisms that could be applied. These have been utilized for previous procurements associated with either ITS or conventional transportation implementations. Looking back to the ITS Cooperative Development Methodology step 400, "Design," and step 410, "Implementation," as shown in Figure 16.3, you will see that these procurement mechanisms represent multiple variations on the division of responsibilities associated with these steps.

By assigning responsibilities in different ways we can significantly affect the cost of the procurement and the probability of a successful outcome. The different mechanisms may also bring different participants into play as roles vary. There are seven primary procurement mechanisms that we have encountered:

- Design, bid, build;
- System manager;
- Design and build;
- Build operate transfer;
- Franchise;
- Shared resources;
- Noncompetitive sole source negotiations.

Each of these is discussed in Sections 16.5–16.11.

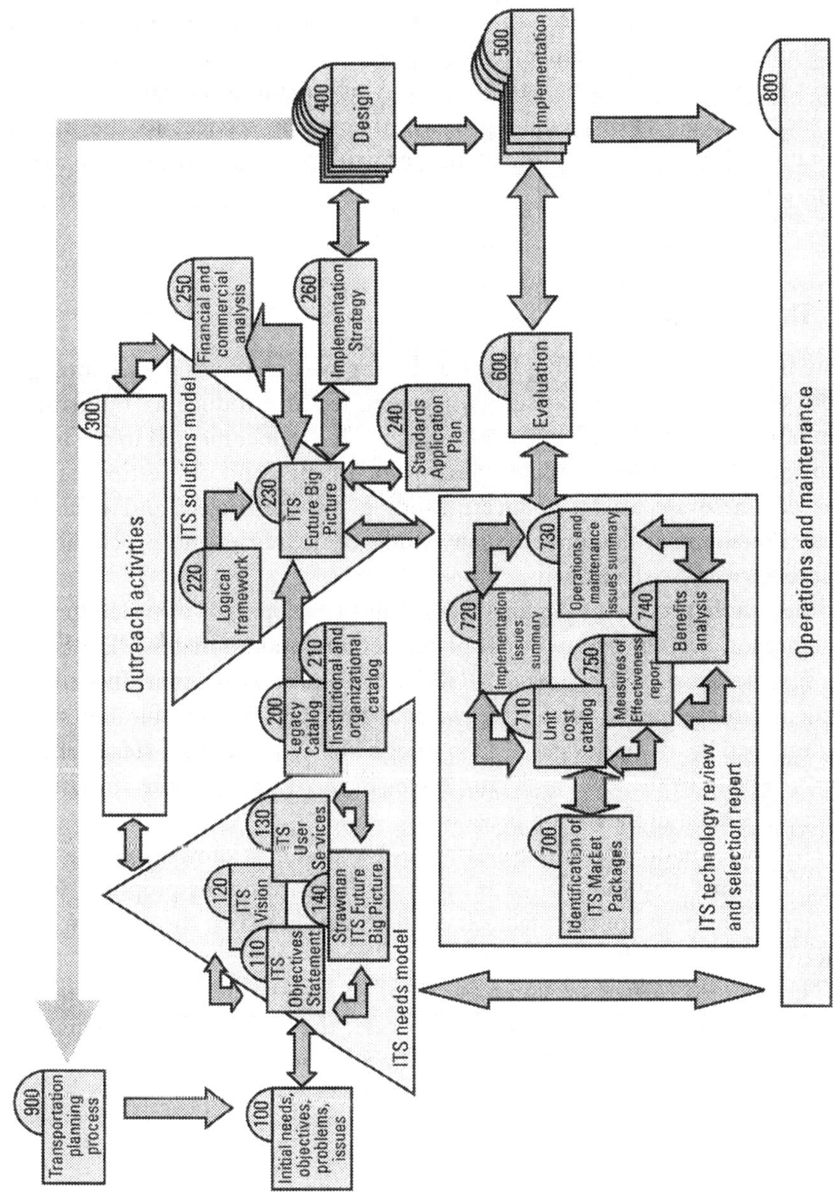


**Figure 16.3** ITS Cooperative Development Methodology steps 411 and 410.

## 16.5 Design, Bid, Build

This is often referred to as the "engineer/contractor" mechanism as the buyer makes use of both an engineer and a contractor to deliver the products and services under separate but concurrent contractual arrangements. Note that the engineer in this case could be part of the buying organization staff or a third party such as a consulting engineer hired to provide services to the buyer organization. Under the auspices of this procurement mechanism, the vendors have the roles of engineer and contractor, described in Sections 16.5.1 and 16.5.2.

### 16.5.1 The Engineer

Starting from the ITS Future Big Picture, the engineer conducts the technology tradeoff studies and further design work required to produce a "buildable" design, often referred to as the "100%" design. This is documented in the form of plans and specifications/detailed design documents, and cost estimates are produced. These are used to allocate funding for the project and establish budgets. Procurement documents such as bills of quantities and procurement specifications are also developed.

This whole process of detailed design and development of procurement documentation is often referred to as plans, designs, and estimates (PD&E) or plans, specifications, and estimates (PS&E). The process is often divided by introducing milestones at which the work is reviewed. These milestones are often described as "30%-design," "75%-design" or "90%-design" stages of the project, referring to the level of detail present in the design and the amount of resources expended in the design process.

This introduction of milestones or review points acknowledges the fact that requirements may still change as more detail regarding the proposed solution emerges and more of the practical context, within which the solution will be placed, is understood.

The engineer delivers a complete detailed design and a set of detailed specifications, enabling the vendor (in this case, the contractor) to accurately assess the cost of doing the work or providing the service and make a robust evaluation of the various risk elements.

### 16.5.2 The Contractor

The main role of the contractor is to develop a detailed proposal for carrying out the work, delivering the products, or delivering the services. In doing so, the contractor makes an assessment not just of the cost of carrying out the work and supplying the products or the services, but also on the degree of risk or liability that would be incurred by the contractor's organization.

The design produced by the engineer is typically released as a request for proposals (RFP). This is publicly advertised and signifies the start of a competitive bidding process under which qualified contractors are entitled to submit proposals or bids to carry out the work, provide the products, or supply the services required. There are also variations on the basic theme of design/bid/build.

One variation introduces a prequalification process where a description of the proposed work, products, or services is published, and contractors are invited to submit documentation describing their relevant qualifications, experience, and expertise. This information is used to develop a list of prequalified contractors, who are then the only ones invited to submit proposals. This prequalification process can be ad hoc and specific to each contracting opportunity or project. It can also be carried out on a systematic basis, where contractors pre-qualify on a regular basis for work, products or services within a defined range. Then, when there is a requirement to contract within that range, the pre-qualified contractors are invited.

### 16.5.3 Advantages

This method has been around for a long time and is one of the traditional mechanisms employed by the transportation profession. Consequently, it is very well understood by both buyers and vendors, and respective roles are well defined. The details involved in setting up contracts under this type of mechanism are well documented as are the various methods employed to settle disputes and resolve conflicts. Many buyer organizations will have a well-developed approach already in place for administering this type of procurement mechanism. On the vendor side of the coin, there are a large number of contractors to choose from as the long track record of utilizing this mechanism has enabled the evolution of the market.

### 16.5.4 Disadvantages

Successful application of this procurement mechanism requires that the design be thoroughly documented before the implementation work begins. This can have several disadvantages, as described in Sections 16.5.4.1–16.5.4.3.

#### 16.5.4.1 Costs and Delays

It can take a long time to get the design work to the point where the engineer feels comfortable in releasing it to a contractor. This is mainly because most procurements of this type will be made on the basis of selecting the contractor with the lowest price. Consequently, the contractor has little or no tolerance to deficiencies in the design and will claim additional unbudgeted compensation from the buyer. This would reflect badly on the engineer, so the engineer tries to avoid this by putting additional effort into the detailed design to the point that ambiguity is at a minimum.

#### 16.5.4.2 Separation of the Design and Implementation Processes

Building on the disadvantage described in Section 16.5.4.1, the fact that the design and implementation are carried out by two different entities can be a major disadvantage in ITS contexts. Innovation in the design and implementation processes is stifled, as the engineer is motivated to minimize any changes from the initial design. The contractor, on the other hand, is actually motivated to find deficiencies and the need for changes to the original design as this provides an opportunity to raise the original price by requesting a formal change order and associated additional compensation.

Skills and experience that the contractor may have acquired in the course of previous implementations may not be applied to the project because of this. It can seem as if a boundary has been established in the middle of a range of activities that would work more effectively as a continuous sequence. The focus may shift from getting the job done as well as possible to transferring blame and responsibility back and forth across the boundary.

#### 16.5.4.3 Dependence on Strong Ability to Specify Initial Requirements

This type of procurement mechanism depends on the engineer's ability to specify requirements precisely. This typically results in a description of "how" the needs are to be addressed and will have a particular approach imbedded as a result. In an area such as ITS where new technologies, products, and services are emerging at an accelerated rate, the engineer and the buyer may not be able to develop a good "how" statement.

## 16.6 System Manager

Using this procurement mechanism, the whole design process for the system is the responsibility of a single organization. Even if the implementation spans several phases and multiple procurements, the system manager remains the common thread or continuity through the whole system implementation. So we still have a separation between the designer and the vendor or implementer, but in this case the designer also takes on roles at the other end of the process.

The system manager, in addition to being the designer of the system also takes on the following roles:

- System engineering;
- System integration;
- Testing and certification;
- Preparation of contract documentation for multiple implementation procurements;
- Coordination of multiple implementation contracts by multiple vendors;
- Interface analysis and management;
- Overall planning and scheduling for the multiple projects that result in full system implementation;
- Training;
- System documentation.

The buyer organization still procures implementation services, products, and subsystems, but the system manager can provide specialist technical support for the procurements.

### 16.6.1 Advantages

#### 16.6.1.1 Single Point of Responsibility for Whole System Design

The system manager has sole responsibility for the system design and the successful integration of all the components. This may include work performed by other contractors and cover testing and certification.

#### 16.6.1.2 Less Scope for Responsibility Transfer and Finger Pointing

Since we have removed the boundary between the engineer and the contractor by giving both jobs to the same organization, the boundary, as far as the buyer organization is concerned, goes away. Of course, in reality it does not really go away; it becomes internalized into the system manager's organization.

#### 16.6.1.3 Innovation Can Be Better Accommodated

As the same organization is responsible for design and integration, it is possible to support more flexibility in the design and more innovation. As the system manager bridges the boundary between designer and implementer, it is possible to adopt a more relaxed approach to design specification, allowing changes to be made in the course of implementation without encountering the problem of change orders and increased price.

#### 16.6.1.4 Complex Systems and Evolving Technologies Can Be Accommodated More Effectively

The system manager with roles at both ends of the procurement can coordinate and optimize the application of advanced technologies through support for the multiple procurements.

#### 16.6.1.5 Legacy Systems Can Be Encompassed More Effectively

The system manager supplies the specialist, technical knowledge required to identify, define, and manage the required interfaces between legacy systems and the one being implemented.

#### 16.6.1.6 Time Saving Due to Use of Single Designer With Multiple Contractors

The system manager has to climb the learning curve only once and can therefore support more efficient liaisons with the multiple contractors than would be possible if there were multiple designers as well.

### 16.6.2 Disadvantages

.There are a number of disadvantages associated with this procurement mechanism; they are described in Sections 16.6.2.1–16.6.2.6.

#### 16.6.2.1 Harder to Find Qualified Vendors

The role of system manager requires a broad range of skills and experience to support the role successfully. There may only be a small number of vendors in the market that possess the necessary capabilities to be a successful system manager. This may limit competition and cause upward price pressure. One solution to this is for several vendors to team up as system manager. This may present its own difficulties, as the system manager organization may not then be seamless.

#### 16.6.2.2 The Transportation Profession Is Less Familiar With This Type of Procurement Mechanism

This procurement mechanism is more typically utilized in the context of large system engineering development projects. The transportation profession is less familiar with it and consequently may not be comfortable with managing and administering it.

#### 16.6.2.3 It Creates a Single Point Failure Mechanism

In simple terms, it puts all your eggs in one basket. If the system manager fails then the whole implementation is in trouble. This raises the stakes on ensuring that a good, qualified, and reliable system manager is selected.

#### 16.6.2.4 Scope for Ambiguity in Requirements

Using this procurement mechanism, the final system or deliverable can be more ambiguous because of the flexibility provided by combining the design and implementation process. Changes in design can be better accommodated because the design and implementation work is closely integrated. This may give the false impression that requirements definition, agreement, and stabilization are less important, and ambiguity is left in the requirements for the project. This can lead to problems when the buyer organization expectations differ from the vendor or system manager's objectives.

#### 16.6.2.5 May Cost More Money

Using this procurement mechanism the buyer organization transfers more of the risk associated with the project to the vendor. This may well lead to higher prices as vendors seek to cover the increased risk with higher reward, or the purchase of insurance.

#### 16.6.2.6 Increased Chance Of Design Error

Under this procurement mechanism there are no independent engineer's reviews on the contractor's design. The removal of this peer review could open up the possibility of undetected design errors unless an alternative review is included in the system manager's approach.

## 16.7 Design and Build

This type of procurement mechanism is often referred to as the "system integrator" approach. It goes one step beyond the system manager approach by allowing a single vendor to take on responsibility for design *and* implementation, or building. The buyer packages design and implementation into a single procurement under which the vendor has to deliver a complete working system that meets the functional requirements specified.

### 16.7.1 Advantages

There are a number of advantages associated with this procurement mechanism.

#### 16.7.1.1 Encourages Innovation in Design

As the vendor will design and build, there is a large incentive to be innovative in design in order to reduce implementation costs. This may also encourage better incorporation of legacy systems and off-the-shelf products as the vendor seeks to provide the most efficient design/build solution.

#### 16.7.1.2 Provides Good Support for Performance-Based or Functional Specification Approaches to Defining the Scope of the Project

As the vendor must deliver a system that meets the specified needs of the buyer, it is necessary to have a well-defined set of requirements to launch this type of procurement mechanism. Therefore, this procurement mechanism lends itself to performance-based approaches to defining requirements. The buyer has the job of defining "what" is required, and the vendor is left to define "how" the requirements should be satisfied.

#### 16.7.1.3 Copes Well With Complex Systems Requiring Complicated Integration Work With Many Undefined Variables at the Start of the Project

As the design and build or implementation activities are tightly integrated and are the responsibility of a single entity, complex systems with multiple unknowns at the start of the project can be managed better. The removal of the boundary or breakpoint between the designer and the implementer provides a stronger environment for dealing with these issues.

#### 16.7.1.4 Saves Time by Integrating the Designer and the Implementer, Avoiding the Need for Management of the Engineer/Contractor Boundary

The removal of the designer/implementer boundary also removes the need to apply resources to managing that boundary, saving both time and money.

#### 16.7.1.5 Supports Consistency and Continuity on Design and Quality Aspects Throughout the Life of the Project by Placing These With one Single Organization

As there is a single responsible entity, it is easier to ensure consistency in design approach and quality levels throughout the life of the project. This, of course, assumes that the vendor is competent and has the resources to support the entire design and build process.

### 16.7.2 Disadvantages

Sections 16.7.2.1–16.7.2.5 discuss the disadvantages to this approach.

#### 16.7.2.1 Needs Well-Developed "How" Specification or Functional Analysis

This mechanism tends to bias the procurement in favor of larger vendor organizations with the full range of expertise and experience required to fulfill both roles. Sales cost of preparing the more complicated proposals may also prevent smaller vendor organizations from bidding.

#### 16.7.2.2 May Cost More

The procurement costs may be higher in terms of the cost to prepare and administer the procurement and the price of the procurement.

#### 16.7.2.3 Makes the Vendor a Critical Point in the Success of the System Implementation

The vendor becomes central to the success of the implementation, thus making the selection of the right vendor crucial. As this also involves the vendor assuming a higher degree of risk, it may also result in the need for the vendor to carry additional insurance or bonding, hence raising the price.

## 16.8 Build Operate Transfer

Using this mechanism, the system is implemented and operated for an agreed upon period after implementation is completed by a single entity, under contract to the client organization. The single entity may also take responsibility for the design of the implementation.

This procurement mechanism works on the basis that the vendor will be able to recover costs and attain a return on investment through some kind of direct subscription from the users over the course of the operating period. The ownership of the implementation is then transferred to the buyer organization, free of charge.

This procurement mechanism is prevalent in the toll road business where there is a clearly identifiable and predictable revenue stream for cost recovery. A variation on this theme is the provision of the revenue stream by the buyer organization on an incremental basis related to benefits attained or usage of the facility. For example, privately financed roadways could be funded by the payment of a fee per user by the buyer, rather than the driver. This is known in some countries as "shadow road pricing."

### 16.8.1 Advantages

There are several advantages associated with this procurement mechanism.

#### 16.8.1.1 Little or no Capital and Operating Cost to the Buyer

The primary advantage of this procurement mechanism is that the initial capital cost of the implementation for the buyer will be very low or zero. The requirement for operations and maintenance expenditure on the part of the buyer would also be minimal. The vendor supplies all the capital; supplies the resources for operations and maintenance, as well as takes much of the risk associated with the implementation.

#### 16.8.1.2 Encourages Use of Private-Sector Funds

The use of private-sector funding is fundamental to this procurement mechanism. This takes full advantage of the private-sector desire to identify and exploit business opportunities in ITS and enable the public sector to achieve policy objectives with the use of funding from other sources.

### 16.8.2 Disadvantages

There are several disadvantages associated with this procurement mechanism.

#### 16.8.2.1 Loss of Buyer Control of Operations

The buyer in this case relinquishes control and management of the implementation for a predefined number of years. This may prevent the buyer from fully integrating the operation of this implementation with others. Taking the toll road scenario again, as an example, if the buyer is a public transportation agency, it may wish to vary tolls in order to achieve a particular demand management goal. However, it would be unable to influence toll levels as these have been set by the design, build, operate, and transfer vendor.

It is possible to establish predefined limits or ranges within which the tolls will lie, and this is common practice for privately financed toll roads, but the buyer would still lose day-to-day influence on operations.

#### 16.8.2.2 May Cost More

The use of this procurement mechanism could increase the overall cost of the implementation. This would be especially true if there is a degree of perceived risk associated with the project on the vendor's side. For example, if there is the possibility that later buyer activities, over which the vendor has no influence, will affect the subject project, then costs may rise. Similarly, if there is a high degree of uncertainty about the stability of the revenue stream that the vendor will rely on to get a return on investment, then overall costs may rise. The cost increases could take the form of higher user fees (tolls for example) or a longer period of operation before transfer back to the buyer.

## 16.9 Franchise

Using this procurement mechanism, a number of vendors compete for the right to offer products and/or services for a predefined period over a defined geographical area. When the subsequent need arises for the products and/or services, the buyer is bound to use the winning vendor. The vendor has, in effect, an exclusive license to offer the products and services.

### 16.9.1 Advantages

The advantage of this procurement mechanism is the flexibility provided to the buyer and the vendor organization within the terms of the franchise. The goods or services to be delivered can be specified in terms of performance, enabling innovative approaches to providing solutions. The cost of procurement can also be contained as the franchise may cover a multiyear period.

### 16.9.2 Disadvantage

The disadvantage is that if the franchise period is a long one, new innovations or services may be stifled by the need to abide by the current agreement.

## 16.10 Shared Resources

This is also referred to as public/private partnership. Under this procurement mechanism, the buyer organization solicits proposal for partnerships from vendors. Responding vendors then enter into negotiations with the buyer regarding fees for goods and services to be provided and "free" goods and services to be provided by each party as part of the partnership. The partnership agreement may also divide potential profits between the buyer and the vendor.

For example, in traveler information systems provision, some vendors may enter into a public/private partnership with the buyer under which the buyer provides raw travel data and the vendor transforms this to useful information for delivery by subscription to travelers. The buyer may receive a proportion of the subscription as part of the agreement.

### 16.10.1 Advantages

There are a number of advantages associated with the use of this procurement mechanism.

#### 16.10.1.1 Takes Advantage of Private-Sector Activities and Investments

This procurement mechanism enables activities and investments made independently of the procurement to be leveraged. For example, if a telecommunications operator has already installed a fiber optic network within a region, then the use of the network would be offered as part of the public/private partnership agreement. Therefore, the buyer would not have to invest in the development of parallel resources and can take advantage of existing investments made by others.

### 16.10.2 Disadvantages

There are several disadvantages associated with this procurement mechanism.

#### 16.10.2.1 Complicated Procurement Process

The procurement process for this type of procurement mechanism may be very complicated and take a long time to complete. The formation of an effective public/private partnership involves the development of a detailed understanding of the needs and objectives of both buyer and vendor organizations.

#### 16.10.2.2 High Procurement Cost

The cost of supporting the development of the relationship between the buyer and the vendor may be high, due to the time and effort required.

## 16.11 Noncompetitive Sole Source Negotiations

This is rarely used but is still a legitimate procurement mechanism. This would typically only be used where there is clear evidence that only the subject single supplier has the essential skills, experience, or qualifications to deliver the goods or services.

### 16.11.1 Advantages

The advantage of this procurement mechanism is that it is very straightforward and cheap to administer.

### 16.11.2 Disadvantage

The disadvantage is that it removes the competitive nature of the procurement process, which may cause upward price pressure and may be in violation of procurement regulations. Success is also highly dependent on making sure that you have identified the best vendor for the job, in advance of the procurement.

## 16.12 Carrying Out an Analysis To Select the Most Appropriate Approach

Now that we have explored the range of procurement options, how do we go about selecting the best approach for the circumstances? The first thing to consider is that there is no single correct choice for all contexts, therefore a thorough analysis of the options should be carried out. Naturally, we think that this should be a structured process acting as a framework for considering all the crucial aspects. Here are some aspects to consider in selecting the most appropriate procurement mechanism for your implementation.

### 16.12.1 Establish the Primary Procurement Objectives

In selecting the procurement mechanism, one of the most important considerations is the desired objective. You need to consider and define what it is you are trying to achieve and what you ideally want to get out of the procurement. This includes a review of the critical elements that you believe should be a part of the delivery. For example, if you believe that the primary procurement objective is rapid implementation, then you may tend more toward the design/build procurement mechanisms. If, on the other hand, price is the primary consideration, then the engineer/contractor, low-bid procurement mechanism may be the best option.

Whatever the chosen procurement mechanism, it should provide positive incentive for the vendor to achieve the objectives of the buyer.

You also need to consider whether the procurement mechanism will actually help to deliver the desired results given a range of other context factors.

### 16.12.2 Market Conditions

You may well decide that, based on your primary objectives alone, one particular procurement mechanism is the most appropriate. You then have to consider the other half of the buyer/vendor system. You have to ask the question "Can the vendors in the market adequately support the chosen procurement mechanism?" There would be no point in selecting a design/build approach, for example, if vendors in the market have no experience or expertise in the use of this procurement mechanism. This should also include consideration of the vendor's experience in project management under the procurement mechanism, as well as technical competence.

It can be useful to compare potential vendors against industry or professional standard yardsticks of performance or competence. In the systems profession, for example, many companies measure themselves against a system or software development maturity model (such as the capability maturity model for software discussed in Chapter 15). This gauges their abilities to develop, manage, and deliver according to a standard rating system.

The availability of off-the-shelf products or solutions should also be considered when reviewing the market.

### 16.12.3 Procurement Regulation Constraints

Many procurement agencies will have to take account of the restrictions and constraints imposed by their own procurement and contracting procedures and regulations, or those of central government agencies providing funding. For example, in some local government agencies it is not legal for public agencies to share revenue or profit with a private organization. In other cases, the public agency supporting the public/private partnership may not be able to receive the profit or revenue as it must be directed to a central fund or to another agency.

### 16.12.4 Buyer Organization Resources and Capability

As well as reviewing what the vendors have to offer, it is essential to consider the resources available within the buying organization. These include the following:

- *Money:* How much is available over what time period? This may also include consideration of funding sources and what conditions have to be satisfied to secure funding. It may be necessary to modify the intended procurement mechanism to take account of constraints or regulations relating to funding.
- *Staff:* Are there sufficient numbers of suitable qualified and experienced staff to successfully support the chosen procurement mechanism?

- *Time:* How much time is available to complete the project? Is timely delivery a critical factor?
- *Requirements:* Does a well-defined, agreed upon set of requirements exist for the procurement? Are the requirements clearly understood and stable?

### 16.12.5 Risks Versus Reward

Procurements usually involve more than just the provision of goods and/or services in return for a fee. Especially in ITS procurements, there is an element of risk transfer accompanying the primary exchange. This makes it essential that the chosen procurement mechanism provide a fair balance between risk and reward for the vendor.

If it does not, then the price is likely to be driven up as the vendor tries to adjust the balance by increasing the reward. It may be possible to avoid the upward price pressure by modifying the procurement mechanism to remove some of the risk, or retain it on the buyer side. Recognition of this risk transfer at an early stage is key to developing a good procurement mechanism.

### 16.12.6 Support of Logical and Creative Procurement Grouping of ITS Future Big Picture Elements

When deciding what parts of the ITS Future Big Picture should be grouped together into procurement packages or projects, it is important to pay particular attention to grouping. Rather than bundle items together according to some historical buying pattern that puts items together because "that's the way you procured them in the past," it would pay to look at how items can be bundled logically or creatively. Logical bundling will put items together that fit with the skills and experience of the vendors in the market. Creative bundling may produce some new groups of items that, while unfamiliar to the vendors, may produce some innovative responses. The grouping selected should also take account of the need for interoperability between components in the system. If the need for interoperability between two components or subsystems is critical to overall system performance, then it may make sense to ensure that they reside in the same procurement package or project.

Grouping should also take account of any need to attain visible, tangible benefits from staged implementations. For example, if you plan to procure a VMS, it may make sense to procure the communications and the ability to generate and display messages on the signs as part of the same procurement. While it may appear logical to procure these independently because of vendor capabilities and expertise, visible benefits will not be achieved until all three elements are procured. Delays between separate procurements could lead to the situation where the traveling public may see the blank signs installed at the side of the road, providing no benefit and developing a negative perception of your ITS efforts. Similarly, the consequential effects of the particular procurement on other ITS and non-ITS projects and programs should be considered.

### 16.12.7 Costs of Procurement Approach

Consideration of the costs associated with different procurement mechanisms should be an important factor in the selection of the procurement mechanism. Again, looking at the issue from the perspective of the buyer/vendor system view, the costs for both the buyer and the vendors should be taken into account. It is also necessary to look at the need to encourage a higher number of realistic, low bids.

Buyer costs may include the costs associated with administering the procurement mechanism, supervision, and management. Vendor costs may include the sales cost involved in developing the proposal and supporting a compliant response to procurement requests or acquiring insurance or bonding to cover risks.

Ideally, you should include all potential vendors in the development process for the ITS Future Big Picture. The ITS vision developed in step 120 of the ITS Cooperative Development Methodology provides a tool to facilitate this. By sharing the vision with potential vendors at an early stage in the development of the ITS Future Big Picture, the vendors are able to identify and define potential opportunities to provide existing products and services or develop new ones. They can also be encouraged to participate as stakeholders. Obviously the participation of vendors in the planning and development of an ITS requires some careful management as it would be undesirable to allow the vendors to lead or unduly influence the requirements analysis process. However, the structure provided by the ITS Cooperative Development Methodology should enable you to manage these issues while obtaining valuable input from the vendors at an early stage.

You may have realized by now that the ITS Cooperative Development Methodology we have been describing in this book is designed to alleviate many of the problems encountered in design and procurement activities. Through the development of a user-driven, requirements-led ITS Future Big Picture, the implementing or buying organization is empowered to consider and select the most appropriate design and implementation options. The approach also lends itself to the use of performance specification approaches to procurement, in which the "what" is specified, leaving the contractor to define the "how."

## 16.13 A Checklist of Activities

To finalize this chapter, here is a suggested checklist of activities to be carried out as you move toward procurement.

- Develop clear requirements;
- Review internal resources;
- Define procurement objectives;
- Review the market;
- Review procurement mechanism options;
- Investigate experiences and lessons learned by other similar organizations;
- Group procurement items logically or creatively;
- Review funding source requirements and procurement regulations;
- Assess risk versus reward factors;
- Develop a financial contingency plan defining how financial settlements will be made if the projects or joint ventures do not work as planned. This should encompass a thorough review of bonding requirements for the selected procurement mechanism.

## 16.14 How Procurement Methods Impact Design and Implementation

Many of the factors for consideration that we have discussed in this chapter will have a direct impact on the implementation strategy for the ITS and the detailed design of each element. Therefore, it may be necessary to either revise the ITS Future Big Picture or make modifications to proposed ITS designs in light of the insight gained in the course of developing the procurement strategy for a particular element.

Further detail on procurement mechanism choices and a list of additional information sources can be found in the report "FHWA Federal-Aid ITS

Procurement Regulations and Contracting Options, FHWA-RD-97-145" available from the Federal Highway Administration or online at the following address:

http://www.tfhrc.gov/qkref/gene/title.htm

# 17

# Financial and Commercial Analysis of Intelligent Transportation Systems

## 17.1 Introduction

This chapter addresses the activities in step 250, "Financial and Commercial Analysis" of the ITS Cooperative Development Methodology. ITS has many potential opportunities for public/private partnerships, leveraged private-sector funding, and commercial business opportunities. A complete ITS Future Big Picture will exploit such opportunities to the fullest. In order to support this, it is necessary to conduct a thorough review of financial and commercial aspects of the proposed ITS. This is what step 250 is all about. To identify, explore, and exploit such opportunities, we have found it useful to develop a financial and commercial analysis report for the ITS Future Big Picture. This primarily describes the third layer of the ITS Future Big Picture, the "commercial layer."

Remember that we discussed in Chapter 4 our use of a slightly different layering for the ITS Future Big Picture than that utilized in the National Architecture for ITS. We choose to separate the commercial aspects of the ITS Future Big Picture from the institutional/organizational to provide the opportunity to treat commercial issues separately from institutional/organizational. This is because commercial issues have a much greater dependence on private-sector involvement than institutional/organizational issues, which are predominately public sector.

In our experience there is a substantial difference in the way the private sector considers and approaches ITS as a series of business opportunities, compared to the public-sector, policy and public service objective-driven approach.

This part of the ITS Cooperative Development Methodology could be viewed as the boundary zone between two very different cultures. As we discuss in Section 17.3.1, the public sector is focused on topics such as social equity, the encouragement of open systems, and maintaining control over management strategies. On the other hand, the private sector has a focus on maximizing revenue and profit and minimizing risks and often has a skeptical view on the reality of potential ITS markets and working with the public sector.

Therefore, a significant feature of a successful financial and commercial analysis will be the identification and definition of the differences in approaches and the development of commercial arrangements that accommodate both approaches. Separating the commercial layer facilitates a focus on the development of private-sector roles within the ITS Future Big Picture, either as public/private partnership opportunities or purely private.

This chapter explores the needs of both public and private organizations participating in ITS developments and deployments, the issues that should be considered in a financial and commercial analysis before providing some thoughts on how to go about developing one.

## 17.2 What Are the Financial and Commercial Aspects of ITS?

Before we get started on describing the issues and discussing ways to deal with them, let's define what we mean by "financial and commercial issues." As we said earlier, these issues relate to layer 3, the commercial layer of the ITS Future Big Picture. The financial and commercial arrangements defined to address the issues must serve to underpin the proposed technical solution and integrate with the institutional/organizational arrangements.

### 17.2.1 Financial

Where is the money going to come from to pay for the initial implementation and subsequent operation and maintenance of ITS? This includes the identification of both public and private funding sources and the mechanisms through which the money will be obtained and applied.

### 17.2.2 Commercial

This involves identifying and describing the revenue flows associated with each part of the overall ITS. These will be positive and negative depending on whether the subsystem generates revenue or requires funding to operate. This also includes the assessment of who will pay for each part of the proposed deployments through the identification and definition of roles and responsibilities associated with project funding. This is also linked to the identification of sources for subsidy or entitlement to profits generated by the system. In an ideal ITS Future Big Picture, the financial and commercial aspects would be configured to ensure that there is a fair balance between responsibility for funding, benefits gained, and distribution of profits.

### 17.2.3 Direct Subscription and Community Funding

When considering the financial and commercial aspects of ITS, we have found it useful to consider two primary mechanisms for funding, which we refer to as *direct subscription* and *community funding*. There is a fundamental choice to be made between the two mechanisms.

#### 17.2.3.1 Direct Subscription

This could also be called "pay per use" as it involves the direct payment for goods and services by the individual, based on personal need and use of the products or services. This requires that there be a consumer demand for the products and services or a clear benefit story that can be used to generate demand through advertising and promotion. It also requires the existence of a practical method for charging subscribers for the use of products and services.

#### 17.2.3.2 Community Funding

This could also be called "public expenditure" as it involves the application of common funding raised through taxation at either a central government or local level. It typically requires the existence of clear, tangible benefits to the community.

One of the important aspects to consider in the financial and commercial analysis is the use of either of these mechanisms. We have found it useful to look for direct subscription opportunities before considering community funding for the following reasons:

- A direct subscription approach, if one is available, ensures that the customer only pays for what he or she really wants. If a suitable opportunity for direct subscription exists, it suggests that the private sector may have a business opportunity and indicates that public funding may not be appropriate.
- If, on the other hand a direct subscription approach is not considered appropriate and there is clear, tangible community benefit to be attained through the deployment of products or services, there may

be a case for community funding. These days, community funding dollars are getting smaller and smaller as governments strive for greater efficiency and lower taxation. Therefore, it may be appropriate to review ITS deployments that do not meet the needs of the direct subscription approach with a view of reconfiguring them to be acceptable. What we are trying to say is that every effort possible should be made to make use of private-sector funding through the direct subscription mechanism.

## 17.3 The Differing Needs of Public Sector and Private Sector

One of the significant features of a financial and commercial analysis for any ITS Future Big Picture is the contrasting needs of the public- and private-sector organizations involved. Any successful approach to the commercialization of ITS involves dealing effectively with this fact. Both sectors have widely differing needs and objectives that should be addressed in any successful ITS Future Big Picture, through the development of commercial arrangements involving the public and private sectors working interactively.

### 17.3.1 Public-Sector Needs

To develop the theme of the contrasting needs, Sections 17.3.1.1–17.3.1.5 discuss a few public-sector needs we have identified and defined in the course of project work on ITS.

#### 17.3.1.1 Fair and Open Competition

Without exception, public agencies strive to establish procurement and deployment environments that fully promote and sustain fair and open competition. This involves the development of a level playing field on which all potential providers are given equal consideration.

#### 17.3.1.2 Lowest Price Bid Acceptance

Many public agencies are obliged under regulation to accept the lowest qualified bid for a project or procurement, even if an alternative bid is technically superior or offers the best value overall. The tendency is to put pressure on service and product providers to be increasingly efficient.

#### 17.3.1.3 Encourage Open Systems

As discussed in Chapter 12, it is not always a good idea to promote the use of standards and the development of open systems. Depending on the stage of the market, open systems may stifle private-sector investment in the development of new products and services, thus removing the fuel necessary to drive and grow the market. However, public agencies have the development and use of standards and the promotion of open systems is a high priority item on their list of ITS actions and initiatives. This is because they support the establishment and maintenance of a high level of competition in the marketplace, encouraging a downward pressure on prices for products and services.

#### 17.3.1.4 Social Equity

As part of their public-service role, public agencies seek to achieve an equitable application of ITS. This means that they look carefully at the benefits and "dis-benefits" of potential ITS applications on different sectors of the community, trying to identify and avoid situations where one sector or group is advantaged to the detriment of another. This could also be viewed as an equal sharing of the potential benefits to be achieved through the application of ITS. From another perspective, it could be viewed as equal sharing of the disbenefits, or negative impacts of any proposed implementation. A good example of the issue of social equity concerns the potential use of variable tolling or congestion pricing for demand management in road networks. It could be argued that the imposition of use related fees advantages the well-off to the detriment of the less well-off.

Due to their need to consider such social equity issues, it becomes very difficult for public-sector organizations to provide differential ITS services. For example, while the traveler information market may be asking for a range of services and varying prices, public agencies may provide a common level of information to all travelers at no cost, or provide no information at all.

#### 17.3.1.5 Control

To safeguard the capability to achieve public-service goals and objectives, there are many aspects of transportation over which public agencies will wish to retain full management and control. These include the operational strategies to be applied to traffic and transportation control systems, the provision of a certain basic level of traveler information, and the establishment of most forms of road and public transit user charging.

### 17.3.2 Private-Sector Needs

The private sector also possesses a range of needs that are unique to this sector and not always held in common with the public sector.

#### 17.3.2.1 Making a Return on Investment

In most cases, private-sector organizations will have invested their own, or someone else's money into the development and production of products or services for the ITS market. To be considered successful, such ventures have to produce a reasonable return on the money invested. The return should typically be the same as that achievable from comparable investments or, more simplistically, larger than the interest that could have been earned if the invested funds had just been deposited in a bank account of some kind. This need is held in common by the public sector also, but there it is designated "getting value for money."

#### 17.3.2.2 Preserving Intellectual Property but Sharing Information

As part of the approach to making a return on invested funds and protecting themselves from early market competition, most providers will attempt to restrict information regarding newly developed products and services. Any advances made in technology, techniques, or procedures are hidden and protected by intellectual property laws. Providers also want to share certain information about products and services as part of the marketing effort to make potential customers aware of the offerings. Therefore, a balance must be struck on the level of detail and the type of information to be disseminated about products and services.

#### 17.3.2.3 Safeguarding Markets, Reassuring Investors

Private-sector organizations have a personal stake in safeguarding the markets for their particular products and services. They need to ensure that the market develops and stabilizes so that they obtain the required customer base to sell enough products or services to make a return on investment. They also need to make sure that they win and retain their desired share of the market by addressing potential competition.

In most cases, the decision to invest in product or service development will be based on a business plan that incorporates market size forecasts, market share estimates, and financial objectives. Having committed to such objectives and bought into the market forecasts, they then do their very best to make the forecasts come true. In order to maximize the protection of their market opportunity and reassure investors or stockholders, many private-sector organizations will seek to develop and deploy proprietary or closed systems as discussed in Chapter 12.

#### 17.3.2.4 Minimizing Risk, Sharing Risk

A major element in the decision to invest resources in product and service development for ITS will be the identification, quantification, and evaluation of risk factors associated with the ITS market. Many product and service providers may trade off market share in favor of a safer, less risky approach to the market. This can involve the development of alliances and coalitions to share the risk over a wider base or reduce the risk being assumed.

#### 17.3.2.5 Stable, Predictable Market Environment

While there are some organizations that are designed for (and therefore seek) dynamic, unpredictable markets, there are many providers who seek a stable, predictable market for their products and services. They need stability in order to develop the business confidence to recruit staff and sink funds into research, development, production, and marketing. This is one of the motivating factors in the private sector's drive for ITS standards in some application areas.

## 17.4 Detailed Business Opportunity Evaluation

Many private-sector product and service providers have developed sophisticated business models that are used to support the development of strategic business plans. These models attempt to capture the salient features of a particular market and develop prediction regarding sales volumes and margins under different scenarios. It would be very unlikely for a private-sector organization to share details of such models outside of their organizations. However, it is important to be aware of the existence of such models and the private sector's need to collect the relevant information as input to them.

The strategic business plans developed based on these model outputs are typically not shared with the outside world either. They seem to vary widely from one organization to another, but we have found that most plans contain some of the main elements discussed in Sections 17.4.1–17.4.7.

### 17.4.1 Situation Analysis

Situation analysis entails a review of the context within which the business venture is being proposed. This includes an assessment of the current and future markets for the product or service and identification and evaluation of current and future competition. This element of the business plan might also review the perceived strengths, weaknesses, opportunities, and threats associated with the market and the products or service.

### 17.4.2 Risk Identification and Evaluation

This includes an analysis of the external factors that might affect the success of the business venture. Risks are typically identified and classified according to the probability of occurrence and the severity of the consequences on the business venture.

### 17.4.3 Nonfinancial Objectives

This element identifies and defines the organization's objectives in participating in the business venture. These are nonfinancial and typically relate to business strategy objectives such as entering new markets or diversification. While there are no financial parameters associated with these objectives, we believe that these are just as important as financial objectives because they provide a good indication of the private-sector organization's motivation for participation in business ventures. An example might be "become the number-one provider of ITS services worldwide."

### 17.4.4 Financial Objectives

The "hard numbers" equivalent of the objectives is described in Section 17.4.3. These would be stated in terms of target sales volumes, margins, staff numbers, and other parameters the organization utilizes to control and manage businesses.

### 17.4.5 Strategies

This would provide descriptions of the strategies to be employed to attain the objectives. Typically, these would define higher level directions to be taken toward achieving the stated objectives. An example would be "improve customer understanding of our products and services."

### 17.4.6 Tactics

This includes the definition and description of more specific actions that will be taken to implement the strategies defined above. For example, a specific tactic related to the strategy example given above would be "develop and deliver customer product and service awareness seminars."

It is also important to note the underlying business philosophy inherent in the development and adoption of strategic business plans. Their use suggests that the private-sector organization has a structured approach to business planning. While it may not be possible to access specific information regarding these plans, it is possible for the public sector to appreciate the motivation and perhaps align accordingly.

There is another level of detail regarding business models for ITS. This level does not reveal detailed financial objectives or parameters for the private sector but defines the relative roles and responsibilities of all essential participants in the delivery of the service or the procurement of the product. This type of business model is usually publicly available and would form a good basis for the commercial layer of the ITS Future Big Picture. Developed in a cooperative manner between public and private partners, this would go a long way in defining workable commercial arrangements.

### 17.4.7 Managing Risk

When the private sector considers an investment or business opportunity, the likely risks are assessed as well as the likely return. Avoidable risks are subject to risk avoidance strategies and unavoidable ones are subject to mitigation strategies under some kind of coherent risk management plan.

## 17.5 How to Approach the Development of a Financial and Commercial Analysis for an ITS Future Big Picture

Now that we have explored some of the issues associated with the development of the ITS Future Big Picture commercial layer and identified the contrasting needs of the public and private sectors, Sections 17.5.1–17.5.6 provide some advice on developing an analysis.

### 17.5.1 Review the Current Version of the ITS Future Big Picture, Identify Potential Private Sector Partners, and Invite Comments and Opinions

The development of the ITS Future Big Picture puts the public sector right into the driving seat with respect to the establishment of financial and commercial arrangements. One of the criticisms we have often heard the private sector utter about the public sector is "they never have their act together." The ITS Future Big Picture ensures that we all "have our act together." Ideally, some private-sector partners may already be involved from an early stage of the process and, obviously, they would be first in line for approach and discussion about potential commercial opportunities. However, this is an appropriate point to take a good look around the region and identify other potential partners that may have an interest or hold some potential for synergy with public-sector objectives.

Use the data flows defined and the subsystems as the basis for a review of commercial opportunities and an assessment of funding sources. Define the factors that determine if a subsystem is a suitable candidate for public/private partnership or purely private operation, and then seek suitable partners with the appropriate qualifications, experience, expertise, and synergistic business directions.

In some cases, private-sector organizations are either not aware of the potential opportunities within an ITS Future Big Picture for a region or do not have enough information to evaluate the opportunities. This means that the public sector is forced into a very unfamiliar role—the role of sales agent. In order to attract private-sector providers and investors to participate, it is necessary to develop a sales story for the ITS Future Big Picture.

Using the communications tools already developed as part of the ITS Cooperative Development Methodology, it is possible to develop a clear, effective set of marketing materials that can inform and attract potential private-sector participants. In particular, the use of the ITS mission and ITS vision as communication aids would help to explain the ITS Future Big Picture objectives, while the use of ITS User Services would provide an effective way to communicate desired results.

### 17.5.2 Involve the Private Sector in the Stakeholder Group

If suitable partners are identified and not already taking part, then they should be involved in active participation in the stakeholder group. Get them to provide input on later generations of the ITS Future Big Picture.

### 17.5.3 Open and Maintain a Meaningful Dialog With Selected Private-Sector Organizations

Whether the private-sector partners take part in the stakeholder group or not, it is important to establish and maintain a meaningful dialog between public and private sectors. Over the time scale upon which the ITS Future Big Picture will be implemented, there may be considerable change in both public and private circumstances and directions. Keep options open for future reviews.

### 17.5.4 Identify, Define, and Understand the Basic Goals, Objectives, and Motivating Factors Within Both Public and Private Organizations

Some private-sector organizations are averse to working with public-sector agencies. This is mainly due to past difficulties in working with the public sector due to significant differences in objectives, culture, and business procedures. Some private-sector organizations view the public sector as slow to react and decide, hard to obtain long term commitment from, and very bureaucratic in nature. On the other side of the coin, the public sector often views the private sector as too motivated by profit and always ready to cut corners to make more margin. Sometimes this can escalate beyond lack of understanding of true goals and motivations to outright distrust of each other.

It is essential to set the scene and create the best environment for leveraging private money toward the achievement of public policy goals and commercial business objectives. Therefore, some effort has to be made to understand both sides of the issues and open a meaningful dialog.

Understanding the different perspectives through good communications supported by the ITS Future Big Picture and the various tools that have been developed in the course of the ITS Cooperative Development Methodology is a key requirement. Outreach between public and private sectors to develop understanding of motivational factors can also help.

This issue has many similarities with the public-public relationship development that is carried out as part of outreach activities discussed in Chapter 13. Conflict resolution may take a different form here though as the conflicts will revolve around public/private interfaces rather than public/public. Consequently, the issues may be of a different nature, with business directions and objectives conflicting with public policy attainment and regulatory constraints rather than "turfism."

One good example of this that we have come across is the issue of driver information. If a public agency provides driver information free as part of its public-service offering, it may undermine the regional market for driver information through direct subscription. Therefore, both the public- and the private-sector organizations involved need a clear agreement on what information will be free and what information will require payment.

Another issue that we have encountered relates to what we call the "Request for Information (RFI) and Request for Proposals (RFP)" dilemma. Most public agencies support fair and open competition through a procurement process involving requests for information and requests for proposals. Qualified providers are invited to describe their products and services as part of this process. Many private-sector approaches to ITS have a great amount of intellectual property or proprietary information associated with them. Consequently, it is very difficult for a private-sector organization with such an approach to respond to a conventional RFI or RFP. This is due to public procurement regulations that make all proposals submitted to a public agency publicly available. Competitors could then have access to the proprietary information, and commercial advantage would be destroyed. It is necessary to address this problem through discussions with the private sector and careful choice of procurement techniques.

We have found that the dialog can be effectively supported utilizing tools such as the ITS User Services, mission, and ITS vision as previously developed in the ITS Cooperative Development Methodology process.

### 17.5.5 Identify and Define Public/Private Partnership and Purely Private Opportunities That Fit Within the ITS Future Big Picture and Address Organizational Objectives

This step entails building on the information already in the ITS Future Big Picture and further defining the commercial arrangements that will be required to support the technical solution and comply with institutional/organizational requirements. The defined opportunities can be agreed on in principle and passed on to the next step for further detail. An issue to be addressed here revolves around public-sector retention of direction and control over the operation of certain elements of ITS. These elements produce benefits and effects that are key to the attainment of public-sector policy objectives. Consequently, the public sector will have a strong desire to retain full control. For example, a city would want to retain control of signal timings and operational management strategies for urban traffic control since these are the basic instruments used to apply transportation policy. Assignment of roles and responsibilities within the commercial layer of the ITS Future Big Picture should take account of the balance between funding and control.

### 17.5.6 Cooperate on the Development of a High-Level Business Plan, Leaving the Private Sector to Address the Detail

As discussed earlier, it will probably not be feasible to define and agree on a detailed business plan including detailed financial parameters for the private sector. However, it is highly desirable to develop a business plan that at least outlines relative roles and responsibilities. This can then act as a coherent framework for later detailed agreements. The detailed content of this business plan will vary according to the needs of the public- and private-sector participants but would typically include topics discussed in Sections 17.5.6.1–17.5.6.5.

#### 17.5.6.1 Definition of Cooperative Mechanisms

Mechanisms such as public-private partnerships need to be identified and defined. The interface between the public and private sectors can be formalized in a number of ways. There are a range of so-called public/private partnership mechanisms that can be employed to support private-sector participation in

ITS, including bartering and cash and service-based transactions. Bartering agreements can be a straight swap of privately provided in-kind services for exclusive access to data or other public property. Cash-type mechanisms would involve simple cash payments by the private sector in return for access to public property. There are a number of other approaches and variations on approaches. At the time of writing, many such mechanisms are being defined and tested as part of telecommunications shared resource projects. Private telecommunications carriers or network developers enter into agreements with public-sector agencies regarding the use of public right of way along highways to accommodate fiber optic communications networks. A very good reference on this subject is "Emerging Trends and Paradigms in Shared Resource Projects" (Nossaman, Guthner, Know & Elliott, Conference Proceedings of the ITS America Annual Meeting, Cobo Hall, Detroit, May 4–7, 1998).

The agreement would have to describe and define the exact details of the cooperative mechanism. Including identification of "who pays for what and when," definition of revenue streams to and from each subsystem, and other details regarding the cooperative mechanisms to support interaction between public and private sectors.

#### 17.5.6.2 Reviewing the ITS Future Big Picture

The financial and commercial analysis, like many other steps in the ITS Cooperative Development Methodology has a two-way link with the ITS Future Big Picture. This indicates that not only will the information about the ITS Future Big Picture have influence on the financial and commercial analysis, but also that the financial and commercial analysis work may have a reciprocal effect on the ITS Future Big Picture. For example, information regarding potential private-sector partners plans within the region may lead to a rethink on some elements of the ITS Future Big Picture in order to maximize leveraging of funds and optimize the ITS Future Big Picture to achieve public-private synergy.

Remember that the ITS Future Big Picture has three levels, and it may be necessary to make changes on all three levels. For example, at an institutional/organizational level, it may be desirable to make changes in regulations, provide tax incentives, or accommodate liability relief to encourage the private sector to venture forth into the regional ITS.

#### 17.5.6.3 Confirm Public Sector Funding as a Complement to the Private-Sector Investment

Ideally, the sources of public-sector funding for the ITS Future Big Picture have been identified earlier in the process. Otherwise, it would be very difficult to get the private sector around the table in the first place. However, it is wise to review public-funding sources at this point, in light of the agreements that are being defined on commercial arrangements.

#### 17.5.6.4 Strike "Win-Win" Sustainable Bargains

The aim should be the development of win-win situations for both the public and private sectors. This ensures that agreements reached are sustainable into the future. Participants should also be prepared to review and revise the agreement in light of significant changes in circumstances.

#### 17.5.6.5 Synchronize Private-Sector Pricing, Product Development, and Promotional Strategies With Public-Sector, Needs, Objectives, Issues, and Problems

Once the ITS Future Big Picture has been revised to take account of the greater amount of detail emerging from the commercial and financial arrangements, it is important to ensure that both public and private implementation strategies are coordinated. Output from the financial and commercial analysis would be supplied as input to step 260, "Developing an ITS Implementation Strategy," as discussed in Chapter 14.

Once you have carried out the activities we suggest above, it would be useful to document the outcome in report format. We do not intend to provide a rigid prescription for the format or content of the report. However, we believe that both of these aspects should be designed specifically to support your objectives in preparing it in the first place. These will typically revolve around the need to review the financial and commercial analysis with a wide range of stakeholders and revise it in the light of these reviews and reviews of the ITS Future Big Picture.

Therefore, we would suggest that the report be written in jargon-free, plain English and with the intended audience firmly in mind throughout the report preparation. In other words, the report should be developed based on a user-focused, needs-driven approach. Does this sound familiar?

As one important part of your target audience is likely to be the private sector, we have found it useful to ask private-sector partners about the content and format of their internal business plans. This helps you to develop a report structure and style to which they can relate more easily.

In any case, an essential component of the documentation for the financial and commercial analysis would be an agreed upon high-level business plan for the ITS Future Big Picture.

# 18

# Considering Economic Evaluation for ITS

## 18.1 Introduction

This chapter addresses the activities covered by step 600 of the ITS Cooperative Development Methodology, "Economic Evaluation." It also pulls together all the other evaluation activity carried out as part of earlier steps in the ITS Cooperative Development Methodology.

This part of the book is all about making choices and taking decisions as you get closer to the detailed final solution to your transportation problems. At this stage in the development of the ITS Future Big Picture we are trying to achieve two objectives. First, we want to narrow choices and make the best decisions about how to invest resources in ITS over the long term. This usually equates to an analysis of costs and benefits associated with proposed deployments. Second, we want to support the "mainstreaming" of ITS into the conventional transportation planning process through the development of common procedures for evaluation. It is becoming increasingly important to be able to assess competing projects within a common evaluation framework.

In our experience, this can be the place in the development cycle where too many resources are expended in search of unnecessary precision or resolution in quantifying benefits and costs. On one project we are aware of, two years of detailed traffic modeling in a major city, resulted in massive data output. When analyzed, it led to the conclusion that the answer to the city's traffic problem was a new bridge over the river. The results were then presented to the mayor of the city, who took the manager of the modeling team aside after the formal presentation and quietly said to him, "If you had asked me, I could have told you that a bridge over the river was the answer two years ago." The lesson we saw in this episode was simple. A lot of resources were used to confirm the obvious. You really want to avoid this, however, if the choice is not obvious, then model enough to make it obvious, but no more than that.

Conventional transportation initiatives are typically evaluated based on a benefit-to-cost ratio (b/c ratio). This involves the determination of benefits that will be attained as a direct result of the initiative. These are then translated into monetary equivalents. Costs for the initiative are then estimated and the b/c ratio is calculated. In most assessment regimes the b/c ratio must equal at least 1.00 (benefits must exceed costs) for the initiative to be progressed to deployment. This approach has the advantage of providing a clearly defined method of economic evaluation but has weaknesses associated with the translation of all benefits into monetary units. Some less tangible, yet important benefits are very difficult to translate into money. For example, how would you put a value on the pain and suffering avoided through reduced accident rates? What is the appropriate value to be placed on the noise and visual intrusion of a new road project avoided through the use of ITS to enhance current capacity?

This is not a uniquely ITS issue; it is common to all transportation projects. One approach to addressing it is to make the b/c ratio just one element in an overall assessment framework. The framework contains quantitative and qualitative criteria and ensures that all issues have been addressed. This then supports a more subjective decision on the best course of action.

Another approach is the use of cost-effectiveness in analysis techniques. This involves the determination of the costs for a range of alternatives that will all produce the same effectiveness. For example, the cost of building new roads could be compared to the costs associated with deploying ITS to give the same capacity increase.

Unlike conventional transportation initiatives, there are no predefined threshold b/c ratios for ITS. We do not have the opportunity to calculate benefits and costs and then compare the ratio of one to another against a predefined yardstick. The precedent has not yet been established, as we do not have a long enough record of accomplishment of ITS deployment from which to draw.

This chapter aims to provide a discussion on the issues to be considered and the ways in which economic evaluation may be carried out for ITS.

## 18.2 The Two Major Types of Evaluation

There are two major types of evaluation that can be carried out for ITS developments. These are traditionally referred to as "formative" and "summative." Formative evaluation is carried out during the course of the development work with the aim of helping you to keep it on course and reach your objectives. It's a kind of "how's it going?" evaluation designed to provide useful short-term feedback into the development process. Summative evaluation is a retrospective look at the whole development effort with the objective of justifying the work and identifying lessons to be learned for the next time. It's a kind of "well, how did it go?" analysis.

Formative evaluation is inherent in the development of the ITS Cooperative Development Methodology. If we look back at the process so far we can identify the following points where formative evaluation has been applied:

- *Requirements analysis:* Identifying and evaluating the needs, problems, objectives, and issues;
- *ITS User Service development:* Developing and evaluating the ITS User Services to be satisfied by the system;
- *ITS technology review:* Reviewing possible technology solutions and evaluating fitness for purpose;
- *ITS Market Package definition:* Grouping technologies into meaningful units, establishing Measures of Effectiveness, and evaluating performance;
- *ITS Future Big Picture development:* Synthesizing ITS Market Packages, evaluating at a high level on technical, commercial, and institutional/organizational criteria;
- *Financial and commercial analysis:* Evaluating the effectiveness of the proposed solution in terms of funding and leveraging of private-sector investment;
- *Institutional/organizational issues analysis:* Identifying institutional/organizational issues and evaluating effectiveness of proposed solution;
- *Implementation strategy developments:* Taking account of multiple criteria, evaluating choice of implementation strategy, and deciding on the optimum.

This formative evaluation is carried out within a framework approach in which a range of criteria are identified and considered to facilitate the comparison of ITS Future Big Picture options.

## 18.3 Why Are You Conducting an Economic Evaluation?

It is important to identify and understand the objectives you wish to achieve in conducting the economic evaluation for ITS. The specific objectives you have in mind may significantly affect the way in which you go about the economic evaluation. As discussed in Section 18.2, there are two major types of evaluation: formative and summative. The choice of approach should be related to what you are trying to do. If your objectives in conducting evaluation activities relate to keeping the current project on track, then formative evaluation approaches should be adopted. If the objective is to learn lessons for subsequent deployments, or gather information to post-justify a decision to invest resources in ITS, then summative methods would be appropriate. We have encountered a few objectives.

### 18.3.1 Learning Lessons for Future Deployments

Future deployments could be later stages in your own deployments or separate deployments by others. One of the important aspects of current ITS deployment is the sharing of lessons learned and case study information. In its current state of maturity, ITS needs to support this type of sharing as a way to share the cost of pioneering and increasing the base of knowledge about ITS planning, development, and deployment. This type of evaluation objective would be best supported through a longer term evaluation cycle aimed at the production of case study information.

### 18.3.2 "Go/No Go" Decisions About ITS Expenditure

If you are trying to justify expenditure in ITS and are supporting a "go/no go" decision on an ITS deployment, program, or project, then you are probably most interested in absolute costs and benefits from previous summative evaluations.

### 18.3.3 Comparison of ITS to Conventional Transportation Alternatives

Maybe your objective is to compare a proposed ITS deployment to a conventional transportation initiative such as a new road, bridge, or transit system. In this case, relative costs and benefits become more important. The use of cost-effectiveness techniques as discussed in Section 18.1 may be appropriate for this objective. Information from previous summative evaluations of ITS and conventional transportation alternatives would also be valuable.

### 18.3.4 Comparison of one ITS Option to Another

When there are several, alternative ITS approaches to provide the same results, you may have to choose between ITS options. This can occur at the ITS Future

Big Picture level or at the detailed design level. At the ITS Future Big Picture level, the choices will be associated with major subsystems; significant technical, institutional/organizational, and commercial directions; and communications strategies. At the detailed design level, there will be many more minor choices regarding the specific design approach, choice of technologies, and methods of operation.

### 18.3.5 Evaluation as Short-Term Feedback for Project Management

This entails the use of evaluation techniques to provide information on how well the development and implementation process is going. This is very useful in the ITS context as there are many more situations where new technologies are being deployed or techniques are being utilized for the first time. In this type of pioneering or groundbreaking situation, it is very useful to have a short-term feedback mechanism to facilitate keeping the project on track. The essential feature of this type of evaluation is the short time from design and deployment to production of evaluation results. The deployment activities would be evaluated on a stage-by-stage basis with rapid evaluation enabling corrective action to later stages, if required.

### 18.3.6 Evaluation of Design Effectiveness

This type of evaluation would be aimed at the assessment of how good one detailed design approach is compared to another. This information would be used to support detailed design choices, sometimes known as tradeoff analysis or trade studies. This is useful for the designer or system developer in the detailed design process.

### 18.3.7 Assessment of Operational and Management Strategies

When we have deployed the ITS, then what is the best way to manage and operate it? What strategies should be deployed to get the best results for the specific context?

## 18.4 A Generic Evaluation Process

We have no intention of laying out a formal set of guidelines or steps for you to follow in the evaluation of your ITS proposals. However, we thought it might be helpful to describe a general approach to the economic evaluation of ITS as a way of giving you some food for thought and conveying some things we have learned. Please treat this as a "brain starter" and develop your own approach based on the thoughts and ideas we provide.

### 18.4.1 Identify and Confirm Evaluation Objectives

Identify, confirm, and agree on what the major objectives are for the proposed deployment. These should be subject to a "sanity" check to ensure that the objectives are well enough defined to be measurable. If you cannot measure it, it will probably have no direct benefit of its own and should not be an objective.

### 18.4.2 Identify and Characterize Potential Solutions

Having identified and confirmed the objectives, the next step is to identify and define the potential solutions. This may be a trial-and-error process that lends itself to a "spiral" development approach—one that entails starting at a high level of abstraction, reducing the number of alternatives then adding detail in successive increments.

### 18.4.3 Establish Measures of Effectiveness, Performance Parameters, and Evaluation Measurements

Establishing Measures of Effectiveness involves identifying appropriate yardsticks and confirming the method to be used for measurement. At this point, it may be decided to adopt sampling or proxy approaches to parameter measurement.

### 18.4.4 Develop an Evaluation Plan

The evaluation plan will outline the activities to be undertaken in order to measure the evaluation parameters. It will also describe the analysis work to be carried out and define the product of the analysis. This may seem trivial, but we have encountered many data collection exercises for evaluation parameters where there was no clear idea about the expected outcome. We find it very difficult to design good data collection approaches unless the result is clearly in sight. The evaluation plan will also identify the need for short- and long-term feedback and the mechanisms required to support such feedback.

### 18.4.5 Collect Evaluation Data and Measure Evaluation Parameters

Conduct the activities required to collect the evaluation data and measure the required parameters. This might be specially commissioned survey work or automated data collection and information processing as direct output from the implemented system. Data may be quantitative, objective data based on direct measurements of parameters or qualitative, subjective data based on travel reaction to proposals or consumer acceptance of products and services.

### 18.4.6 Provide Short-Term Formative Evaluation Feedback

Communicate the short-term formative evaluation feedback to the appropriate team members. This should be interactive with information supplied to the team members and reaction and comments received back from the recipients of the information. This will typically be utilized for project management and design guidance purposes.

### 18.4.7 Provide Medium-Term Summative Feedback

At predefined and agreed on intervals, or major milestones in the planning, development, and deployment schedule, provide medium-term summative evaluation information. This information should also be communicated in an interactive fashion. Medium-term summative information could also be treated as interim staged output from the long-term summative evaluation process.

### 18.4.8 Provide Long-Term Summative Feedback

This would typically be in the form of a retrospective report, documenting the events and summarizing the evaluation results. Recommendations for revisions in procedures, lessons learned, and other case study material would be included in a "recommendations" section of the report.

## 18.5 The ITS Measurement Paradox

One of the issues that confront economic evaluation attempts is the difficulty of measuring evaluation parameters directly. We just do not have the sensors or communications capability to measure and utilize many of the parameters required for thorough analysis. This leads us to the adoption of "proxy" measurement or sampling techniques. Proxy measurements involve the establishment of a relationship between the parameter we need and a parameter that is easier to measure and monitor (the proxy parameter). We then measure the proxy parameter and infer the value of the required parameter using the established relationship.

The sampling approach involves the planning and execution of ad hoc survey and data collection exercises aimed at collecting a small sample of the parameters we require. This involves the deployment of temporary sensors and the use of manual survey techniques over a short time period and perhaps over a small sample of the overall geographic area. The samples are then processed to infer the required parameter values over longer periods and wider geographical areas.

This type of sampling also takes place when the benefits achieved in smaller field and operational tests are scaled up to represent expected benefits from full deployment.

Both proxy and sampling techniques introduce errors and uncertainties into the parameter data. These include those due to sampling and scaling error and those due to assumptions regarding the transferability of results from one region to another, or from one project to another.

An interesting aspect of the measurement problem is the way in which our view of evaluation has been molded to fit currently available methods of collecting data. Very few existing evaluation methodologies or guidelines take account of the capabilities of new data collection possibilities. This is another "chicken and egg" problem associated with ITS implementation. The deployment of ITS provides the sensor and communications capability necessary to support more effective data collection.

## 18.6 What Can You Measure?

The first step in evaluation is to decide what to measure. Sections 18.6.1 and 18.6.2 provide a list of potential ITS benefits and some suggestions on what to measure to gauge these benefits. Note that some of the suggestions may require the application of ITS as a data collection and analysis tool. Our objective is to provide you with a number of items to consider and to support lateral thinking when it comes to data collection and analysis for economic evaluation.

While there is a wide range of potential benefits to be derived from the application and deployment of ITS, almost all will fall into one of four main categories—safety, efficiency, productivity, and environmental. Most benefits relate to improving the effectiveness of the transportation process, or minimizing the undesirable side effects of the transportation process. Sections 18.6.1 and 18.6.2 list ITS benefits under these headings and describes each benefit.

### 18.6.1 Minimizing Undesirable Side Effects

Sections 18.6.1.1–18.6.1.2 describe ways to minimize undesirable side effects of the transportation process.

#### 18.6.1.1 Safety

Safety benefits are mainly associated with the reduction of transportation-related accidents or a decrease in the severity of accidents. Unfortunately, data collection regarding accidents involving automobiles is not a perfect science. Information about the root causes of accidents and even details describing the accident itself are approximate and incomplete. In most jurisdictions, police notification is not required if there is no injury associated with an accident. Therefore, statistical information on noninjury accidents is difficult to find. At the current time, there are no sensors or data storage devices fitted to vehicles to record quantitative data regarding accidents. Consequently, we have to rely on anecdotal information and laboratory tests to establish causes and effects. This often drives us to a higher level of abstraction when determining safety benefits or leads us to use proxies for the data required.

*Accident Rate Reductions*

This constitutes one of the most direct methods of measuring safety improvements. Improvements are typically measured based on a comparison of "before" and "after" accident studies. An important issue in using this Measure of Effectiveness is the determination of the actual safety effect generated. Have you improved the overall situation in the region, or just moved the accidents from the location where the ITS has been installed to another location on the network?

It is also critical to isolate other external influences from the measurement. Accident rates over the whole region may have dropped because of other factors such as reduction in kilometers traveled or better law enforcement. This is usually handled by establishing a "control" or accident-monitoring point away from the location where the improvement is being made. Accident statistics associated with the control point are used to determine the effect of these external factors so that they can be discounted from the comparison. As well as measuring the overall reduction in accidents, it is also possible to assess the severity of accidents to determine if safety improvements have reduced the severity of accidents by reducing the number of fatalities.

There are several proxy measurements that can be made to assist in the evaluation of safety effects of ITS. One is to determine the reduction in the overall number of recorded incidents that take place on a stretch of highway.

An incident is an event that causes some unusual change in the traffic pattern, leading to unexpected changes in the level of congestion and travel speeds along the highway. These can be measured by conventional traffic sensors. If a reduction in the number of incidents will lead to a proportionate reduction in the number of accidents, the before and after incident rate could be used as a Measure of Effectiveness. Note that incidents cause congestion and even secondary accidents but do not involve bodily injuries. Thus, incident response times may have less impact on safety than accident response times.

Another proxy is the measurement of incident response times. This is becoming increasingly popular as a means of measuring the effectiveness of incident, accident, and emergency response. The underlying assumption is that faster response to an accident by traffic management and emergency services will lead to an increase in the probability of survival for those injured in the accident. The response time includes the time required to detect, verify, and respond to the accident with the appropriate resources. This has the virtue of being easy to measure since the response process is well documented in most cases.

Another safety proxy measurement would relate to insurance industry statistics. Published information regarding the variation in insurance premiums and total expenditure in vehicle repair can be used as a proxy for the determination of safety effects. This may be most effective for large area improvements, as information may only be available in aggregate form.

#### 18.6.1.2 Environmental

ITS can provide a number of environmental benefits as follows.

##### *Reduced Emissions*

This can be measured as a regional parameter in terms of the total weight of pollutants produced by the transportation process. It can also be measured as a local value based on air quality measures. In many cases, proxy parameters are used. For example, the total fuel consumed for transportation use could be measured and used to derive an estimated total weight of pollutants. Average vehicle speeds, accelerations, and decelerations could be measured or modeled and used to estimate and predict air quality parameters. Direct air quality measurements can also be made for localized values.

##### *Reduced Noise*

This can be measured by noise meters and predicted by mathematical simulation models relating total traffic flow or network usage to noise volumes and frequencies.

#### *Reduced Visual Intrusion*

This is not typically directly measured but determined from laboratory experiments. These involve subjecting users to various transportation scenarios and asking them how they perceive the threat, risk, or nuisance value of the scenario. The presence and magnitude of the visual intrusion represents a degradation to the environment just like noise or air pollution.

### 18.6.2 Direct Improvements in the Effectiveness of the Transportation Process

ITS can produce direct improvements in the effectiveness of the transportation process by supporting the following benefits:

#### 18.6.2.1 Efficiency

#### *Increased Capacity*

This can be measured in terms of the number of vehicles that can be accommodated in a given time period, or the number of people that can be transported from origin to destination. Capacity can vary over time. In fact, the variation of capacity over time can have many causal factors. For transit systems, it may relate to the time taken to board the vehicle, number of operators available, operational vehicles, and the headway between them. For road-based transportation, it may relate to the number of road closures for construction and maintenance or just the volume of traffic on the road network. For road-based freight transportation, it could be affected by the number of drivers and trucks available. In most analyses, you assume that capacity is fixed over time. In some cases, account has been taken of time varying capacity by treating the morning and evening peak hours separately, determining different fixed capacities for these periods.

We believe another factor may be useful in the evaluation of capacity effects. We call it "capacity availability." This would indicate the proportion of the total time over which the stated capacity would be available. For example, in electronic toll collection, the downtime required for maintaining and repairing each lane would be taken into account, producing an availability factor of say 85%. In other words, the maximum capacity of the facility would be available 85% of the time. Use of this factor enables us to favor transportation deployments with higher reliability and lower maintenance and repair needs.

#### *Faster Vehicle Speeds*

There are several ways to measure the speed of travel within a transportation network. One measure is the total journey time from door to door for a specific trip. Another is the journey time along specific links or stretches of road in the network. Such speeds can be measured as either time or space mean speeds. Time mean speed averages the speed of vehicles passing a fixed point over a predetermined time. Space mean speed averages the instantaneous speeds of vehicles within a defined section of roadway. Average instantaneous vehicle speeds can also be used as a measure, but it is debatable if this is really an efficiency measure since wide variations in speed are not taken into account. Increased instantaneous speeds can also lead to a reduction in safety. For these reasons, journey speeds are often used as a measure of efficiency for transportation networks. Journey speed is the time taken to complete the journey, divided by the length of the journey measured in distance.

#### *Increased Utilization of Transportation Capacity*

This measures how much of the available capacity is actually utilized. This would be a useful measure of how balanced the use of the whole transportation network is. For example, if the utilization of the road network was running at 70% while the utilization of the train network was running at 30%, then there would be considerable scope for balancing the load to improve transportation conditions.

#### *Optimal Route, Mode, Timing Choice*

Measuring this would give an indication of how well the user is employing the transportation system. If you think of the whole regional transportation system as a tool being used by the traveler to assist in a task (making a journey), then the results will depend on how good the tool is and how well the tool is used. Suboptimal use of the tool will negate some of the capabilities of the tool. Within the transportation context, many choices are presented to the traveler. If these choices are not made carefully, it leads to suboptimal use of the transportation facility. In an ideal world, the whole transportation capability within a region would be used optimally. This never happens, of course, but any progress towards this ideal will help the transportation situation.

This requires the measurement of a set of parameters that are difficult to measure directly, leading to the use of proxy and sample techniques as discussed in Section 18.5.

#### *Reduced Congestion, Delays, Stops*

This entails before and after studies to measure the congestion, delay, and the number of stops vehicles and travelers have to make in the course of their journeys.

#### *Improved Journey Time Reliability*

It is interesting to note that many travelers seek more reliable journey times before reduced journey times. Increasing the reliability of journey times may also unlock what we call the "reliability bonus." Many transportation system users, especially freight carriers operating under *just-in-time* delivery contracts will "buy" journey time reliability by incorporating a time buffer into the journey time. If the load has to be there at 4:00 p.m., the driver will schedule to arrive at 3:00 p.m. and wait at the door for the hour. This provides a contingency for unexpected delays and longer than predicted journey times. If the reliability of journey time predictions could be improved, this buffer could be reduced, freeing the vehicle and operator for another assignment, hence increasing the production volume per vehicle and operator.

### 18.6.2.2 Productivity

#### *Reduced Freight Operating Costs*

Cheaper freight movement—through the reduction of congestion and the more effective use of fleet resources—would be measured in terms of production volume per vehicle.

#### *Accessibility to Jobs*

The difficulty encountered by staff in getting to and from places of employment has several effects on regional productivity. In markets where skilled labor is in short supply, employers may be in competition for limited labor resources. Those employers situated within efficient transportation networks will have a competitive advantage and have a better chance of avoiding the staff shortage constraint on productivity. Transportation efficiency will also support faster and more reliable imports of raw materials and exports of manufactured goods. Measurement of this parameter would have to be by the use of proxy parameters relating to industrial output and efficiency.

## 18.7 Some Ideas on How to Approach the Economic Evaluation of ITS

Evaluation actually takes place at several points in the ITS Cooperative Development Methodology. In fact, one way to look at the ITS Cooperative Development Methodology is to think of it as a structured approach to identifying and assessing or evaluating alternatives for ITS planning, development and deployment. As illustrated in Figure 18.1, our overall philosophy in the

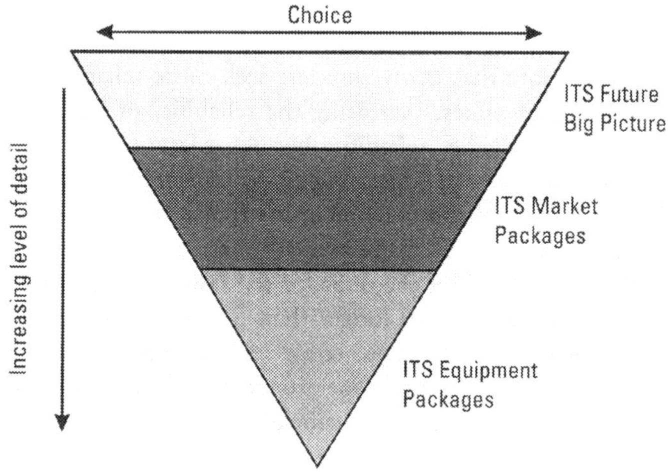


**Figure 18.1** From coarse to fine evaluation.

economic evaluation of ITS involves starting at a coarse level of detail, making early choices on a broad basis, adding more detail to remaining options, then continuing to refine the evaluation process.

As the number of options diminishes, the amount of detail in the evaluation increases. This is our way of managing the resources required to conduct the economic evaluation, while attempting to provide as rigorous an evaluation as possible.

This is supported by the ITS Cooperative Development Methodology, shown in Figure 18.2.

In the early stages leading up to the development of the ITS Future Big Picture level, only significant objectives, needs, issues, problems, major benefit types, and approximate costs are considered. A number of options for the ITS Future Big Picture are considered by comparing functionality in the ITS Future Big Picture with the needs, issues, problems, and objectives encapsulated in the ITS User Services. The support activities providing input into the definition of each layer of the ITS Future Big Picture are utilized to assist in the decision process, ensuring that the single ITS Future Big Picture option that is eventually selected is fully balanced.

ITS Market Packages are identified and selected to match the ITS User Services and in this way a broad evaluation is conducted. The evaluation is carried out based on what to include and what not to include, with minimal consideration of performance or effectiveness. An exception to this will occur if there are multiple ITS Market Package options available to address the ITS User Services. Subsequently, some basic assessment of the relative effectiveness

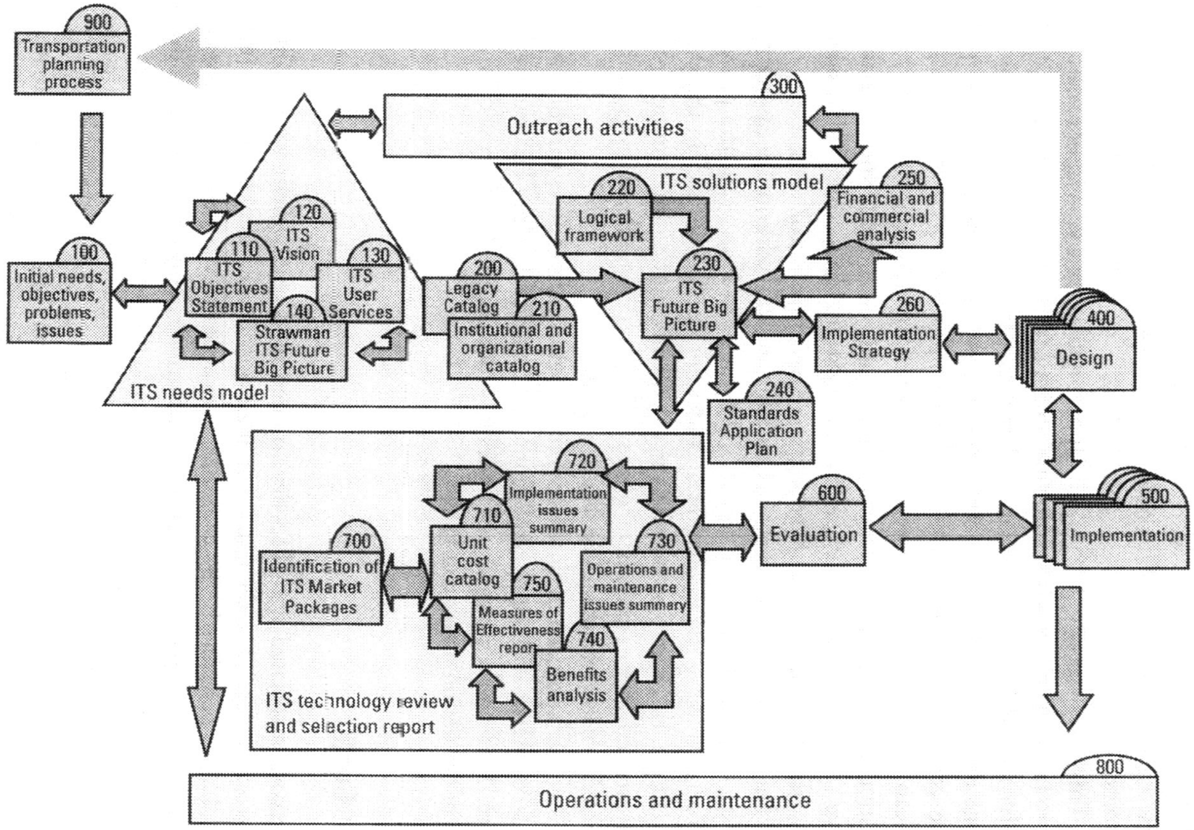


**Figure 18.2** ITS Cooperative Development Methodology.

of the ITS Market Packages will have to be made. This would consider technical, institutional/organizational, and commercial evaluation factors derived from the various activities in the ITS Cooperative Development Methodology. It may be necessary to work with a range of anticipated values for evaluation parameters or Measures of Effectiveness due to the absence of design detail at this point.

This raises an issue that we call "appropriate accuracy." On one hand, we want to uncover the very best information about the economics of the options available so that we can support good decisions. On the other hand, we have to manage the expenditure on analysis so that it stays in reasonable proportion to the cost of deployment and provides "appropriate accuracy." What do we mean by "appropriate accuracy?" We mean the adoption of accuracy and resolution levels that are appropriate considering the errors and estimations involved. This also includes consideration of external factors that could significantly affect the outcome but over which we have no control.

This requires us to think of evaluation and the selection of appropriate accuracy levels in terms of a "system." We need to look at all the processes involved in making the evaluation and identify the "weak link" in the chain. For example, there may be no point in calculating traffic signal timings to the nearest second, if the traffic data we are using to calculate the timings has been estimated to a lower accuracy of say, one minute.

Another approach to this issue is the development of an evaluatory design that is consistent with the overall framework specified in the ITS Future Big Picture. This would incorporate the design work required to narrow down the range of evaluation parameter values.

## 18.8 Costs

Some costs associated with ITS development and deployment vary in direct proportion to use (market uptake or market penetration of consumer products). This is particularly the case in the advanced travel information application area, where the traveling public are the primary consumers of the products and services. This contrasts to advanced traffic management systems where the local transportation agency or operator is the primary customer for products and services.

This makes it essential to develop some kind of business model that can be used to predict likely market conditions and determine the appropriate level of costs and benefits.

## 18.9 Determining Benefits for ITS

The first step in determining the benefits of ITS is to establish performance parameters, or Measures of Effectiveness. We have already carried out this task as part of the identification and definition of ITS User Services and ITS Market Packages earlier in the ITS Cooperative Development Methodology. Ideally, we would now be able to directly measure the Measures of Effectiveness and determine the exact level of benefit that might be obtained. However, there are some elements of ITS that are very difficult to measure in a quantitative manner—things like traveler reaction to new ITS facilities, for example. In these cases, it may be necessary to adopt qualitative benefits measures.

## 18.10 Private-Sector Leverage Possibilities as Defined in the Commercial Layer

In addition to the purely public-sector funding aspects of ITS, it is important to take account of the private funding possibilities that may manifest themselves in the course of the development of the ITS Future Big Picture. During the financial and commercial analysis activities, for example, possibilities for public/private partnership may be identified and defined. The existence of these opportunities to leverage private funding toward the achievement of public policy goals, while offering real business opportunities to the private sector, may influence the economic evaluation of ITS. In a conventional transportation initiative such as a new highway or transit system, the balance of costs and benefits on the public side may dominate the decision to invest or not. Where there are significant public/private partnership opportunities, the economic evaluation will have to take account of the existence of the private funding possibilities, influencing the final decision to invest or not. In many cases the availability of private-sector funding may depend on a short window of opportunity, making it necessary to consider timing of deployments as well as total investment costs.

## 18.11 Designs

As we move through the development of an implementation strategy toward detailed design, there are fewer major choices but many detailed options and tradeoffs to be considered. At this detailed design and implementation level, more resolution is applied to the economic evaluation to assist in the design and implementation choices. Indeed, much of the detail is only possible at this stage as design choices are starting to be made, enabling accurate unit cost estimates to be developed and exact functionality and operational parameters to be determined.

## 18.12 Mathematical Simulation Modeling

In many cases, an important component in the assessment of the economics of an ITS deployment proposal will relate to the use of mathematical simulation models to identify and quantify likely impacts and benefits of the proposed deployment. Simulation modeling will also be necessary for benefit assessment at various market penetration rates for ITS products and services.

We cover this in more detail in Chapter 19. These tools are extremely useful in supporting "what if?" analyses and supporting a better understanding of the potential effects of the ITS on the transportation network.

However, we should recognize the constraints that we work with in making use of such tools. We need to understand and appreciate the fact that models may sometimes provide an imperfect reflection of reality, either due to inherent variations in the modeling philosophy, or due to errors and inaccuracies in the data used to provide input to the model.

This leads us to the belief that it may be necessary to build more flexibility into our plans and designs for ITS. It should be recognized that there is a degree of uncertainty in our approach and that adaptive capability to cope with a range of values may be an important element of future transportation systems.

## 18.13 Information Sources for ITS Benefits and Costs

As an additional check on your evaluation, or maybe even as a low-cost alternative to conducting your own evaluation, you might wish to review benefits and costs from other previous implementations. Although traveler behavior varies from one region to another and there may be other location specific factors not taken into account, this information can be very helpful in forming a coarse picture of costs and benefits, or checking that your own evaluation is making sense.

There are many examples of previous ITS implementations around the world. These have generated a wealth of information on both benefits and costs. To help you find information on what people have learned in the past, please refer to the following reference section. For additional information on evaluation methods, please refer to [6].

## References

[1] "Intelligent Transportation Systems Action Guide–Realizing the Benefits," eds. Roger W. Gilroy, Craig Roberts, and Oliver Yandle, Intelligent Transportation Society of America publication, 1996.

[2] Leviäkangas, P., "Some New Aspects and Ideas for Economic Evaluation of ITS Investments," Proceedings of the 5th World Congress on Intelligent Transportation Systems, Seoul, Korea, Oct. 12–16, 1998.

[3] Mizutani, H., et al., "AHS Evaluation System," Advanced Cruise-Assist Highway System Research Association (AHSRA).

[4] Baum, H., W. H., Schulz, "Economic Evaluation of CHAUFFEUR," Proceedings of the 4th World Congress on ITS, Berlin, 1997.

[5] Johnson, Dr. William F., et al., "Benefit-Cost Assessment of Intelligent Transportation Systems (ITS) Implementation in Canada," Paper No.: 1044, Transport Canada, Transportation Development Centre, 1997.

[6] Weiss, C. H., *Evaluation: Methods for Studying Programs and Policies,* Prentice Hall, December 1997.

# 19

# ITS Simulation Modeling Issues

## 19.1 Introduction

Mathematical simulation modeling techniques have become an important part of the evaluation process for transportation initiatives. The use of abstract models to represent real-life situations can have considerable benefit in terms of managing the cost of analysis and evaluation. These techniques also open possibilities for analysis that simply were not available previously. In particular, the ability to create, evaluate, and modify designs without the need to actually implement them provides the capability to conduct "what if?" analyses that can be of significant value in the design process. In the ITS context, simulation models can be especially attractive as many of the products and services intended for deployment are relatively new and we want to develop a detailed understanding of the likely impacts and effects.

There is not a specific separate activity for simulation modeling in the ITS Cooperative Development Methodology. Simulation models would be utilized in support of the activities incorporated into step 500, "Evaluation," of the ITS Cooperative Development Methodology.

In this chapter, we explore some of the issues associated with the use of simulation models for evaluating ITS designs and implementations and developing operational and management strategies for ITS. Transportation has many modes and there are a great number of mathematical simulation models addressing many of the modes. We focus on traffic- and highway-related simulation models in this chapter, as these are most prevalent in the transportation community. Consequently, they should be of most interest and concern to the target audience for the book. It should be noted that there are many other mathematical simulation models available for use in aviation, rail, sea, and multimodal transportation. Much of what we have to say about simulation modeling issues will be relevant to these other models as well. We have to be honest and tell you that we had traffic simulation models in mind when we wrote this chapter. We could probably write a whole book just on the subject of simulation models alone, if we were smart enough.

We expect that in the future there will be a higher degree of integration between these mode-specific models, reflecting the growing trend toward a more integrated approach to transportation.

Our objective in writing this chapter is not to provide a state-of-the-art review of mathematical simulation models for ITS. Instead, we are attempting to provide some thoughts and insights that may be helpful when you address the job of selecting and utilizing such models. We especially hope to provide some information and practical guidance on the "balanced" use of such models. By this we mean the attainment of a balance between the resources deployed on the simulation process and the resources deployed on actual deployment. This also relates to the achievement of the appropriate degree of accuracy in the simulation. As in many things associated with ITS, there is a cost versus performance balance to be struck in data collection and data accuracy for ITS simulation models and evaluation.

## 19.2 What Is Simulation Modeling?

There are many definitions, but we will give you the "McQueen version." Simulation modeling is the art of observing some sort of real phenomenon, understanding the underlying mechanisms, then developing a computer software program based on a representation of those mechanisms that replicates real life. The items being simulated and the context within which the simulation takes place are given a numerical representation that can be fed into the simulation model as input. Output from the model is also in numerical form, requiring interpretation to highlight the real-life effects.

For example, in the case of modeling for traffic management applications, the road network is typically represented as a series of nodes and links. A link is a length of road with essentially homogeneous characteristics such as total width and number of lanes. A node is where two or more links meet. The vehicles are represented by classifying them according to type. For example, there may be three vehicle types—private cars, trucks, and buses. Each type would have its own set of characteristics such as acceleration, maximum speed, and amount of road space taken up.

The translation of real physical objects and contexts into parameters that can be represented numerically lies at the heart of mathematical simulation modeling. This is also the point at which many of the errors and inaccuracies are introduced.

Of course, many real-life mechanisms are so sophisticated and complex that we have to summarize and approximate in order to work within our constraints. These constraints can be resource constraints in terms of time and money, or intellectual constraints, in terms of an incomplete understanding of what is happening. Therefore, in our view, simulation modeling is typically a tradeoff between accuracy and realism on one side of the balance and level of resources required on the other side.

There is also a type of simulation that involves the use of human subjects in simulated ITS environments just like the cockpit simulators used to train pilots. This is especially important for vehicle safety testing using either fixed or motion-based simulators. Such simulations have been used to assess the effectiveness of advanced braking systems, adaptive cruise controls, and in-vehicle information systems. They will be important in assessing the effects of in-vehicle systems on driver behavior.

## 19.3 "To Simulate or Not To Simulate?" That Is the Question

While we recognize the value of simulation models, we do not believe that the use of simulation modeling for ITS evaluation is an automatic choice. Options should be considered as part of an overall decision to model or not to model.

Sections 19.3.1–19.3.3 describe the alternatives to the use of a simulation model that, in our experience, are usually available.

### 19.3.1 "Just Build It"

Instead of using a simulation model, you could just go ahead, implement the design or strategy, and see what happens. This may seem like a strange option, but it is a real possibility and one that many implementers choose. If you are convinced that the benefits of the proposed implementation are so overwhelming and the risks of unexpected side effects or impacts is low, then you may want to go ahead and do it.

We suppose this goes deeper than the choice to use a simulation model or not and relates to the choice to evaluate effects prior to implementation or not. The main point here is that it is not compulsory to use a simulation model, or even carry out an evaluation for that matter. You should only conduct these activities if they will produce worthwhile results and information that will justify the proposed investment of resources or shape the design or implementation.

### 19.3.2 "Piggyback"

You could save a great deal of resources by identifying a similar evaluation that someone has already carried out. Of course, you have to be careful that the results and conclusions are equally applicable to your context, but why reinvent the wheel? Even if the context within which the existing evaluation was conducted differs from yours, it may be possible to draw broad conclusions regarding likely effects and benefits. Depending on how mature the products and services that you plan to evaluate and deploy are, there may be a substantial body of existing literature that you can review. Such reviews often yield valuable information concerning lessons learned in the course of evaluations and deployments as well as results. This could be a great way to manage the cost of evaluation and simulation and avoid falling into traps that others have already encountered.

### 19.3.3 "Dip a Toe in the Water"

Another option is to conduct a small-scale experimental or pilot deployment to see what happens. This is not a simple choice since there are many factors to be weighed. You may not have the option to conduct a small-scale deployment, experiment, or pilot project since it may not be technically or economically feasible. Sometimes simulation lets you try things that would not be feasible in real life, like trying 50 different variations to see which one is best. Try to get committee or board approval to do an experiment or pilot on that scale and see how far you get. On the other hand, the act of conducting the pilot may be just the kind of cooperative activity that you need to bring your stakeholder group together and start them thinking about sharing and working together.

This should start you thinking about why you are trying to simulate ITS in the first place. Is it because you want to try so many options that you cannot afford to do it for every option for real? Is it because you cannot really produce realistic conditions in a pilot or an experiment?

We cannot say what the best option is for you. It may be simulate or experiment. It may even be a combination of both. However, we just want to make you aware of the options and give you some food for thought as you approach the job. Perhaps when all is said and done, it comes down to a comparison of cost against performance for experimentation compared to simulation. It might also relate to how much we really know about the effects of our ITS products and services. If we have no idea what will happen when we deploy them, then maybe an experiment should be conducted before any simulation work is carried out. Remember that some of the ITS products and services you plan to deploy may well have data collection capabilities that can be utilized to support the evaluation of the pilot.

In any event, we would encourage you to think through the various options with a questioning attitude, keeping your objective firmly in sight.

## 19.4 What Kinds of ITS Can You Simulate?

Remember that in Chapter 2, we introduced you to the seven areas of ITS, as follows:

- Traveler information;
- Public transportation management;
- Emergency management;
- Advanced vehicle control and safety systems;
- Electronic payment;
- Commercial vehicle operations;
- Traffic management.

There are simulation models available to support analysis in all of these areas. For the purposes of explanation and discussion, we have focused on two areas, traffic management and traveler information, discussed in Sections 19.4.1 and 19.4.2.

### 19.4.1 Traffic Management

By representing both the vehicles and the road network in a simulation model, it would be possible to investigate the effects of changes to the network, changes to the vehicles, and changes in driver behavior. Note that there are not too many models around that are good at doing all three of these together. Our current approach to transportation, while effective enough, has resulted in the development of a number of "polarized" interest groups. For example, there are vehicle manufacturers, transit operating companies, traffic signal engineers, and commercial vehicle operators. Each of these groups has specific goals and cultures and has little dialog with the other groups. Models are no different to other products and services in the market; they are developed and marketed according to the structure of the market (what the customer wants). Therefore, there are a number of mode-specific models around.

We came across a classic example of this in some work we were involved in with a major automobile manufacturer a few years ago. The project involved the integration of an engine simulation model with a traffic simulation model to produce a new model capable of predicting engine emissions under various traffic conditions.

At one of the early meetings for the project, there was an information exchange session in which the users of the two models were to exchange information on the capabilities and features of the models. The traffic modelers started by explaining that the traffic model required a number of parameters to describe the road network in terms of how wide each road was, the distance between intersections, and how the various roads were interconnected. One of the engine modelers interrupted at this point and asked, "How many parameters do you use to describe the vehicle?" to which the modeler replied, "Just one; we assume that the vehicle will be represented as a block of road space."

We could all see the surprise on the engine modeler's face, and it all became clear as she went on to explain that the engine simulation model required at least 40 different parameters to describe the vehicle in terms of engine size, transmission type, and engine management arrangements. This really highlighted the polarization between the two groups. The traffic modelers, used to working for public agencies responsible for the road network, quite rightly had a focus on the network. The engine modelers, employed by an automobile manufacturer, had the same focus on the vehicle. The transportation community has an abundance of polarized interest groups.

### 19.4.2 Traveler Information

Modeling the effects of traveler information involves modeling human behavior. This typically corresponds to the modeling of reaction to information, or choices made between alternatives. In many cases, the behavior is represented by a cost function that determines the "cost" of each alternative and assumes that the traveler will select the lowest "cost" option. The cost function is usually developed based on survey work in which travelers in the regions are asked to state a preference for different options. This is sometime known as the "stated-preference" approach. There is another approach that involves survey work that monitors and observes actual traveler behavior such as choices. This is referred to as the "revealed preference" approach, as the travelers reveal their preference through actual choices rather than hypothetical ones.

Traveler information can be provided to the traveler through a number of delivery channels including kiosks, the Internet, telephones, pagers, and roadside based VMS. This last delivery channel, VMS, can be modeled using traffic models.

## 19.5 The Range of Mathematical Simulation Models Available

We want to provide you with some information on the range of mathematical simulation models for traffic management and traveler information that is available around the world. When we started to carry out the research for this part of the book, we were confronted with a daunting task. Despite the fact that we had narrowed the focus of the discussion to traffic management and traveler information application modeling, we were aware that there were a large number of such models around the world. Fortunately, some initial research uncovered a European Union research project doing this exact task. It is called the SMARTEST project. The project addresses mathematical simulation modeling for dynamic traffic management problems caused by incidents, heavy traffic, accidents, road works, and events. It covers incident management, intersection control, motorway flow control, dynamic route guidance, and regional traffic information. The project's objectives are to review existing models, identify opportunities to develop or enhance models to fill gaps in the current offering, and develop a "best practice" manual for the application of simulation modeling techniques to road transportation.

As part of its early activities, the SMARTEST team has carried out a thorough review of traffic management and traveler information models worldwide (see the technical paper entitled "Review of Micro-Simulation Models," by Staffan Algers, Eric Bernauer, Marco Boero, Laurent Breheret, Carlo Di Taranto, Mark Dougherty, Ken Fox, and Jean-Francois Gabard, SMARTEST Project Deliverable 3). Micro- and macro-simulation models are discussed in Section 19.6.5. Table 19.1 shows a summary of the findings as extracted from the team's deliverable. The team identified a total of 58 micro-simulation models and studied the 32 listed in Table 19.1.

Table 19.1 gives you some indication of the depth and breadth of the range of simulation models available. They address both urban and freeway contexts and provide a variety of modeling options. As we stated earlier, it is not our intention to provide a state-of-the-art review of simulation models in this chapter. There are many other publications that do this job well, especially

**Table 19.1**
List of Analyzed Microsimulation Models

| Model Name | Developing Organization |
|---|---|
| AIMSUN2 | Universitat Politècnica de Catalunya, Barcelona |
| ANATOLL | ISIS and Centre d'Etudes Techniques de l'Equipement |
| AUTOBAHN | Benz Consult—GmbH |
| CASIMIR | Institut National de Recherche sur les Transports et la Sécurité |
| CORSIM | Federal Highway Administration |
| DRACULA | Institute for Transport Studies, University of Leeds |
| FLEXSYT II | Ministry of Transport |
| FREEVU | University of Waterloo, Department of Civil Engineering |
| FRESIM | Federal Highway Administration |
| HUTSIM | Helsinki University of Technology |
| INTEGRATION | Queen's University, Transportation Research Group |
| MELROSE | Mitsubishi Electric Corporation |
| MICROSIM | Centre of Parallel Computing (ZPR), University of Cologne |
| MICSTRAN | National Research Institute of Police Science |
| MITSIM | Massachusetts Institute of Technology |
| MIXIC | Netherlands Organisation for Applied Scientific Research—TNO |
| NEMIS | Mizar Automazione, Turin |
| NETSIM | Federal Highway Administration |
| PADSIM | Nottingham Trent University—NTU |
| PARAMICS | The Edinburgh Parallel Computing Centre and Quadstone Ltd |
| PHAROS | Institute for Simulation and Training |
| PLANSIM-T | Centre of Parallel Computing (ZPR), University of Cologne |
| SHIVA | Robotics Institute—CMU |
| SIGSIM | University of Newcastle |
| SIMDAC | ONERA—Centre d'Etudes et de Recherche de Toulouse |
| SIMNET | Technical University, Berlin |
| SISTM | Transport Research Laboratory, Crowthorne |
| SITRA-B+ | ONERA—Centre d'Etudes et de Recherche de Toulouse |
| SITRAS | University of New South Wales, School of Civil Engineering |
| TRANSIMS | Los Alamos National Laboratory |
| THOREAU | The MITRE Corporation |
| VISSIM | PTV System Software and Consulting GMBH |

the SMARTEST deliverable we have drawn from here. Just be aware of the range and choice that is available.

## 19.6 Things to Consider in ITS Simulation Modeling

In Section 19.5, we explored simulation models and addressed the decision to make use of them. In this section, we will discuss some aspects that we believe you should consider when making use of the models. These include defining your requirements, assessing the total cost associated with acquiring and using the model, balancing cost against performance, and learning how to really use the model.

### 19.6.1 Defining Your Requirements

Requirements for mathematical simulation models can cover a wide range of items including the model's capabilities, the type of hardware platform it will require to operate, and the computer programming language the model utilizes.

#### 19.6.1.1 Modeling Abilities

This is an important aspect in modeling selection. Consideration of this factor requires that you first establish your exact objectives in using the model. This would include an assessment of the ITS services and products you wish to evaluate and the Measures of Effectiveness to be utilized in performance evaluation. The accuracy of the modeling technique employed by the model should also be considered. This would involve going beyond the "black box" approach to the model through the identification and questioning of the underlying modeling philosophy. This is particularly important for models that attempt to emulate human behavior.

#### 19.6.1.2 Hardware Platform

You should consider the platform the model requires to be able to operate within your operating needs. This includes the type of hardware such as computers, data storage, display, plotting, and printing. You should also consider the software aspects of the platform required. This will address operating systems: Does it operate under Windows 95, Windows NT, UNIX, or some other operating system?

#### 19.6.1.3 Computer Programming Language

You should also consider the native language used to support the software itself. You may not need this information as you begin use of the model, but it will become an important factor when you later need to enhance and improve the model's original capabilities.

### 19.6.2 Evaluating the Total Costs

We recommend that you consider a "whole life cost" approach to the selection of the model(s) you intend to use. What we mean by this is that you should identify and weigh all the cost components associated with acquiring, operating, and maintaining the model and not just the initial capital costs of acquisition. We have found that these cost components typically fall into the categories described in Sections 19.6.2.1–19.6.2.9.

#### 19.6.2.1 Cost of Acquisition

This covers the evaluation of the cost of buying, leasing, and licensing the model software. This may require decisions on how many people will use the model and at what locations. Many software vendors have varying fees according to the number of users ($/seat) or number of installations (use locations) for the model software.

#### 19.6.2.2 Cost of Installation and Testing

Once you have acquired the model software, it has to be installed and tested. This may be a simple "plug-and-play" operation, or it may require a considerable investment of resources in the installation of new hardware and software. Ideally, the model supplier would provide information from previous experiences in installing the model.

#### 19.6.2.3 Cost of Environment

This is related to the platform costs but covers the wider cost of housing the hardware and software and providing an optimum operating environment. These days, many models are based on PC hardware and software with minimum needs for special environments; however, it is worth reviewing and including in the overall cost assessment.

#### 19.6.2.4 Training

This falls into several categories. The first is the training required to use the model. It may be the case that your organization already uses simulation models and may have staff experienced in the use and operation of particular models or model types. If you decide to adopt a new model, there may be a significant cost of retraining and a loss of previous investment. You should also consider the intended working life of the new model within this context. It is important to balance the investment in learning to use the model, with the expected life of the model. You would not want to invest several person years in learning all aspects of the model, if it has a short working life of one or two years. The degree to which the model is future-proof should be taken into account as far as possible.

#### 19.6.2.5 Operation

Actually running or operating the model may consume staff, hardware, and software resources. Complex models may tie up shared hardware and software for long periods, inducing an opportunity cost covering the loss of access to other users, or the cost of acquiring additional hardware and software to maintain service levels. Operation of the model and the collection of traffic parameters from the model output is an important and time-consuming activity. Output parameters such as average journey times, traffic volumes, predicted incident rates, and emissions can be used as input to the cost-and-benefit evaluation.

#### 19.6.2.6 Data Collection

This could be a significant investment requirement in terms of the effort required to collect and prepare data for input to the model. Mathematical simulation models are a lot like traffic management and traveler information systems. The quality of the information that comes out is directly proportional to the quality of the data that you put in. Systems developers have recognized this and have coined the term *garbage in/garbage out* (GIGO).

There are several questions you should ask as part of the review of this aspect:

- Does it need new data that you do not have now?
- If you have the data, does the data required exist in electronic form or does it need preprocessing?
- Does the model provide data preprocessing support tools?
- What input screens are utilized and are they user-friendly?
- What location referencing requirements does the model have with respect to data?

#### 19.6.2.7 Calibrating the Model

Even after the operating platform is installed, the data has been input, and the model is up and running, there may well be further work required to calibrate the model for your specific local conditions. This typically requires field data to be collected using specially commissioned surveys. Data to be collected can include actual traffic volumes, traffic speeds, and journey times on important links in the network. The model is then used to make predictions about the same traffic conditions as the survey and the results from the model are compared to the measured results. In a traffic modeling exercise we know of, this procedure was carried out and the modelers were presented with an inexplicable difference between observed traffic volumes and those predicted by the model. The observed results were all much higher than the model predicted. Further investigation revealed the cause of the discrepancy—during the peak periods, when congestion was very bad, drivers were using the hard shoulder on the highway as a running lane. This use of unauthorized road space was providing more capacity than expected.

#### 19.6.2.8 Maintaining the Model

Unless the model is to be used for a very short-term, limited-scope application, developing a simulation model is typically more than a one-shot deal. The model will need to be updated and maintained over the course of its working life. This may include updates to both the simulation modeling software and the data used as input to the model.

#### 19.6.2.9 Upgrading the Model

Part of the maintenance of the model may involve the addition of new functionality or new features. When reviewing available models, the ease with which the model can be enhanced or expanded in the future should be taken into account. This is particularly relevant to the modeling and representation of ITS products and services due to the dynamic nature of the market.

### 19.6.3 Cost Versus Performance

We do not have a magic formula for making cost versus performance decisions with regard to simulation against experimentation. We would suggest the use of a multicriteria framework analysis approach, as discussed in Section 18.2. We hope that you will identify an approach that balances the full cost of model acquisition, operation, and maintenance against the results that can be attained. We think of the balance in terms of a middle ground between the extremes discussed in Sections 19.6.4 and 19.6.5.

### 19.6.4 Extreme Simulation

At the simulation extreme there is a high investment of resources in simulation modeling, perhaps leading to the situation where the model is costing as much as or more than actually carrying out the implementation. The model is so complicated and requires so much data that it takes as long to carry out the modeling as it would to actually deploy it. When the results of the modeling work become available, options suggested by the modeling results are intuitively obvious and could have been deduced without modeling.

### 19.6.5 Extreme Experimentation

The experimentation extreme makes use of no simulation modeling and uses deployments to learn lessons. Trial-and-error techniques are used to determine the best approaches, learning through costly mistakes and long, hard evolution. Many false starts may be a feature of this extreme.

You should also be aware of the different levels of simulation modeling that can be achieved with various models. Models are typically described as macroscopic, mesoscopic, or microscopic, depending on the level of detail at which they model the subjects. Macroscopic models will typically deal with major traffic patterns or trends, while microscopic ones may model the vehicle, the road network, and the behavior of the driver in detail. Some macroscopic models make use of analytical approaches rather than true simulation, where the items to be modeled have been represented by an analytical approximation instead of a faithful replication of behavior.

The term *mesoscopic* tends to be applied to models that can operate in either macroscopic or microscopic mode depending on the level of detail available in the input data. Using such models, it may be possible to model an entire road network at the macroscopic or more abstract level of detail and specific locations or facilities within the network at the microscopic or detailed level.

### 19.6.6 Learning How to Really Use it

We have a big concern about how simulation models are used. We call it the "black box" syndrome. It revolves around the naïve use of simulation models as "black boxes." What we mean by this is that users with no knowledge or experience in the use of the model and consequently little or no idea of the internal processes of the model ( inside the "black box"), can make use of the model fairly easily. Sophisticated software with user-friendly interfaces have made life easier for the professional modeler but also support this kind of usage that we think poses a significant problem. In particular, the ability to subject model output to critical review, drawing on experience accumulated from a few previous assignments, is of great value in terms of quality control. The value of an experienced team supporting the model should not be underestimated. In fact, this is something that should be taken into account when selecting the model in the first place. If the model is relatively new, there may be a limited pool of experienced resources available to support it.

An experienced modeling team can also provide very valuable support when it comes to interpreting the output from the model and turning it into something that will be meaningful to the stakeholder group. We assume that one of the major destinations of the results from the model work will be some sort of presentation communicating choices to this group.

### 19.6.7 Presenting the Results

Last, but not least in the list of activities involved in the application and use of simulation models, is the final step of making sense of it all and communicating the results to the intended audience. We would approach this in the same manner as described in Chapter 13 by first identifying the audience, their characteristics, and backgrounds and then deciding on the content, format, and structure of the presentation. You should bear in mind that there is no point in doing an excellent job of evaluating and simulating the proposed ITS deployments only to fail to communicate this to the target audience.

# 20

# Experiences

## 20.1 Introduction

We have now completed our excursion through the various aspects of the ITS Cooperative Development Methodology. In the course of the book, we have introduced many new terms and concepts within an overall philosophical framework that may be strange and unfamiliar to many readers. At this point, it would be helpful to look back on some of the key concepts and perhaps bring them to life a bit. Accordingly, this chapter shares stories from some of our projects, highlighting experiences with specific concepts and tools. The stories come from three projects with which we have been associated over the past three years. Two of the projects were carried out in the United States, and one in Southeast Asia. The latter project is described in more detail than the other two, giving us the opportunity to illustrate a few of the major concepts in action on a single project.

### 20.1.1 "Application of ITS to Recreational Travel, Cape Cod"

The first story comes from a project that was known as the "Application of ITS to Recreational Travel, Cape Cod" project carried out in 1994 in the United States on behalf of FHWA, the Massachusetts Highway Department (MHD), and the Cape Cod Commission. The project was aimed at investigating the potential application of ITS products and services to the transportation needs, issues, problems, and objectives on the Cape Cod peninsula. As much of the traffic on the Cape is associated with leisure and recreational travel, this meant that the project would address the use of ITS to improve economic development, tourism, public transit, and recreational vehicle travel.

This was Bob's first assignment in the United States, giving him the optimistic impression that all assignments would be set in such beautiful surroundings.

Cape Cod has some unique traffic and transportation problems relating to the cyclical peaks caused by tourist and recreational travel and to the physical access arrangements to the Cape. The "ship canal" effectively turns the Cape into an island. There are two crossing points across the ship canal funneling all road traffic entering and leaving the Cape. The project called for the use of the FHWA ITS early deployment planning process to identify and define some potential ITS applications to the Cape Cod region.

One of the interesting aspects of this project was the development of a full range of Measures of Effectiveness for potential ITS applications. You will recall that we discuss these in Chapter 4, where they are introduced as the first step in the assessment and evaluation of benefits for ITS. The Measures of Effectiveness developed for the Cape Cod project provide a useful example of what we mean by the term. Table 20.1 shows a summary of the Measures of Effectiveness defined in the course of the project.

Another interesting aspect of the project was the Strawman ITS Future Big Picture used to communicate the emerging technical, institutional/organizational, and commercial solution to the stakeholder group. This group consisted of traffic and transportation professionals, representatives of the leisure and recreation industry, and representatives of the local economic development agency. The Strawman is illustrated in Figures 20.1–20.3, which show the technical, institutional/organizational, and commercial solutions, respectively.

The Strawman was defined as four primary ITS Market Packages:

- The Cape Cod smart card;
- Cape Cod congestion (C3) index;
- Recreational traveler information system (RTIS);
- Enhanced demand actuated transit (EDAT).

These ITS Market Packages are described in Sections 20.1.2.1–20.1.2.4.

#### 20.1.1.1 Cape Cod Smart Card

The Cape Cod smart card is defined as a chip card, with an integrated circuit containing a programmable microprocessor for management and control of stored data. A transaction may be completed by swiping the card past, or

**Table 20.1**
Cape Cod ITS Measures of Effectiveness

| General Area | Performance Measures |
|---|---|
| Economic development | Retail sales ($/year)<br>Number of new businesses (number/year)<br>Employment growth (%/year) |
| Tourism | Per capita income of tourists ($/year)<br>Dollars spent per day per tourist ($)<br>Measure of tourist satisfaction<br>Measure of service quality<br>Number of tourists (tourists/year)<br>Hotel occupancy rates (%) |
| Public transit | Number of transit rides per year (number rides/year)<br>Transit vehicle occupancy (number of people/transit vehicle)<br>Revenue ($)<br>Number of routes served<br>Hours of operation<br>Modal split |
| Vehicle travel (private) | Travel times (minutes)<br>Vehicle kilometers traveled (kilometers)<br>Number of accidents (accidents)<br>Level of service criteria on routes<br>Person throughput (persons/hour)<br>Queue lengths (number of vehicles)<br>Traveler stress (subjective measurement) |

inserting the card into, a reader. The smart card is capable of two-way communications, which will allow a number of transactions related to traveler services using a financial clearinghouse, transportation management center (TMC), transit management center, or traveler information center. The complexity and extent of the operating software required for smart cards are functions of the application.

The development of the Cape Cod smart card network is considered an individual subsystem; however, the functions of the smart card will support subsequent project subsystems. The Cape Cod smart card is an "open" system that allows transactions to be made both at multiple locations and for multiple applications, as opposed to a "closed" system, which restricts the number of card uses.

The primary function of the Cape Cod smart card will be the support of financial transactions made by recreational travelers. The financial functions

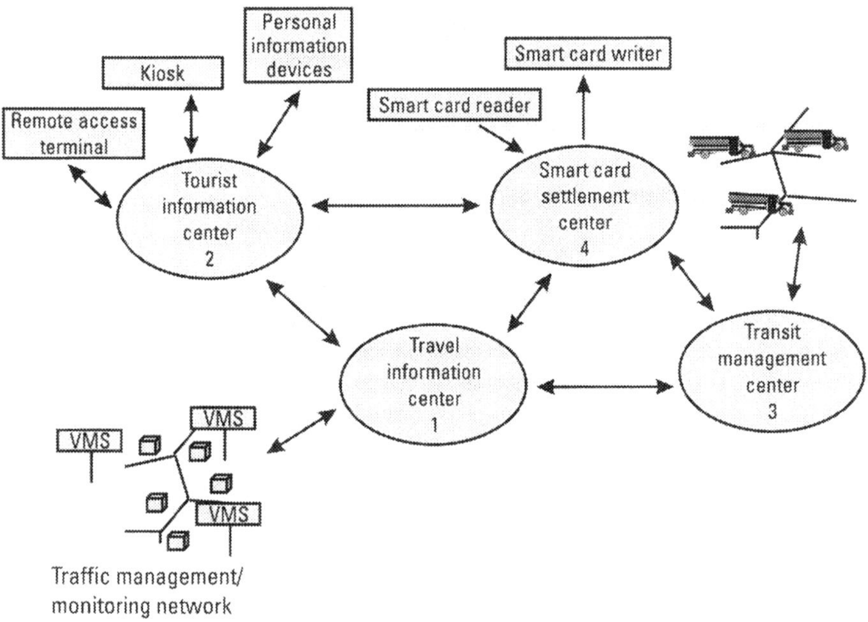


**Figure 20.1** Cape Cod Strawman ITS Future Big Picture—technical solution.

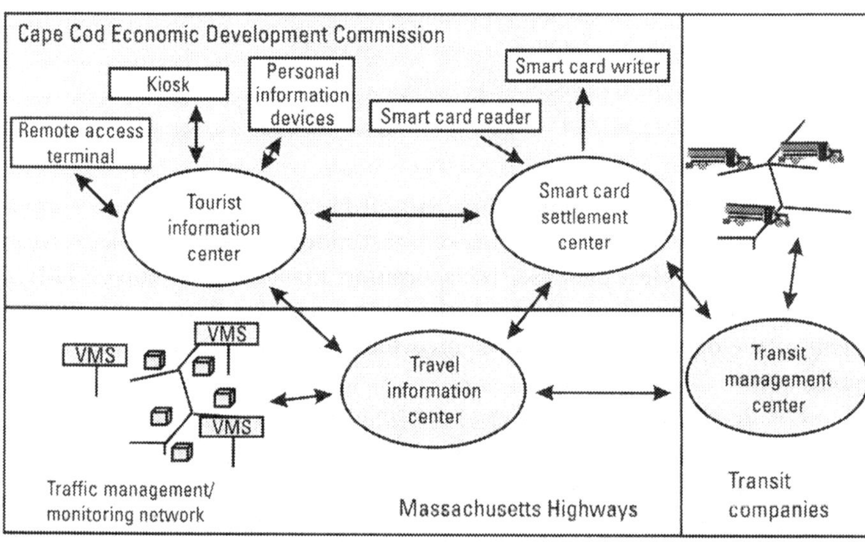


**Figure 20.2** Cape Cod Strawman ITS Future Big Picture—institutional/organizational solution.

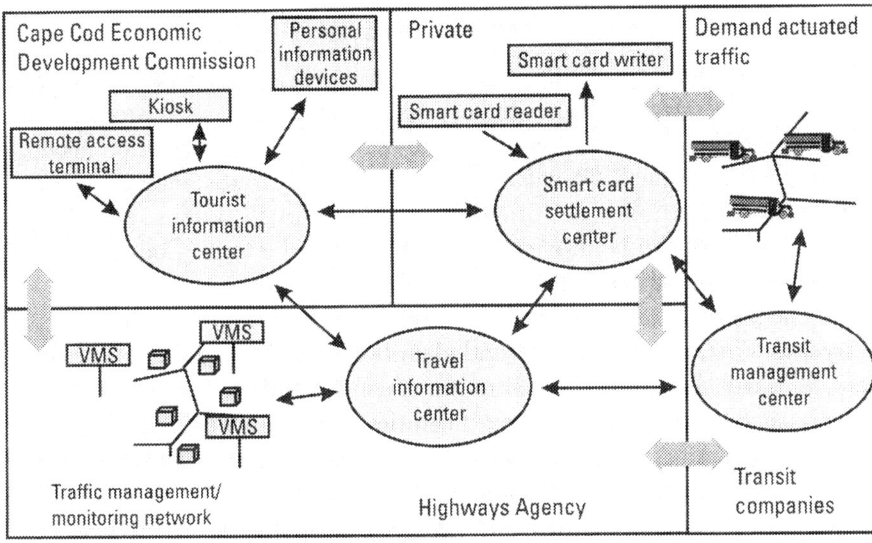


**Figure 20.3** Cape Cod Strawman ITS Future Big Picture—commercial solution.

of the Cape Cod smart card subsystem will include a financial clearinghouse that will act as the data fusion center for merchants accepting the low-value smart card transactions. Information stored on the card is collected and compiled at the smart card settlement center. The center will handle payment activities, while monetary value recording will occur at the point of the transaction.

#### 20.1.1.2 Cape Cod Congestion (C3) Index

The major elements contained in this subsystem will be the establishment of a travel information center, the development of the C3 algorithm, and recreational traveler demand management. The traveler information center will be used to provide transportation network information to travelers. The information, which is collected, used, and disseminated by the travel information center, will deal primarily with vehicular travel. Processed information can be disseminated to travelers through information centers, VMS, HAR, PDAs, telephone lines, and kiosks. Information provided to the traveler by the C3 index might include travel times, queue lengths, incidents and delays, weather conditions, and parking conditions. The traffic predictions will be used to stagger upper-, mid-, and lower-Cape arrival/departure times as a means of effectively managing recreational traveler demands.

#### 20.1.1.3 Recreational Traveler Information System (RTIS)

The development of the RTIS on the Cape will provide a service that matches the interests of travelers to specific features and amenities unique to the Cape Cod region. The RTIS is supported by the existing visitor information network system. Data from the RTIS will be adapted to provide tourist interest information. Information may be provided locally through kiosks and PDAs. Information may also be provided nationally using the Cape Cod smart card, the Internet, or other remote access technologies.

The information that is stored on the Cape Cod smart card has two primary uses: First, a traveler profile and identification will be stored on the smart card. This will allow the processor to provide individualized information to the traveler such as points of interest, amenities, and individualized intermodal transit schedules based upon information that is provided by the traveler and stored on the smart card. The traveler profile will be determined through information that is voluntarily provided by travelers; it may include travel patterns, including origins and destinations.

The traveler profile and travel pattern data will also be useful for improving the Cape's traveler and tourism services and for developing economic indicators for use in development of economic growth-oriented development strategies.

#### 20.1.1.4 Enhanced Demand Actuated Transit (EDAT)

The primary function of this subsystem will be the development of a TMC. An important feature of the TMC will be real-time automatic vehicle location on buses. This subsystem may also encourage transit ridership through traveler information. The TMC will also implement the demand-actuated transit strategy. Demand-actuated transit involves requesting transportation service and receiving transit information via communications methods such as modified telephone services or kiosks. Smart cards may also be used for electronic toll collection.

This subsystem aims to optimize transit system use by optimizing existing routes using traffic data for avoidance of incidents. Transit service could be made more economically viable by implementing arrangements with local taxi companies to facilitate demand-actuated transit or paratransit service.

This project resulted in the development of an ITS Future Big Picture for the application of ITS to recreational travel applications in Cape Cod. The stakeholder group endorsed the final results, which were presented to the Federal Highway Administration as the basis for future project definition and deployments.

## 20.2 Communication Alternatives for Scenic Byways

The second story comes from a project known as the "Communication Alternatives for Scenic Byways" project carried out in 1994 for FHWA in the United States. This project involved an interesting stakeholder group composed of transportation professionals and representatives from the outdoor advertising agencies and the leisure/recreation industries. The specific objective of the project was to explore the potential for applying ITS products and services to the support of advertising on scenic byways. This issue represented a microcosm of the dilemma faced by transportation professionals. On one hand, the economic well-being of industries and service businesses along the scenic routes depended on the use of billboard-style advertising for creating and stimulating interest in their offerings. On the other hand, the visual intrusion related to the billboards was a cause for concern in the environmental lobby. The application of ITS technologies held out the promise of providing the desired information to the traveler, while removing the undesirable side effects. The part of this project we want to focus on is the development of the ITS vision. In this case, we made use of a graphical approach to the description of the vision as shown in Figure 20.4. This enabled us to communicate the essential elements of the vision with minimum recourse to jargon and terminology.

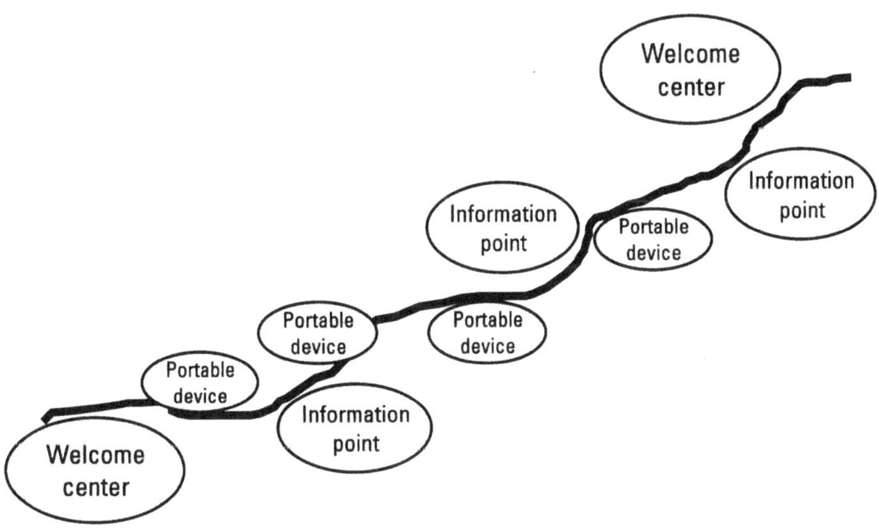


**Figure 20.4** Communications alternatives for scenic byways vision.

### 20.2.1 ITS Vision

The vision for the project was relatively compact, involving the application of an advanced traveler information system along a defined scenic byways corridor as shown in Figure 20.4.

There was no formal, written ITS vision used on this project. The project team relied entirely on the graphics with supporting verbal explanations to the stakeholder group. The essence of the vision was captured in the graphics: the establishment of a traveler information source that would collate and distribute current information on things to see and do along the corridor. Travelers along the corridor would be able to use their own personal information devices, such as in-vehicle information systems, or rent or borrow a device for use along the corridor. This involved the integration of the information system operation with the operation of conventional tourist welcome centers at the gateways to the corridor.

This project work reinforced the value of the ITS vision within the overall context of a structured system engineering approach to ITS planning and development. In this case, the vision was used to communicate important technical, institutional/organizational, and commercial features of the proposed solution to an audience comprised of essentially nontechnical stakeholders. The experience also taught us that the vision does not have to take the form of a text narrative. In the case of this project, the vision was a series of simple graphics. This indicates that perhaps in the future we should make use of sophisticated visualization and animation tools to support the creation and evolution of an ITS vision. This might enable the vision to be more readily adopted in the light of comments from the stakeholders and new insights from the technologists and system designers. The experience also highlighted the need to keep the vision current as the project progresses. This ensures that the vision correctly reflects the direction of the technical ITS Future Big Picture development work.

The outcome of this project was the identification and definition of a range of communications technologies that could be applied to the delivery of commercial and traveler information along scenic byways in the United States. The technologies were defined within an overall framework or ITS Future Big Picture that indicated how they might work at technical, institutional/organizational, and commercial levels. This was endorsed by the stakeholder group and presented to the Federal Highway Administration for consideration as part of the overall scenic byways program.

## 20.3 "Integrated ITS for the Klang Valley"

Our third and final story comes from the project carried out in 1995 in Kuala Lumpur, Malaysia, for a private-sector client. This project was interesting from a couple of perspectives. First, it was in Southeast Asia in a developing country with strong economic growth. Second, it involved a diverse group of stakeholders with multiple divergent ideas about the objectives for the work. The context diagram, shown in Figure 20.5, representing the highest level of the logical ITS Future Big Picture for the project, illustrates the range of stakeholders that were involved in the project. Table 20.2 summarizes the organizations and their primary roles in the transportation context in Malaysia.

The client for the study was a multinational industrial conglomerate with origins and major current business in Malaysia. The focus was on the application of advanced traffic management systems along a corridor from the Kuala Lumpur metropolitan area to the Port of Klang, some 20 miles to the west of the city.

We want to focus on four aspects of the project. The ITS Objectives Statement, the ITS User Services, Measures of Effectiveness, and the ITS Market Packages adopted for the project.

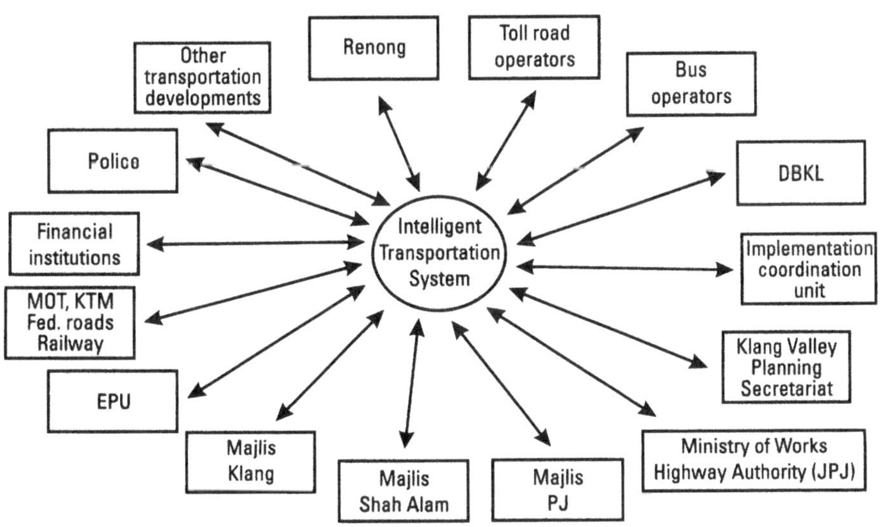


**Figure 20.5** Context diagram for the "Integrated ITS for the Klang Valley" project.

**Table 20.2**
Klang Valley Institutions and Their Transport-Related Functions

| No. | Name | Function |
|---|---|---|
| 1 | Economic Planning Unit (EPU) | Advises the prime minister on, amongst other things, privatization initiatives |
| 2 | Ministry of Transport (headed by the Minister of Transport) | Regulatory, policy-drafting, and administrative function |
| 3 | DBKL (town hall or city government; it falls under the prime minister's office) | Planning, traffic control, maintenance, operational function |
| 4 | MPPJ (development agency for the new city of Putra Jaya) | Planning, traffic control, maintenance, operational function |
| 5 | Municipality of Shah Alam | Planning, traffic control, maintenance, operational function |
| 6 | Municipality of Klang | Planning, traffic control, maintenance, operational function |
| 7 | JKR (Ministry of Public Works) | Road planning, construction, and maintenance |
| 8 | Park May | Bus operator |
| 9 | Interakota | Bus operator |
| 10 | PLUS | Toll road operator |
| 11 | KTM | Commuter rail services |

### 20.3.1 ITS Objectives Statement

The ITS Objectives Statement was developed as the product of one of the early consensus workshops held during the formative days of the project. It goes like this:

> To provide clearly identifiable solutions that have the potential to provide tangible improvements to the transportation context in the Klang Valley, while forming the basis for the commercialization of related opportunities.

You will notice that the ITS Objectives Statement does not really say very much about the overall project. What it does do is to provide a statement that all stakeholders bought into, enabling us to have some chance of a successful outcome to the project. Without it, we would have been doomed to failure as no one agreed on the overall objective for the work.

### 20.3.2 ITS User Services

The second aspect was the ITS User Services developed and adopted for the project work. These were based on the ITS User Services developed during the National Architecture for ITS program in the United States, but with additions and modifications specially for the Kuala Lumpur transportation context. The ITS User Services were developed based on an inventory of transportation needs, issues, problems, and objectives that were identified in collaboration with the stakeholder group and other local transportation agencies. We want to show you how the ITS User Services were derived from the needs, issues, problems, and objectives identified and confirmed through interaction with the stakeholder group. Therefore, we will start by describing the needs, issues, objectives, and problems. Subsequently, we will show you a mapping or relationship between these and selected ITS User Services.

### 20.3.3 Needs, Issues, Problems and Objectives

These were developed in close cooperation with client staff and members of the stakeholder group. The initial input was collated and structured under six headings, discussed in Sections 20.3.3.1–20.3.3.6.

#### 20.3.3.1 Incident Control and Management

There are a number of problems and issues associated with the management of traffic incidents and the control of road traffic. These include the following:

- There are not enough special task forces units to deal with these adequately.
- There are not enough tow-in facilities. This is a service that must be coordinated by the emergency services control center, and it should be part of the overall traffic control and system management concept.
- Response times to incidents are too long because of the obstructions being created by the traffic congestion or the ensuing congestion after the occurrence of an accident or incident. Control and management of the emergency services are also needed if this problem is to be addressed.
- The traffic density increases after an accident, and emergency vehicles find it difficult to reach the scene of the accident quickly enough.
- The concerned authorities are not equipped adequately for special situation management, such as chemical spills.

- The emergency services are not properly coordinated and deployed. The only efficient use of resources can be obtained when this function is integrated and centralized. This makes it possible for the proper recording of incidents and the optimization of labor and resources by central dispatching and control.
- Flooding is not dealt with adequately, and no automatic warning system exists. As this problem occurs frequently in the Klang Valley, the particular places where it occurs must be known by now. It would therefore be a relatively simple exercise to overcome most of the congestion caused by flooding through early warning systems.

A number of needs, issues, problems, and objectives relate to the lack of traffic and transportation information in the region. These include the following:

- Street surface markings are invisible under heavy traffic conditions, and overhead signage is required.
- Road signs are often found to be inconsistent and, in many instances, poorly located.
- Road signs in Malay will be very difficult to understand by the foreigners during the 1998 Commonwealth Games.
- The demarcation of the important routes for the 1998 games must be planned and catered for in advance as an uninformed driver can do unpredictable things in an unfamiliar situation.

#### 20.3.3.2 Financing Implementation, Control, Management and Maintenance

There are a number of problems associated with financing implementation, control, management, and maintenance. These include the following:

- Inadequate funds to develop and maintain the traffic control system in the Klang Valley;
- Inadequate funds for construction and maintenance of new road infrastructure;
- Inadequate funds to provide a high standard of law-enforcement.

#### 20.3.3.3 Legal, Organizational, and Institutional

There are a number of legal, organizational, and institutional issues identified as follows:

- Fragmented services provided by several authorities or departments all operating under different legislation;
- Insufficient law enforcement because of the deployment of manpower;
- Division of financial responsibilities;
- Spatial distribution of services;
- Lack of proper planning.

#### 20.3.3.4 Public Transport

The following needs, issues, problems and objectives relate to public transport in the region.

- Multiple tickets and systems are being used.
- Unpredictable schedules for buses exist.
- There is an imbalance between the supply and demand of taxi services.
- There is inadequate information on services, routes, frequencies, costs, and transfer opportunities.

#### 20.3.3.5 Special Situation Management

Issues and problems associated with the management of traffic during special events were identified as follows:

- No facilities to handle VIP routes adequately;
- Traffic congestion caused by major sporting events;
- Ceremonial parades;
- Requirement of special traffic management arrangements for 1998 Commonwealth Games.

These problems, needs, issues, and objectives were summarized and numbered for future reference as shown in Table 20.3

Table 20.3 provides an overview of the raw input from the stakeholder group. It has some duplication and overlap in many of the needs, issues, problems, and objectives but was used in this format to provide feedback to the stakeholder group. This confirmed that we had listened carefully and recorded faithfully by retaining the format of their input and not disguising their perceptions and initial ideas.

**Table 20.3**
Klang Valley Problems, Needs, Objectives, and Issues

| Number | Description |
|---|---|
| P20 | Coordination of traffic signals |
| P21 | Emergency services not coordinated |
| P22 | Emergency services not optimally deployed |
| P23 | Emergency services resource deployment |
| P24 | Fragmented services |
| P25 | Imbalance between demand and supply in taxi services |
| P26 | Inadequate funds |
| P27 | Inadequate road signs |
| P28 | Inadequate street surface markings |
| P29 | Insufficient incident management resources |
| P30 | Insufficient manpower—police |
| P31 | Insufficient staffing for traffic management |
| P32 | Lack of information on transportation system status |
| P33 | Lack of cohesive and integrated planning |
| P34 | Lack of traffic demand management |
| P35 | Motorcycles on freeways |
| P36 | Moving violations |
| P37 | Multiple ticketing systems |
| P38 | Nonconformation to traffic act |
| P39 | Overloading of vehicles |
| P40 | Police controlling intersections |
| P41 | Post-flash flood traffic congestion |
| P42 | Red light jumping |
| P43 | Replace traffic circles or signalize |
| P44 | Signalizing more intersections |
| P45 | Slow reaction times responding to incidents |
| P46 | Speeding |
| P47 | Traffic signal settings |

Once we had achieved this initial confirmation, we carried out some further analysis work on the needs, issues, problems, and objectives in order to explore the importance that members of the stakeholder group placed on each. It should be noted that our objective in conducting the ranking process was to identify which problems should be solved first, rather than remove problems from consideration altogether. As discussed in Chapter 3, the most effective way to develop a complete, sustainable ITS solution is to develop an ITS Future Big Picture that addresses all requirements, then develop a staged implementation plan or strategy.

### 20.3.4 Ranking of the Most Immediate Needs, Issues, Problems, and Objectives

The ranking of these needs, issues, problems, and objectives was achieved by using a delphi process to seek consensus on the order of the needs, issues, problems, and objectives, starting with the most critical problem and going down to the least important one. A two-stage process was followed whereby consensus was first sought on the weighting of the needs, issues, problems, and objectives in each of the five categories. Then the rating of each of the needs, issues, problems, and objectives was determined. The rating and weighting process essentially involved the allocation of weights to the various categories of needs, issues, problems, and objectives. The first goal was to distinguish their relative importance among the categories. Second, the members in the study team received the opportunity to rate each of the needs, issues, problems, and objectives on a scale of one to ten in terms of their severity (one indicating low severity and 10 indicating high). The rates and the weights are subsequently multiplied to provide the final score that can be ranked by sorting in a descending order.

The weighting of the needs, issues, problems, and objectives was done on a category basis. The following weights had been selected by the study team (with the highest weight meaning that the category is considered to be of a higher priority then any other category):

1. Congestion management = 5;
2. Safety = 4;
3. Transportation demand management = 3;
4. Funding = 2;
5. Law enforcement = 1.

During one of the stakeholder group meetings, the members were given an overview of the needs, issues, problems, and objectives. They were then asked to rank the needs, issues, problems, and objectives by scoring them according to the relative importance each individual attached to it. The questionnaire used to support this activity is shown in Figure 20.6.

Fourteen questionnaires were completed. There were 28 needs, issues, problems and objectives that had to be scored in terms of their level of severity ranging from a low importance (0 value) to a high importance (100).

Table 20.4 provides the results of the process, which was held within the study team and later confirmed in the liaison group.

#### 20.3.4.1 Mapping Transportation Needs, Issues, Problems, and Objectives to ITS User Services

This comprised an initial review of the needs, issues, problems, and objectives defined in the preceding section and a comparison with the 29 U.S National Architecture development program ITS User Services. Note that at the time the project was conducted there were 29 ITS User Services; there are now 30. This revealed the need to define additional ITS User Services specifically for the Klang Valley context. These related to the following needs, issues, problems, and objectives:

- Red light violation;
- Motorcycles on freeways;

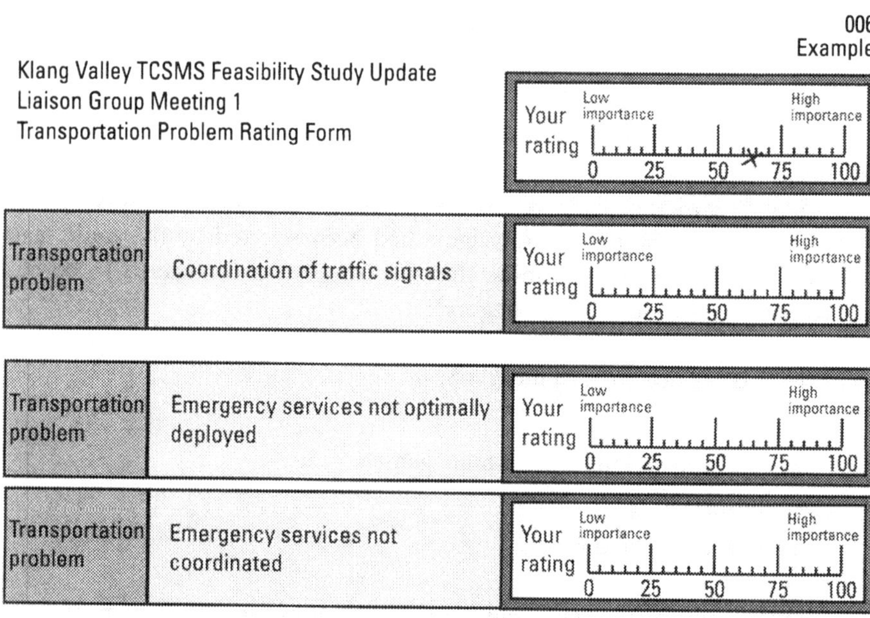


**Figure 20.6** Klang Valley questionnaire.

**Table 20.4**
Results of Ranking Process

| Ranking Number | Problem Number | Description | Weighting | Rating | Score |
|---|---|---|---|---|---|
| 1 | P20 | Coordination of traffic signals | 5 | 10 | 50 |
| 2 | P47 | Traffic signal settings | 5 | 10 | 50 |
| 3 | P37 | Multiple ticketing systems | 5 | 9 | 45 |
| 4 | P31 | Insufficient staffing for TCSMS | 5 | 8 | 40 |
| 5 | P43 | Replace traffic circles or signalize | 5 | 8 | 40 |
| 6 | P45 | Slow reaction times responding to incidents | 5 | 8 | 40 |
| 7 | P27 | Inadequate road signs | 5 | 7 | 35 |
| 8 | P28 | Inadequate street surface markings | 5 | 7 | 35 |
| 9 | P40 | Police controlling intersections | 5 | 7 | 35 |
| 10 | P44 | Signalizing more intersections | 5 | 7 | 35 |
| 11 | P21 | Emergency services not coordinated | 4 | 8 | 32 |
| 12 | P48 | Unpredictable public transport services | 3 | 10 | 30 |
| 13 | P22 | Emergency services not optimally deployed | 4 | 7 | 28 |
| 14 | P32 | Lack of information on transportation system status | 3 | 9 | 27 |
| 15 | P26 | Inadequate funds | 3 | 8 | 24 |
| 16 | P29 | Insufficient incident management resources | 4 | 6 | 24 |
| 17 | P33 | Lack of cohesive and integrated planning | 3 | 8 | 24 |
| 18 | P34 | Lack of traffic demand management | 3 | 8 | 24 |
| 19 | P41 | Post-flash flood traffic congestion | 4 | 6 | 24 |
| 20 | P35 | Motorcycles on freeways | 4 | 5 | 20 |
| 21 | P23 | Emergency Services Resource Deployment | 3 | 6 | 18 |
| 22 | P25 | Imbalance between demand and supply in taxi services | 3 | 6 | 18 |
| 23 | P24 | Fragmented services | 3 | 5 | 15 |
| 24 | P30 | Insufficient manpower—police | 1 | 6 | 6 |
| 25 | P36 | Moving violations | 1 | 6 | 6 |
| 26 | P38 | Nonconformation to traffic act | 1 | 4 | 4 |
| 27 | P39 | Overloading of vehicles | 1 | 4 | 4 |
| 28 | P42 | Red light jumping | 1 | 4 | 4 |
| 29 | P46 | Speeding | 1 | 4 | 4 |

- Speeding;
- Moving violations;
- Nonconformation to traffic act;
- Overloading of vehicles.

These needs, issues, problems, and objectives require the provision of an ITS User Service addressing enforcement, which could have been assumed to be included in ITS service 40, "Traffic Control," but were best addressed separately. Therefore, a new ITS User Service U45, "Enforcement," was defined.

Another significant problem related to the lack of cohesive, integrated transportation planning in the Klang Valley. This can be addressed through the application of conventional transportation initiatives and organizational approaches. It can also be supported through the provision of travel demand pattern information, trip distribution data, and the use of the ITS to provide travel survey data on a comprehensive and continuous basis. Therefore, another new ITS User Service, U135, "Trip Survey and Travel Data," was defined. The definitions of these two new ITS User Services are described in Sections 20.3.5 and 20.3.6.

### 20.3.5 U45: "Provide Enforcement Support"

This provides support for police and traffic authorities in the enforcement of traffic regulations and the application of highway law. Facilities will be provided for the following:

- Automatic detection of speeding violators and reporting;
- Automatic detection and reporting of commercial vehicles operating outside of legal requirements with regard to speed, weight, height, and width;
- Automatic detection and reporting of the operation of prohibited vehicles on freeways (e.g., motorcycles and farm and construction equipment);
- Automatic detection and reporting of other moving traffic offenses;
- Automatic enforcement of car parking regulations through the use of electronic meters, both on- and off-vehicle;
- Automatic enforcement of traffic signal compliance (i.e., red light jumping detection and reporting).

### 20.3.6 U135: "Provide Trip Survey and Travel Data Services"

This provides services to the transportation planner and responsible authorities in support of coordinated transportation planning and policy formulation. This includes the use of the various sensors and data collection capabilities of an ITS in order to determine the following:

- Travel demand patterns for public transport;
- Travel demand patterns for private cars;
- Travel demand patterns for commercial vehicles;
- Variation of total traffic flow overtime;
- Variation of total transport demand overtime;
- Generation of coordinated transportation management strategies;
- Effects of weather on transport system operation.

### 20.3.7 ITS User Services Finally Selected

These new ITS User Services were added to the initial list and a mapping of transportation needs, issues, problems, and objectives to ITS User Services was developed as shown in Table 20.5.

The results of the mapping exercise clearly show the need for an integrated approach to ITS development and implementation for the Klang Valley. There is an even spread of ITS User Services required to address the transportation needs, issues, problems, and objectives, indicating that an integrated ITS with capability to provide multiple ITS User Services is required.

As the following services were not mapped to transportation needs, issues, problems, and objectives, they were assumed to be required so late in the Implementation Strategy time line that corresponding deployments would be beyond the scope of the current project. They were therefore discarded at this point and not considered for further system development work.

- U230: "Longitudinal Collision Avoidance;"
- U240: "Lateral Collision Avoidance;"
- U260: "Vision Enhancement for Collision Avoidance;"
- U270: "Safety Readiness;"
- U280: "Precollision Restraint Deployment;"
- U290: "Automated Highway Systems."

**Table 20.5**
Mapping Transportation Needs, Issues, Problems, and Objectives to ITS User Services

| Transportation Problems / ITS User Services | 10 En Route Driver Information | 20 Route Guidance | 30 Traveler Services Information | 40 Traffic Control | 45 Enforcement | 50 Incident Management | 60 Emissions Testing and Mitigation | 70 Pre-Trip Travel Information | 80 Ride Matching and Reservation | 90 Demand Management and Operations | 100 Public Transportation Management | 110 En Route Public Transport Information | 120 Personalized Public Transport | 130 Public Transport Security | 135 Trip Survey and Travel Data | 140 Electronic Payment Services | 150 Commercial Vehicle Electronic Clearance | 160 Automated Roadside Safety Inspection | 170 Onboard Safety Monitoring | 180 Commercial Vehicle Administrative Processes | 190 Hazardous Material Incident Response | 200 Commercial Fleet Management | 210 Emergency Notification and Personal Security | 220 Emergency Vehicle Management | 230 Longitudinal Collision Avoidance | 240 Lateral Collision Avoidance | 250 Intersection Collision Avoidance | 260 Vision Enhancement for Crash Avoidance | 270 Safety Readiness | 280 Pre-Crash Restraint Deployment | 290 Automated Highway Systems |
|---|---|---|---|---|---|---|---|---|---|---|---|---|---|---|---|---|---|---|---|---|---|---|---|---|---|---|---|---|---|---|---|
| Coordination of traffic signals | | | | • | | | • | | | • | • | | | | | | | | | | | | | • | | | | | | | |
| Emergency services not coordinated | | | | | | | | | | | | | | | | | | | | | | | | • | | | | | | | |
| Emergency services not optimally deployed | | | | | | | | | | | | | | | | | | | | | | | | • | | | | | | | |
| Emergency Services Resource Deployment | | | | | | • | | | | | | | | | • | | | | | | • | | | • | | | | | | | |
| Fragmented services | | | | • | | | | | | • | • | | | | • | | | | | | | • | | • | | | | | | | |
| Imbalance between demand and supply in taxi services | • | | | | | | | | • | • | • | • | • | | | | | | | | | | | | | | | | | | |
| Inadequate funds | | | | | | | | | | | | | | | | | | | | | | | | | | | | | | | |
| Inadequate road signs | • | • | • | | | | | • | | | | | | | | | | | | | | | | | | | | | | | |
| Inadequate street surface markings | | | | | | | | | | | | | | | | | | | | | | | | | | | | | | | |
| Insufficient incident management resources | | • | | | | • | | | | | | | | | | | | | • | | • | | • | • | | | | | | | |
| Insufficient manpower—authorities | • | • | • | • | | • | • | • | | • | • | • | | | | • | | | | | | | | | | | | | | | |
| Insufficient staffing TCSMS | • | • | • | • | | • | • | • | | • | • | • | | | | • | • | • | • | • | • | • | • | | | | | | | | |
| Lack of appropiate info. on transportation system status | • | • | • | | | | | • | • | • | • | • | | | • | | | | | | | • | | | | | | | | | |
| Lack of cohesive and intergrated planning | | | | | | | | | | | | | | | • | | | | | | | | | | | | | | | | |
| Lack of traffic demand management | • | • | • | | | | • | • | • | • | • | • | | | | • | | | | | | • | | | | | | | | | |
| Motorcycles on freeways | | | | | • | | | | | | | | | | | | | | | | | | | | | | | | | | |
| Moving offences | | | | | • | | | | | | | | | | | | | | | | | | | | | | | | | | |
| Multiple ticketing systems | | | | | | | | • | | | • | • | | | • | • | | | | | | | | | | | | | | | |
| Nonconformation to traffic act | | | | | • | | | | | | | | | | | | | • | • | | | | | | | | | | | | |
| Overloading of vehicles | | | | | • | | | | | | | | | | | | | • | • | • | | | | | | | | | | | |
| Police controlling intersections | • | • | • | • | | • | | • | | | • | • | | | | | | | | | | • | | • | | | • | | | | |
| Post flash flood traffic congestion | • | • | • | | | • | | • | | | • | • | | | | | | | | | | • | | • | | | | | | | |
| Red light jumping | | | | | • | | | | | | | | | | | | | | | | | | | | | | | | | | |
| Replace traffic circles or signalize | | | | • | | | | | | | • | | | | | | | | | | | | | • | | | • | | | | |
| Signalizing more intersections | | | | • | | | • | | | • | • | | | | | | | | | | | | | • | | | • | | | | |
| Slow reaction times responding to incidents | • | • | • | • | | • | | • | | • | • | • | | • | | • | | | | | • | • | • | • | | | | | | | |
| Speeding | | | | | • | | | | | | | | | | | | | | | | | | | | | | | | | | |
| Traffic signal settings | | | | • | | | | | | | • | | | | | | | | | | | | | • | | | • | | | | |
| Unpredictable public transport services | | | | • | | • | | | | • | • | • | | | | | | | | | | | | | | | | | | | |

Therefore, including the two newly defined ITS User Services; there are a total of 25 ITS User Services for the Klang Valley TCSMS study. These are listed in Table 20.6

There were also some needs, issues, problems, and objectives that were not fully addressable by the ITS. This was anticipated in the study methodology, and these needs, issues, problems, and objectives were addressed in a separate set of recommendations for basic solutions that was reported as part of the overall study. The needs, issues, problems, and objectives not fully addressable by ITS were the following:

**Table 20.6**
Klang Valley ITS User Services

| | |
|---|---|
| Travel and transportation management | U10 En Route Driver Information<br>U20 Route Guidance<br>U30 Traveler Services Information<br>U40 Traffic Control<br>U50 Incident Management<br>U60 Emissions Testing and Mitigation |
| Travel demand management | U70 Pretrip Travel Information<br>U80 Ride Matching and Reservation<br>U90 Demand Management and Operations |
| Public transportation operations | U100 Public Transport Management<br>U110 En Route Public Transport Information<br>U120 Personalized Public Transport<br>U130 Public Transport Security |
| Electronic payment services | U140 Electronic Payment Services |
| Commercial vehicle operations | U150 Commercial Vehicle Electronic Clearance<br>U160 Automated Roadside Safety Inspection<br>U170 Onboard Safety Monitoring<br>U180 Commercial Vehicle Administrative Processes<br>U190 Hazardous Material Incident Response<br>U200 Commercial Fleet Management |
| Emergency management | U210 Emergency Notification and Personal Security<br>U220 Emergency Vehicle Management |
| Advanced vehicle control and safety systems | U250 Intersection Collision Avoidance |
| Enforcement | U45 Enforcement |
| Transportation planning | U135 Trip Survey and Travel Data |

- Inadequate street surface markings;
- Inadequate funds;
- Lack of cohesive and integrated planning.

## 20.4 Measures of Effectiveness (MOEs)

In parallel with the development of the ITS User Services, a range of performance parameters, or Measures of Effectiveness (MOEs) were identified and defined. As discussed in Chapter 14, the identification and definition of MOEs at this early stage in the project helps to ensure that ITS User Services are valid. In the case of this study, MOEs fall into two distinct categories related to the primary objectives of the work, transportation MOEs, described in Section 20.4.1, and commercial MOEs, described in Section 20.4.2.

### 20.4.1 Transportation MOEs

Potential MOEs can be defined for each transportation problem identified in the problem statement, providing an initial view of the "yardsticks" that can be used for measuring system effectiveness. Once the proposed technical solution has been defined in some detail, it will be necessary to review these MOEs, since modification and enhancement may be required to take account of the specific nature of the proposed solution. This is important, as it will be the performance of the final technical solution and its elements that will be measured directly.

The initial set of transportation MOEs identified for the study are shown in Table 20.7 and defined in Sections 20.4.1.1–20.4.1.23.

#### 20.4.1.1 Accident Rates

Annual accident rates for the area covered by the implementation are measured in accidents per million vehicle kilometers traveled per year.

#### 20.4.1.2 Additional Delay Attributable to Incidents

This comprises the total delay directly attributable to incidents occurring on the network. It is measured in person hours per delay per year. This MOE will require derivation from traffic volumes, average vehicle occupancy, and historical traffic flow data.

**Table 20.7**

Mapping of Transportation Measures of Effectiveness to Needs, Issues, Problems, and Objectives

| | Needs, Issues, Problems, Objectives | | | | | | | | | | | | | | | | | | | | | | | | | | | | |
|---|---|---|---|---|---|---|---|---|---|---|---|---|---|---|---|---|---|---|---|---|---|---|---|---|---|---|---|---|
| **Transportation Measures of Effectiveness** | Coordination of traffic signals | Emergency services not coordinated | Emergency services not optimally deployed | Emergency services resource deployment | Fragmented services | Imbalance between demand and supply in taxi services | Inadequate funds | Inadequate road signs | Inadequate street surface markings | Insufficient incident management resources | Insufficient manpower—authorities | Insufficient staffing TCSMS | Lack of appropriate information on transportation system status | Lack of cohesive and integrated planning | Lack of traffic demand management | Motorcycles on freeways | Moving offenses | Multiple ticketing systems | Nonconformation to traffic act | Overloading of vehicles | Police controlling intersections | Post-flash flood traffic congestion | Red light jumping | Replace traffic circles or signalize | Signalizing more intersections | Slow reaction times responding to incidents | Speeding | Traffic signal settings | Unpredictable public transport services |
| Accident rates | | | | | | | | • | • | | | | | | | • | • | | | • | | | • | | | | • | | |
| Additional delay attributable to incidents | • | | | | | | | | | • | | | | • | • | | | | | | | • | | | | | | | |
| Average vehicle speeds through network | • | | | | | | | | • | | | | • | • | | | | | | | | | | | | | | • | |
| Bus schedule adherence | • | | | | | | | | • | | | | • | • | | | | • | | | | | | | | | | • | • |
| Cost of police traffic management operations | | | | | | | | | | | | | | | | • | • | | • | • | • | | • | | | | • | | |
| Emergency vehicle response times | • | • | • | | | | | | • | | | | • | • | | | | | | | | | | | | • | | • | |
| Journey times through the road network | • | | | | | | | • | • | | | | • | • | | | | | | | | | | | | | | • | |
| Modal split | • | | | | • | | | | | | | | • | • | • | | | • | | | | | | | | | | | |
| Network recovery time after incident | • | | | | | | | | | • | | | | • | | | | | | | | • | | | | | | | |
| Number of licensed taxis | | | | | | • | | | | | | | | | | | | | | | | | | | | | | | |
| Number of stops at traffic signals | • | | | | | | | | | | | | | • | • | | | | | | | | | | | | | • | |
| Number of traffic violations | | | | | | | | | | | | | | | | • | • | | • | • | | | • | | | | • | | |
| Operating cost of bus services | • | | | | | | | | • | | | | • | • | | | | • | | | | | | | | | | • | • |
| Operating cost of emergency services | • | • | • | | | | | | • | | | | • | • | | | | | | | | | | | | | | • | |
| Operating cost of transportation planning | | | | | | | | | | | | | | • | | | | | | | | | | | | | | | |
| Queue lengths at major intersections | • | | | | | | | • | • | | | | • | • | | | | | | | | | | | | | | • | |
| Revenue from traffic offense fines | | | | | | | | | | | | | | | | • | • | | • | • | | | • | | | | • | | |
| Staffing levels required for local authorities | | | | | • | | | | | | • | | | • | | | | | | | | | | | | | | | |
| Staffing required for ATMS | • | | | | | | | | | | • | • | | • | | | | | | | | | | | | | | | |
| Taxi company revenue | | | | | | • | | | | | | | | | | | | | | | | | | | | | | | |
| Total delay incurred | • | | | | • | | | • | • | | | | • | • | | | | | | | | | | | | | | • | |
| Total network throughput | • | | | | • | | | • | • | | | | | • | • | | | • | | | | | | | | | | • | |
| Traveler satisfaction | | | | | | • | | | | | | | | | | | | | | | | | | | | | | | |

#### 20.4.1.3 Average Vehicle Speeds Through Network

This describes the average journey speed through the network, measured over the course of a year. It is expressed as an annual average in kilometers per hour.

#### 20.4.1.4 Bus Schedule Adherence

This is a measure of how closely bus services follow the published service schedules. It is expressed as a percentage of total services meeting operational goals; these can vary according to specific targets, but a typical goal would be to have services arriving at each bus stop within five minutes of the published arrival time.

#### 20.4.1.5 Cost of Police Traffic Management Operations

This is the total cost of operating police traffic management and enforcement of traffic-related offenses. It includes administration and staff costs and is expressed in ringgits per annum.

#### 20.4.1.6 Emergency Vehicle Response Times

This is a measure of time elapsed between receipt of request for service and arrival of unit. It is measured in minutes and averaged over the year.

#### 20.4.1.7 Journey Times Through the Road Network

These are the average journey times to and from major origins and destinations on the transportation network. They are measured and averaged over one year.

#### 20.4.1.8 Modal Split

This is the proportion of total number of travelers choosing to travel by various modes. It is expressed as a percentage and averaged over the course of one year for each of the major travel modes:

- Private vehicle;
- Bus;
- Taxi;
- Train;
- Light rapid transit.

The modal split figures would be compared to target values set as part of transportation policy for the region. In the Kuala Lumpur context the overall objective would be to increase use of transit modes.

#### 20.4.1.9 Network Recovery Time After Incident

This is the time in hours from the clearance of the incident to re-establishment of normal operating conditions. This measures the incident management capability of the system and is expressed in hours. It should be averaged over all incidents occurring over a one-year period.

#### 20.4.1.10 Number of Licensed Taxis

This is the total number of registered licensed taxi vehicles for an annual period. An increase in the number of licensed taxis would indicate an improvement in the enforcement process and improved access to noncar and multiperson modes of travel.

#### 20.4.1.11 Number of Stops at Traffic Signals

This would be an output from a traffic model and is not measured directly. Before and after figures would be compared.

#### 20.4.1.12 Number of Traffic Violations

This is the total number of traffic-related offenses recorded by the police over a one-year period.

#### 20.4.1.13 Operating Cost of Bus Services

This is the total cost in ringgits for operating bus services in the Klang Valley. It includes staff time, administration, equipment, fuel, and maintenance and is totaled for each one-year period.

#### 20.4.1.14 Operating Cost of Emergency Services

This is the total cost in ringgits for operating emergency services in the Klang Valley. It includes staff time, administration, equipment, fuel, and maintenance. It is totaled for each one-year period.

#### 20.4.1.15 Operating Cost of Transportation Planning and Operations

This is the total cost in ringgits for transportation planning and operations in the Klang Valley. It includes staff time, administration, and equipment and is totaled for each one-year period

#### 20.4.1.16 Queue Lengths at Major Intersections

This figure, the average number of vehicles, or length of queue in meters, is established by survey or from a traffic model. It should be determined for a sample of significant intersections in the area and averaged over a year.

#### 20.4.1.17 Revenue From Traffic Offense Fines

This constitutes the total annual revenue from traffic-related offenses, expressed in ringgits.

#### 20.4.1.18 Staffing Levels Required for Local Authorities

This is the total number of staff required to operate and support all transportation-related public services in the Klang Valley area. It is expressed as numbers of people and total salary cost averaged over a year.

#### 20.4.1.19 Staffing Required for Traffic Management

This is the total number of staff required to operate and support the ATMS facility in the Klang Valley area. It is expressed as numbers of people and total salary cost averaged over a year.

#### 20.4.1.20 Taxi Company Revenue

The total annual revenue from the provision of taxi services, expressed in ringgits, may have to be derived from tax returns rather than measured directly.

#### 20.4.1.21 Total Delay Incurred

Established through survey or modeling techniques, this is the total number of hours wasted every year due to congestion. It is the difference between the time taken to make journeys at free flow speeds and the time taken during congested conditions. It is expressed in person hours per year.

#### 20.4.1.22 Total Network Throughput

This figure is expressed in terms of total number of trips people made on the transportation network in the course of an average working day. It should be averaged out over a year.

#### 20.4.1.23 Traveler Satisfaction

This index is measured using market research techniques such as stated preference and other interview surveys. It is expressed in terms of percentage of travelers who are happy/unhappy with various aspects of the transportation system.

### 20.4.2 Commercial MOEs

These represent the business or commercial view of successful system implementation and consist of a combination of financial statistics used to measure the economic health of the business aspects of the proposed implementation. These include measures such as the following:

- Total capital investment required;
- Cash flow;
- Contribution;
- Internal rate of return.

These are the same sort of measures as used in any business enterprise. The definition of commercial MOEs was addressed through the development of a number of business models as part of the development of the commercial layer of the ITS Future Big Picture for this project.

## 20.5 ITS Market Packages

Based on work carried out for the National Architecture for the ITS development program in the United States, the ITS Market Packages can be defined and mapped against ITS User Services as shown in Table 20.8.

These represent the basic units that will be selected and combined to form the ITS architecture, or framework. This will then form the basis for the development of an ITS design.

### 20.5.1 ITS Market Package Descriptions

The following ITS Market Packages were defined.

#### 20.5.1.1 M10: "Public Transport Passenger and Fare Management"

These are technology packages that provide functions for in-vehicle and central management of passenger loading and fare management.

The following facilities are provided:

1. *Fare payment:* This allows the vehicle operator to detect improper fare payment and other information about the passenger and if necessary prevent boarding. Fare information is recorded in real-time onboard the vehicle and is either relayed to a public transport management center in real time or stored on the vehicle for downloading at the end of each day. Real-time data transfer would be achieved using a one-way communications link from the vehicle to the public transport management center. Fare payment is achieved using a debit card or smart card and is handled electronically.

**Table 20.8**
Klang Valley ITS Market Packages to User Services Mapping

| ITS Market Packages / ITS User Services | 10 En Route Driver Information | 20 Route Guidance | 30 Traveller Services Information | 40 Traffic Control | 45 Enforcement | 50 Incident Management | 60 Emissions Testing and Mitigation | 70 Pretrip Travel Information | 80 Ride Matching and Reservation | 90 Demand Management and Operations | 100 Public Transportation Management | 110 En-Route Public Transport Information | 120 Personalized Public Transport | 130 Public Transport Security | 135 Trip Survey and Travel Data | 140 Electronic Payment Services | 150 Commercial Vehicle Electronic Clearance | 160 Automated Roadside Safety Inspection | 170 Onboard Safety Monitoring | 180 Commercial Vehicle Administrative Processes | 190 Hazardous Material Incident Response | 200 Commercial Fleet Management | 210 Emergency Notification and Personal Security | 220 Emergency Vehicle Management | 230 Longitudinal Collision Avoidance | 240 Lateral Collision Avoidance | 250 Intersection Collision Avoidance | 260 Vision Enhancement for Crash Avoidance | 270 Safety Readiness | 280 Pre-Crash Restraint Deployment | 290 Automated Highway Systems |
|---|---|---|---|---|---|---|---|---|---|---|---|---|---|---|---|---|---|---|---|---|---|---|---|---|---|---|---|---|---|---|---|
| M10 Passenger & Fare Management | | | | | | | | • | | | • | • | • | | • | • | | | | | | | | | | | | | | | |
| M20 Public Transport Vehicle Tracking | | | | | | | | • | | | • | • | • | • | • | | | | | | | | • | | | | | | | | |
| M30 Public Transport Security | | | | | • | | | • | | | • | • | • | • | | | | | | | | | | | | | | | | | |
| M40 Public Transport Operations Planning | | | | | | | | | | • | • | | • | | • | | | | | | | | | | | | | | | | |
| M50 Public Transport Maintenance | | | | | | | • | • | | | • | • | • | • | | | | | | | | | | | | | | | | | |
| M60 Broadcast-Based ATIS | • | • | • | • | | • | | • | • | • | • | • | • | • | • | | | | | | | | | • | | | | | | | |
| M70 Interactive ATIS With Infrastructure-Based Route Planning | • | • | • | • | | • | | • | • | • | • | • | • | • | • | • | | | | | | | | • | | | | | | | |
| M80 Interactive ATIS With GIS, GPS | • | • | • | • | | • | | • | • | • | • | • | • | • | • | | • | | | • | | | | • | | | | | | | |
| M90 Mayday Support | • | • | • | • | | • | | • | • | • | • | • | • | | | | • | | | • | • | | • | • | | | | | | | |
| M100 Parking Management | • | | • | | • | | | • | | • | • | | | | • | • | | | | | | | | | | | | | | | |
| M110 Billing Support | | | • | | • | | • | | | • | | | | | | • | | | | | | | | | | | | | | | |
| M120 In-Vehicle Signing | • | • | | • | • | | | | | • | | | • | | | • | | | | | • | | | • | | | | | | | |
| M130 Basic Traffic Monitoring/Management | • | | | • | • | • | • | • | | • | | • | | | • | | | | | | | | | • | | | | | | | |
| M140 Area Wide Traffic Management | • | | | • | • | • | • | • | | • | | • | | | • | | | | | | | | | • | | | | | | | |
| M150 Traffic Prediction | • | • | | • | | • | | • | | • | | • | | | • | | | | | | | | | • | | | | | | | |
| M160 Travel Demand Management | • | • | | • | | • | | • | | • | | • | | | • | • | | | | | | | | • | | | | | | | |
| M170 Automated Toll Plaza | | • | | • | • | | | | | • | | | | | • | • | | | | | | | | | | | | | | | |
| M180 HOV Lane Management | • | • | | • | | • | | • | | • | | • | | | | | | | | | | | | • | | | | | | | |
| M190 Reversible Lane Management | • | | | • | | • | | • | | • | | • | | | | | | | | | | | | • | | | | | | | |
| M200 Probe Vehicle Instrumentation | • | • | • | • | • | • | • | • | • | • | • | • | • | • | • | | • | | | • | | | | • | | | | | | | |
| M210 Onboard Safety | | | | | • | | | | | | | | | | | | | | • | | • | | | | | | • | | | | |
| M220 Private Vehicle Safety | | | | | • | | | | | | | | | | | | | | • | | • | | | | • | • | | | | | |
| M230 Driver Visibility Improvement | | | | | | | | | | | | | | | | | | | | | | | | | • | • | • | | | | |
| M240 Advanced Vehicle Longitudinal Controls | | | | | | | | | | | | | | | | | | | • | | • | | | | • | • | • | | | | |
| M250 Automated Highway Systems | • | | | • | | • | | • | | • | | • | | | | | | | • | | • | | | • | • | • | • | | | | |
| M260 Intersection Collision Avoidance | | | | • | | • | | | | • | | | | | | | | | • | | | | | • | • | • | • | | | | |
| M270 Vehicle Tracking And Dispatch | | | | | | | | | | | | | | | • | | • | | • | • | | • | | | • | • | | | | | |
| M280 Material Tracking | | | | | | | | | | | | | | | | | • | • | • | • | • | • | | | | | | | | | |
| M290 Credentials | | | | | • | | | | | | | | | | | • | • | • | • | • | | • | | | | | | | | | |
| M300 Safety For Commercial Vehicles | | | | | • | | | | | | | | | | | | • | • | • | • | • | • | | | | | | | | | |
| M310 Emergency Response | | | | | | | | | | | | | | • | | | • | | | • | • | | • | • | | | | | | | |
| M320 Emergency Vehicle Maintenance | | | | | | | • | | | | | | | | | | | | | | | | | • | | | | | | | |

2. *Passenger counting:* Passenger loading on individual vehicles is determined using sensors fitted to doorways on the vehicles. Both vehicle operator and central operation staff can have access to passenger volume information in real time. This information can be used to review deployment of the vehicle fleet in real time and adjust to suit changes in the demand pattern as manifested by the actual passenger loading. Passenger loading can also be stored as historical data and used as the basis for both short-term and long-term demand predictions. These can be utilized to improve the planning process for public transport operations.

Another use of this facility revolves around the allocation of public transport subsidy funds. In certain contexts, local or central government provide a subsidy to a privately operated transport company based on number of passengers carried. This is usually part of a wider traffic demand management strategy aimed at encouraging use of public transport. In these cases, the passenger count information can be used to ensure accurate disbursement of the subsidy.

#### 20.5.1.2 M20: "Public Transport Vehicle Tracking"

This provides automatic vehicle location capabilities to track vehicle movements and update public transport schedules. Either terrestrial or satellite-based technologies can be deployed to provide accurate data on current geographical locations of all public transport vehicles in the fleet.

Equipment onboard the vehicle carries out the position determination, which is then reported to a public TMC utilizing a radio communications link from vehicle to management center.

At the public transport management center, the data is collected and consolidated by a software system that provides operations staff with up-to-date information on vehicle schedule and route adherence.

The software system can also be linked to a traveler information system that provides current traffic condition information to the public transport management center. This can be used to adapt public transport operations to real-time conditions.

The software system could in fact be an expert system utilizing AI techniques to generate advice on central operation or vehicle operator actions to optimize the performance of the public transport service.

Another possible extension is to provide an interface to an advanced traffic management or area traffic control system, enabling the public TMS to request priority treatment for vehicles that have fallen behind schedule.

Typical technologies utilized in this ITS Market Package include the following:

- *GPS:* A GPS receiver fitted to the public transport vehicle can be used to determine location.
- *Vehicle-to-roadside communications (VRC):* This uses short-range radio or infrared communications techniques. Specific vehicles are located by identification as they pass each beacon.
- *Dead reckoning:* Periodic dead reckoning vehicle positioning estimates are made with the aid of wheel sensors and an electronic compass. The assumed vehicle position is continuously updated by utilizing a map-matching technique. This involves comparison of the actual vehicle turning movements with map data stored on disc.

Using the capabilities of CD-I technology, 600 Mbytes of digital data can be stored on a single 120-millimeter diameter disc. Methods have been developed for conveniently partitioning the map data into "parcels" and arranging the corresponding data blocks in the most efficient sequence on the disc.

An important data-management aspect is the efficient way in which the information is structured in each block: The minimum amounts of information, which correspond to the nodes and "chains" on the map, are linked via pointer addresses.

The components fitted into a vehicle typically comprise the following:

- An on-board computer for controlling and processing information;
- A CD-ROM or DVD disc player;
- Sensors for determining the position of the vehicle.

#### 20.5.1.3 M30: "Public Transport Security"

Systems to improve the security and safety of all public transport users include operators and passengers. Safety-critical public areas are monitored using this ITS Market Package. These may include the following:

- Onboard the vehicle;
- At bus stops;
- At bus stations or depots;
- In parking areas.

Onboard video cameras, audio, or other sensing technologies can be deployed to allow the operator to monitor the safety and security of passengers onboard the vehicle. In an emergency, an alarm is generated either manually by the operator or automatically by the monitoring system. Sensors and communications techniques such as packet data radio and cellular radio can be incorporated in these ITS Market Packages.

Public transport-related areas such as bus stations, light rapid transit stations, and park and ride locations can also be monitored by static monitoring systems connected by wireline back to a control center.

#### 20.5.1.4 M40: "Public Transport Operations Planning"

These technology packages automate the planning work required for successful public transport operation. The following functions could be provided:

1. *Vehicle operator assignment and monitoring:* Vehicle operators are assigned to particular vehicles and route itineraries based on availability, eligibility, and record of accomplishment. A central vehicle operator database is utilized for this activity, combined with vehicle location information from an automatic vehicle location system. This database can also be used for wages and personnel operations.
2. *Vehicle routing and scheduling:* This takes advantage of automatic vehicle location technology to develop and disseminate schedules based on actual demand and operating parameters. Specially developed software determines the scheduling and routing required for optimum system performance, and the information is disseminated through wireless communication techniques to the vehicles.

Note that if vehicle routes are to be changed in real time, then some form of in-vehicle navigation system may also be required to guide the operator efficiently to the new route and perhaps assist in keeping to the new route for the first few journeys. Although this facility can be created, it is sometimes of little practical use as the standard agreements stipulate that strict adherence to routes must be observed at all instances.

#### 20.5.1.5 M50: "Public Transport Maintenance"

These provide vehicle condition monitoring and maintenance scheduling functions to public transport operators. These include both real-time and "off-line" processes such as work task and technician scheduling and work monitoring. Sensors onboard the vehicle can include transducers, temperature gauges, door mechanism sensors, tire pressure gauges, and engine management systems (including emission monitoring).

An onboard microprocessor polls each of these sensors and determines if there are any irregularities. These are then transmitted back to the public transport management center for action.

The onboard processor can also prepare summary reports of historical sensor data that can be stored onboard the vehicle for later retrieval and analysis. This analysis can be performed by an expert system that provides predictions of failure rates and anticipates the need to repair essential components. This is interfaced with the work scheduling activities to effectively plan repair shop workloads. Linked to an inventory control system, it can also reduce the need for parts inventory by providing more accurate estimates of parts requirements.

#### 20.5.1.6 M60: "Broadcast-Based Advanced Traveler Information Service"

This package involves the collection and dissemination of travel-related data and information. The core of this package is the capability to collect data and process it into information suitable for dissemination to the traveler. The package will typically consist of the following elements:

- *Data collection network:* This can be an advanced telemetry system, with communications networks linking an array of vehicle sensors strategically placed on the transport network. It could also be an organizational arrangement, under which data is supplied to a central point by a variety of methods and by many data suppliers. For example, road traffic data could be collected from the police by telephone and fax and fused with traffic sensor data originating from DBKL.
- *Data fusion facility:* This would be a computer-based facility for inputting, storing, formatting, and retrieving transport-related data from a large variety of sources. A sophisticated software suite would manage the data and carry out the necessary reformatting for data fusion and compatibility. This suite would also interface with the communications network and drive the devices required to disseminate the information to the travelers.
- *Information dissemination network:* In the case of broadcast-based ATIS, the information dissemination network could take the following forms:
    - *Voice radio broadcasts:* This would be an enhanced version of the current service offered by TIME radio for example. A wider range of transport data could be provided by accessing the control center.
    - *Radio broadcast using FM sideband techniques:* Instead of voice bulletins, the desired information is broadcast as a data stream on top of

an existing FM stereo radio station. Travelers would require a specially modified radio in order to receive the information.

- *Roadside VMS:* The information from the control center is used to drive displays on VMS along the highway. The VMS are linked to the control center via a communications network consisting of either copper wire or fiber optic telephone cable.
- *In-vehicle signing utilizing roadside beacons:* Another variation on the broadcast theme is to install roadside beacons connected on a wire network back to the control center. These beacons or transceivers can establish a vehicle to roadside communications link with vehicles equipped with an in-vehicle unit. This receives messages from the beacons and displays them to the driver.

#### 20.5.1.7 M70: "Interactive ATIS With Infrastructure-Based Route Planning"

This package includes the features of broadcast-based ATIS but provides the traveler with an enhanced set of travel information features. The primary enhancement is the provision of a route guidance service, in addition to the travel information. The traveler has a two-way communications device that allows origin and destination to be specified to a central point at which the best route is determined. The route determination is based on real-time traffic information and current journey times along links in the network. This route is then communicated to the traveler via two-way communications.

The device owned by the traveler can be one of a variety of electronic and communications devices as long as the device will send a user-specified set of parameters as required by the central control to calculate the best route.

#### 20.5.1.8 M80: "Interactive ATIS With GIS and GPS"

In this package, the user (i.e., the traveler) will need to own or lease a sophisticated integrated electronics package incorporating a route guidance algorithm, GPS, and GIS. In other words, the in-vehicle unit will be capable of determining its current location on a map and plotting the best route on a screen in the vehicle. Two-way communications with a control center is also required in order to transfer real-time information on journey times along selected links of the network. The journey times are used by the on-board system when determining the best route for the current traffic conditions.

It would be possible to phase the introduction of this ITS Market Package by installing the in-vehicle units and having them operate in autonomous mode until the real-time journey time information is available at a later date. Routes would then be calculated by the onboard unit based on historic traffic information rather than dynamic information. This phasing would allow the establishment of the two-way mobile communications network to be delayed until market penetration had been achieved with the in-vehicle equipment.

Two-way communications with the vehicle can be achieved in the following ways:

- Roadside beacons utilizing microwave communications;
- Roadside beacons utilizing infrared communications;
- Packet data radio;
- Cellular radio.

#### 20.5.1.9 M90: "Mayday Support"

This package allows the user (i.e., the driver) to initiate a request for emergency assistance. The request may be either manually or automatically initiated. A simple device in the vehicle or a modification to a cellular telephone would enable the traveler to access this service. The request for assistance could be directed straight to the emergency services or through a separate control center depending on the organizational arrangements. The mayday message is carried via a wide-area communications technology such as cellular radio or packet data radio.

A more sophisticated in-vehicle device with vehicle tracking capability could also report vehicle location automatically in case of an emergency.

#### 20.5.1.10 M100: "Parking Management"

This package includes monitoring car park space availability and reporting this information to subscribing travelers. This can be integrated with a route guidance/navigation facility that then directs drivers to the available space.

It can also include support for electronic transaction management enabling the car-parking operator to charge automatically for the use of the car parking space.

#### 20.5.1.11 M110: "Billing Support"

This would use existing wireline communications and financial transaction infrastructure such as the bank clearinghouse network and MasterCard/Visa network, etc. Processing hardware and specialized software is also used to support a range of billing activities associated with ITS. These include the following:

- Toll payment;
- Area road pricing;

- Parking payment;
- Public transport fare payment.

Other nontransport related low-value financial transactions can also be addressed by this ITS Market Package.

#### 20.5.1.12 M120: "In-Vehicle Driver Information"

This provides in-vehicle display of locally relevant information on current road conditions. It requires the installation of roadside and in-vehicle communications hardware. Roadside hardware is typically in the form of roadside transceivers or beacons with short-range wireless communications capabilities to passing equipped vehicles. The beacons can be independent, carrying only a local message about the immediate driving context such as advisory speed limit on approach to a sharp bend. They can also be networked by wireline system back to a control center enabling more flexible messaging.

This is also known as low-power advanced HAR, as it is possible to use the car radio as the communications device inside the vehicle. In this case, the beacon would be broadcasting its message on a commercial radio frequency but at low power in order to avoid interference with commercial radio operations.

Another technology that may be adopted is FM sideband, or RDS-TMC. This utilizes commercial FM stereo radio broadcasts but superimposes the data signal on the edge of the voice/music signal where it is inaudible to the human ear but detectable by specially modified radios. The broadcast can come from the beacon or from a central radio station if location coding has also been implemented.

#### 20.5.1.13 M130: "Basic Traffic Monitoring/Management"

This provides traffic engineers with the ability to monitor and manage traffic flow at major intersections and along major highways. Functions addressed include the following:

- Traffic flow monitoring;
- Identification and verification of incidents;
- Data collection and analysis;
- Dissemination of information to users and other service providers.

It includes road sensors, communications networks, data fusion, and analysis software.

### 20.5.1.14 M140: "Area-Wide Traffic Management"

The core of this ITS Market Package is a TMC, where monitoring and management of the traffic flows within the area under control takes place. This center also houses the software, hardware, and operations staff required to control freeway ramp meters, VMS, and surface street signalized intersection control equipment and traffic signals. CCTV cameras covering important parts of the network are also controlled and monitored from the TMC.

### 20.5.1.15 M150: "Traffic Prediction"

This ITS Market Package facilitates the prediction of travel demand patterns to support better route choice and journey timing planning for travelers. Specially developed software is utilized to make future predictions of traffic conditions based on current and historical traffic patterns. A large database of historical traffic patterns is also required to support this activity.

### 20.5.1.16 M160: "Travel Demand Management"

This supports the implementation of travel demand management programs and policies. Information on vehicle emission levels, parking availability, usage levels, and vehicle occupancy are collected by sensors. The information is stored and analyzed by a central computer facility that reports on the current situation compared to that required by policy and regulation. This enables policies to be developed and implemented. Special hardware for sensing air quality and other relevant parameters are provided.

### 20.5.1.17 M170: "Automated Toll Plaza"

This includes the capabilities required to electronically collect tolls, detect and process violators. Two-way short-range communications with equipped vehicles enable toll collection and processing. Wireline interfaces with traffic management and billing centers may be required depending on the choice of technology employed for toll collection.

### 20.5.1.18 M180: "High-Occupancy Vehicle Lane Management"

Where dedicated facilities have been provided for high occupancy vehicles such as buses, taxis, and high-occupancy private cars, this package provides the management capability. This is executed through the use of coordinated ramp meters and lane usage indicator VMS. Enforcement is achieved using video-based occupancy detectors and vehicle classification detectors.

### 20.5.1.19 M190: "Reversible Lane Management"

Where reversible or contraflow lanes have been implemented, this package provides the management capability. This is achieved using communications links to instruct VMS to change displays. Contra software is also required. This would typically form part of a more comprehensive area traffic control system.

### 20.5.1.20 M200: "Probe Vehicle Infrastructure"

This package includes the necessary capability to use equipped vehicles as traffic probes. Probe vehicles can report a range of traffic condition information, as experienced by the vehicle, back to a control center. This is achieved using a communications uplink such as vehicle to roadside short-range radio communication, packet data radio, or cellular telephone.

Depending on the sophistication of the in-vehicle equipment, this can include one or more of the following:

- Journey times along links traversed by the probe vehicle;
- Road surface conditions experienced by the probe vehicle;
- Traffic speeds;
- Weather conditions by monitoring use of probe vehicle windshield wipers and headlights.

Probe vehicle reporting can be carried out on a "by exception" basis under which the probe vehicle only reports conditions that lie outside a predefined range or tolerance band. This is particularly useful in cutting down the volume of messages to be handled by the communications network and control center. Probes can also be constrained to different reporting frequencies depending on geographical location, so that in areas with many probes present, reporting is less frequent. Conversely, in areas with few probes, the vehicle can be asked to transmit more frequently. Data from individual vehicles is aggregated to summary form to protect anonymity.

An entire vehicle population in a city could act as a probe, or a defined user fleet such as official cars, taxis, or rental cars could be designated as probes. Current research on the subject indicates that a sample size of approximately 1% of the entire vehicle population may be enough to obtain reliable statistics.

Another approach is to make the equipment in the vehicle very simple by utilizing automatic vehicle identification technology. This involves the installation of a simple tag on the windscreen of the vehicle. This would be the same type as that currently used for electronic toll collection. Vehicle "ID" is then established by reading the tags with roadside beacons at essential locations. Vehicles are then "tracked" through the network, and the data is statistically summarized over a number of probes to obtain current network operating condition information.

#### 20.5.1.21 M210: "Onboard Safety"

This ITS Market Package provides the functionality required to monitor and report the status of important vehicle components and the driver. This is an onboard system to detect and diagnose critical situations involving vehicle components and driver condition. Included on-board sensors monitor the following:

- Steering;
- Brakes;
- Acceleration;
- Emissions;
- Fuel consumption;
- Engine management functions.

A microprocessor-based safety management system onboard the vehicle monitors these sensor readings and generates warnings to the driver if any components become critical. In some situations, the warning may be accompanied by a control action such as automatic application of brakes.

Driver condition sensors include alertness monitors and optional "fitness to drive" testing equipment.

Implementations are expected to be autonomous—self-contained within the vehicle—although it is feasible to set up an automatic roadside monitoring system to read the status of these parameters for individual vehicles. This would require a one-way communications link from vehicle to roadside.

#### 20.5.1.22 M220: "Private Vehicle Safety"

This includes technologies that monitor the vehicles' immediate environment and deploy safety systems in response to the detection of potential hazards. This involves the use of on-board sensors to sense local conditions, determine collision probabilities, and deploy precrash safety systems.

The sensors would determine the location of and range to adjacent vehicles or roadway obstacles. These would be supplemented by additional sensors measuring existing local road surface conditions, weather conditions, and roadway geometry. An on-board microprocessor-based safety management system would accept and process this information, generate warnings, and initiate the deployment of safety systems such as supplemental restraints if deemed appropriate.

Sensor technologies that will be deployed include infrared ranging and vehicle radar technologies.

### 20.5.1.23 M230: "Driver Vision Enhancement"

This includes technologies that enhance driver vision onboard the vehicle. It uses sensor technologies such as infrared and ultraviolet vision systems and head-up displays to provide the driver with an enhanced view of the road ahead.

### 20.5.1.24 M240: "Advanced Vehicle Longitudinal Controls"

This ITS Market Package provides the functionality to automate vehicle speed and headway control. Utilizing radar, or infrared ranging technologies to detect the distance to the vehicle in front, a microprocessor-based control system will assess vehicle speed, process sensor inputs, and determine appropriate control actions. These include automatic braking, acceleration, or speed maintenance through the application of automated vehicle control techniques.

This is sometimes referred to as intelligent cruise control as it extends the functionality of currently available cruise control systems by determining appropriate headways.

### 20.5.1.25 M250: "Automated Highway Systems"

These technologies provide the capability to automate some or all of the driving function, utilizing a combination of the following ITS Market Packages.

- M240 advanced vehicle longitudinal controls;
- M230 driver vision enhancement;
- M220 private vehicle safety;
- M210 onboard safety;
- M260 intersection collision avoidance.

These technologies are accompanied by enhanced capabilities such as automated lateral control to automatically steer the vehicle and keep it in lane or on the appropriate route.

This is the longer-term future vision for automated highway systems and represents the culmination of development and market entry for the ITS Market Packages.

#### 20.5.1.26 M260: "Intersection Collision Avoidance"

This includes technologies for intersection condition monitoring and the generation of warnings and control actions. Roadside sensors monitor vehicle positions and closing rates on approach to intersections. A control system then accepts this data and determines the probability of a collision at the intersection. It subsequently generates appropriate warnings and control actions that are transmitted to the affected vehicles. Short-range vehicles to roadside communications techniques are utilized to dialog with the vehicles.

The whole package acts as a sort of smart intersection manager by monitoring the intersection and issuing coordinating instructions.

#### 20.5.1.27 M270: "Vehicle Tracking and Dispatch"

This ITS Market Package provides functions that enable the tracking of goods vehicle locations, development and maintenance of delivery itineraries, and monitoring of fuel usage. The package provides fleet managers with functions to automatically track and dispatch vehicles. The core of this package is a fleet management center that accommodates the automatic vehicle location and fleet management hardware and software. This will consist of workstations and software required to capture and process vehicle location data and perform fleet management processes.

In-vehicle equipment will range from relatively crude devices to measure distance traveled and fuel used to more sophisticated units using GPS, dead reckoning, and/or map matching techniques.

A communications medium is also required to transmit position data from the equipped vehicle back to the fleet management center.

Beacon-based vehicle to roadside communication (DSRC) technologies can also be deployed to provide vehicle location.

#### 20.5.1.28 M280: "Material Tracking"

This is an ITS Market Package that tracks cargo location and progress through a multimodal trip chain and monitors cargo condition. This is similar to vehicle tracking, but with the sensors fitted to the cargo containers. The sensors also monitor the condition of the cargo by measuring parameters such as temperature, pressure, shock loading, and vibration. Location and condition information is transmitted back to a fleet management center where the data is processed and compared with desired values and actions formulated.

A typical application of this ITS Market Package would deploy simple electronic article surveillance tags on cargo containers. These utilize the same principles as electronic toll collection tags but are "ruggedized" for goods handling use.

#### 20.5.1.29 M290: "Credentials"

This is an ITS Market Package that provides automated license application and purchase, reporting of license and credential status, and preclearance at roadside inspection stations. Credentials such as licenses and inspection certificates can be purchased electronically and stored in the same format on the vehicle. This allows the goods vehicle operator to buy the credentials in bulk and remotely by logging in to a computer/communication-based credential acquisition and monitoring system.

A typical implementation would utilize automatic vehicle identification (AVI) tags with two-way read/write capability and high-speed vehicle to roadside communications facilities. Infrastructure at the roadside consists of transceivers linked to a wireline communications network and on to an interface with the relevant regulatory agency.

Roadside inspection stations would also be equipped with weigh-in motion sensors and wireline links back to a central database.

#### 20.5.1.30 M300: "Safety for Commercial Vehicles"

This ITS Market Package provides functions for on-board safety monitoring for commercial vehicles and automated reporting of safety status. It also provides roadside support for reading on-board safety data and facilities for automated inspection at the roadside.

Sensors onboard the vehicle monitor safety critical components such as brakes, headlights, steering, indicators, and engine management system functions. Sensors on the loading area also monitor the stability and equilibrium of the cargo. The number of consecutive driving hours performed by the operator can also be measured by integrating a tachograph into the sensor package. This information is stored electronically onboard the vehicle and transmitted to both the fleet management center and local roadside inspection stations as required.

For hazardous material transport applications, information concerning safe handling techniques for the cargo would also be stored electronically onboard the vehicle. In the fleet management center, the data is accepted and processed by dedicated hardware and software with action items determined and reported to the management center-operating staff.

At the roadside inspection station, the data from the vehicles can be read from a transceiver and processed using a microprocessor-based workstation that is linked by wireline to the regulatory agencies database.

#### 20.5.1.31 M310: "Emergency Response"

This ITS Market Package provides the necessary functions to automate the notification of emergency vehicles once the incident has been verified, the location has been established, and the nature of the incident decided. The package will include a workstation with fleet management and automatic vehicle location software, typically located in an emergency management center.

In-vehicle equipment will range from relatively crude devices to measure distance traveled and fuel used to more sophisticated units using GPS, dead reckoning, and/or map matching techniques.

A communications medium is also required to transmit position data from the equipped vehicle back to the emergency management center. In addition, communications links may be required between the emergency management center operators, other emergency response agencies, and traffic and traveler information centers. These links may be either wireline or wireless—probably cellular radio techniques.

#### 20.5.1.32 M320: "Emergency Vehicle Maintenance"

This ITS Market Package provides the functionality to support automated vehicle maintenance planning through the use of vehicle mileage, vehicle condition monitoring, and maintenance schedule planning. It would typically be provided as an extension to emergency fleet management functions and utilizes the same in-vehicle and communications network equipment. Additional sensors would monitor parameters including the following:

- Vehicle speed;
- Brake wear;
- Tire wear;
- Engine management functions;
- Total miles traveled;
- Fuel consumption;
- Oil consumption.

An additional software suite is also required in the emergency management center to accept, process, and summarize this data.

These represent the basic units that will be selected and combined to form the ITS architecture or framework. This will then form the basis for the development of an ITS design.

## 20.6 ITS Future Big Picture and Conceptual Design

As discussed in Chapter 4, the maximum benefit will be attained when the selected ITS products and services are interoperable and compatible. Instead of leaving this to chance or a long evolutionary process with many associated risks, the approach chosen involves the definition of an ITS Future Big Picture, framework, or system architecture that describes the overall system operation and defines each major subsystem of the system and what data needs to be exchanged between each subsystem. This ITS Future Big Picture concept provides an initial view on such a framework.

The overall ITS Future Big Picture concept has been designed as a modular approach in order to provide flexibility for the following reasons:

- To accommodate institutional and organizational requirements;
- To facilitate the introduction of new technologies as they develop;
- To enable new system capabilities to be added as new needs are identified.

Within this ITS Future Big Picture framework it will be possible to accommodate a range of detailed design choices. While some of these design choices have been illustrated, it was outside the scope of the study to comprehensively address the definition of a detailed system design.

Having decided on the ITS Future Big Picture to be adopted for the implementation, it is possible to take the next step toward detailed design. This involves the creation of a conceptual design that describes the proposed system in sufficient detail to enable it to be described to stakeholders and client groups. This level of detail will be technology-specific in that the technologies, or ITS Market Packages, required will be described in some detail, but not technology-dependent as choices regarding the use of particular equipment and particular vendors will not be made at this point.

Section 20.5 described a set of "building blocks" or ITS Market Packages that could be combined in various configurations to create an ITS that addresses all the transportation problems identified for the Klang Valley area. What is now required is a physical representation of the proposed ITS implementation. This requires that the various ITS Market Packages be assembled into subsystems that can be implemented as a physical unit and cost estimates determined accordingly. Table 20.9 shows the mapping of ITS Market Packages to subsystems.

**Table 20.9**
ITS Market Packages to Subsystem Mapping

| ITS Market Packages | ITS Modules | Emergency Management | Travel Information | Commercial Fleet Management | Billing Support | Enforcement | Traffic Management | Parking Management | Toll Plaza | Public Transport management |
|---|---|---|---|---|---|---|---|---|---|---|
| M10 Passenger & Fare Management | | | | | | | | | | • |
| M20 Public Transport Vehicle Tracking | | | | | | | | | | • |
| M30 Public Transport Security | | | | | | | | | | • |
| M40 Public Transport Operations Planning | | | | | | | | | | • |
| M50 Public Transport Maintenance | | | | | | | | | | • |
| M60 Broadcast-Based ATIS | | | • | | | | | | | |
| M70 Interactive ATIS With Infrastructure-Based Route Planning | | | • | | | | | | | |
| M80 Interactive ATIS With GIS, GPS | | | • | | | | | | | |
| M90 Mayday Support | | | • | | | | | | | |
| M100 Parking Management | | | | | | | | • | | |
| M110 Billing Support | | | | | • | | | | | |
| M120 In-Vehicle Signing | | | • | | | | | | | |
| M130 Basic Traffic Monitoring/Management | | | | | | | • | | | |
| M140 Area Wide Traffic Management | | | | | | | • | | | |
| M150 Traffic Prediction | | | | | | | • | | | |
| M160 Travel Demand Management | | | | | | | • | | | |
| M170 Automated Toll Plaza | | | | | | | | | • | |
| M180 HOV Lane Management | | | | | | | • | | | |
| M190 Reversible Lane Management | | | | | | | • | | | |
| M200 Probe Vehicle Instrumentation | | | • | | | | | | | |
| M270 Vehicle Tracking And Dispatch | | • | | • | | | | | | |
| M280 Material Tracking | | | | • | | | | | | |
| M290 Credentials | | | | • | | | | | | |
| M300 Safety For Commercial Vehicles | | | | • | | | | | | |
| M310 Emergency Response | | • | | | | | | | | |
| M320 Emergency Vehicle Maintenance | | • | | | | | | | | |

ITS Market Packages have been grouped together by considering the following criteria:

- Sharing substantial volumes of data;
- Addressing the same customer group;
- Requiring close coordination in order to operate successfully;
- Using the same physical equipment.

Note that only those ITS Market Packages that were considered appropriate for inclusion in the system development work have been utilized. In particular, those ITS Market Packages associated with automated highway systems have been ignored as these were considered to be in the long-term future and beyond the scope of the current study-planning horizon.

Thus, we have the basis for a physical representation of the proposed system architecture, which is composed of nine subsystems. In order to complete the conceptual design task, it is now necessary to develop some detailed descriptions for each of the nine subsystems. These descriptions will be utilized to confirm that the conceptual design can support the vision.They will also be used as input to the institutional/organizational and commercial analysis activities. The conceptual design is illustrated in Figure 20.7.

It is worth noting at this stage that the conceptual design, while providing a higher resolution picture of the proposed implementation, will not be sufficient to provide the basis for RFPs and other tendering processes. Further work

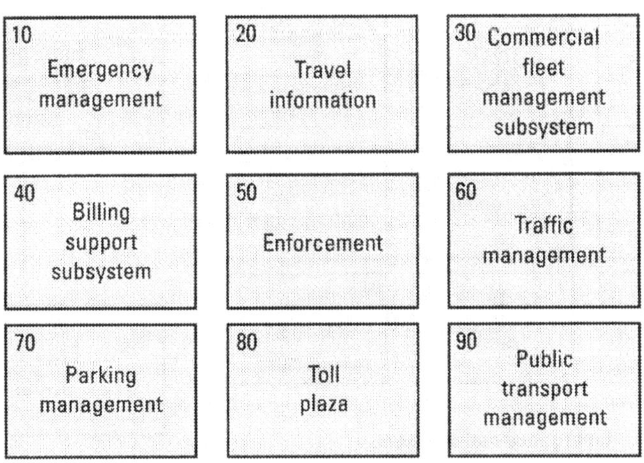


**Figure 20.7** Conceptual design subsystems.

will be required in order to take and confirm detailed design choices such as choice of communications media and detailed design of in-vehicle and portable/personal units. In order to provide a wider range of illustrative material, the conceptual description has been extended to include a range of possible design choices in selected areas. These, along with the subsystem descriptions are contained in Sections 20.7– 20.15.

## 20.7 Subsystem 10—Emergency Management

This subsystem addresses the needs of fire, police, and ambulance services regarding effective fleet management and efficient incident response times. The subsystem consists of two primary elements, an in-vehicle subsystem and a control center subsystem. The two elements work together to provide the functionality required by the emergency management operators.

Figure 20.8 illustrates the two primary subsystems. The emergency in-vehicle subsystem consists of the communications hardware and software necessary to support two-way wireless communications with the emergency management center, an interface to allow communications with the vehicle operator, and a vehicle location, or positioning system. The emergency management center subsystem consists of another operator interface, hardware, and software required to support the task of deploying and dispatching emergency vehicles. Detailed specification of the hardware and software will require

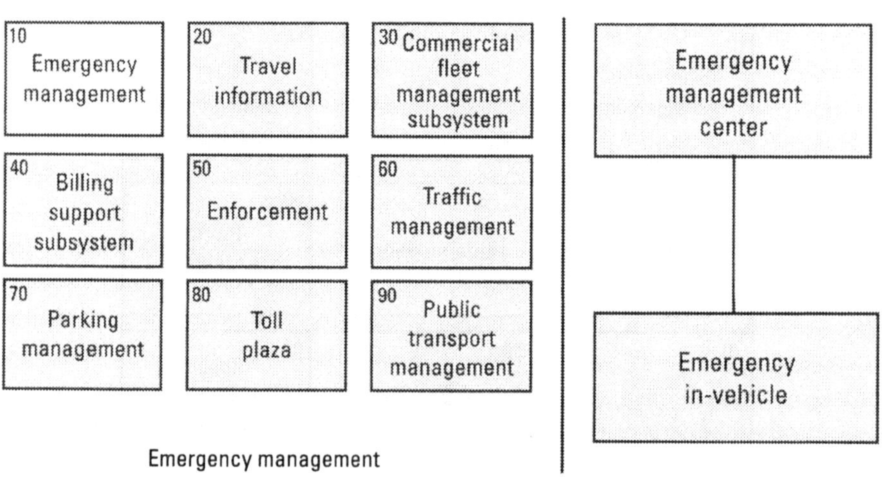


**Figure 20.8** Emergency management primary subsystems.

further requirements analysis and system design activities. Figures 20.9–20.12 show some possible design choices that could be adopted.

The ITS Market Packages incorporated into this subsystem are illustrated in Figure 20.13.

## 20.8 Subsystem 20—Travel Information

This subsystem provides the ability to collect data regarding the current status of the transportation network, collate and synthesize that data, and disseminate it to travelers through multiple communications channels and media. The object is to present the traveler with better and more comprehensive information regarding transport choices, enabling better choices to be made. Figure 20.14 illustrates the composition of this subsystem.

The operation of the subsystem pivots around a traffic information center that, in a similar manner to the emergency management center described above, receives, collates, synthesizes, and stores transportation network data. This data is then processed to provide travel information in various formats and is transmitted to travelers through a variety of techniques and devices.

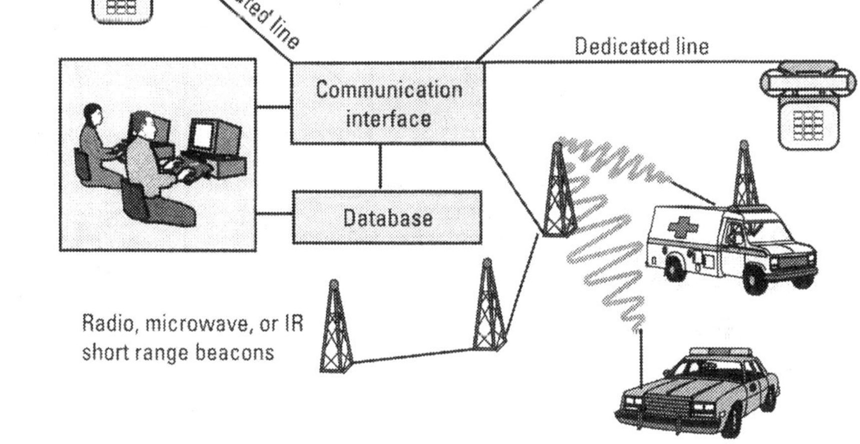


**Figure 20.9** Emergency management center—design choice 1.

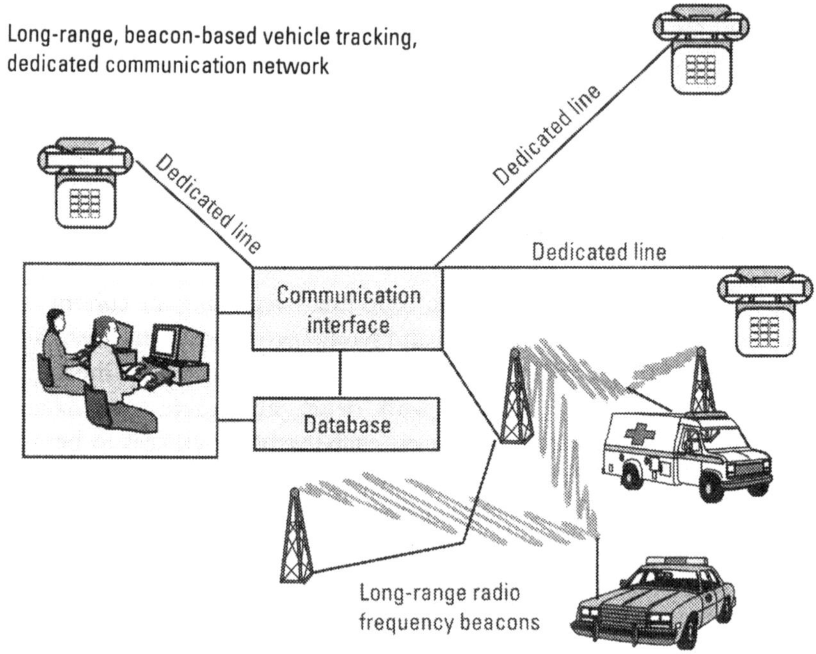


**Figure 20.10** Emergency management center—design choice 2.

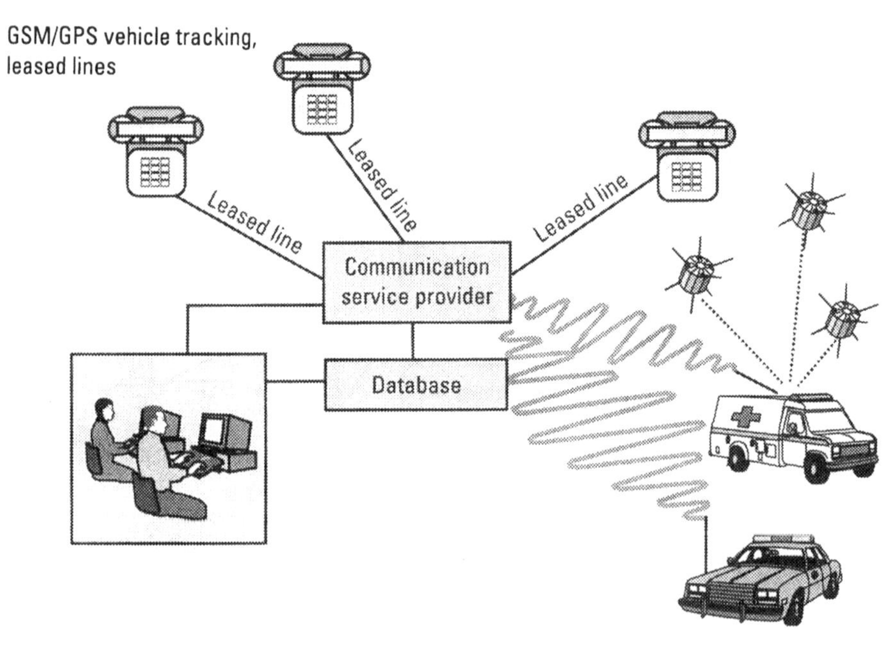


**Figure 20.11** Emergency management center—design choice 3.

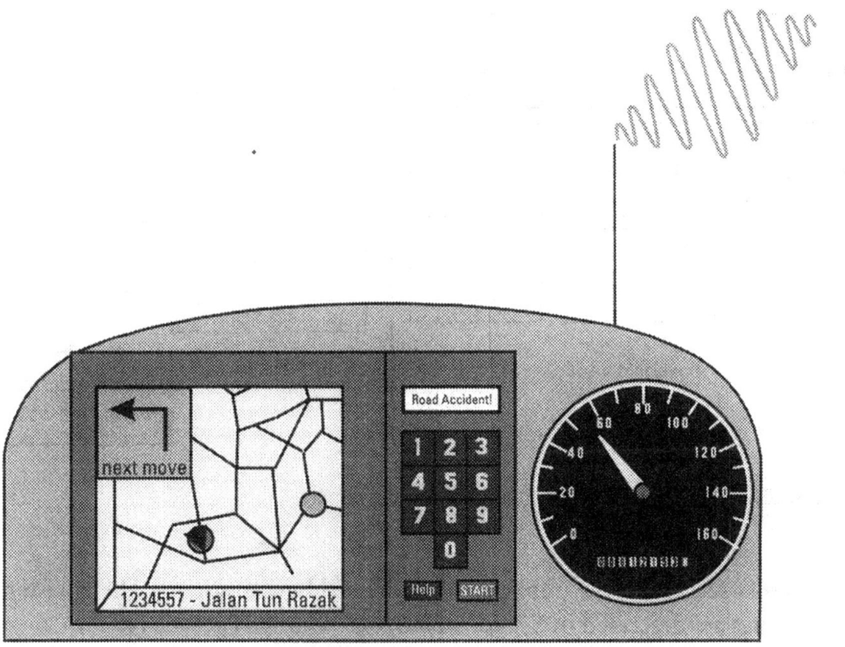


**Figure 20.12** Emergency vehicle subsystem design choice.

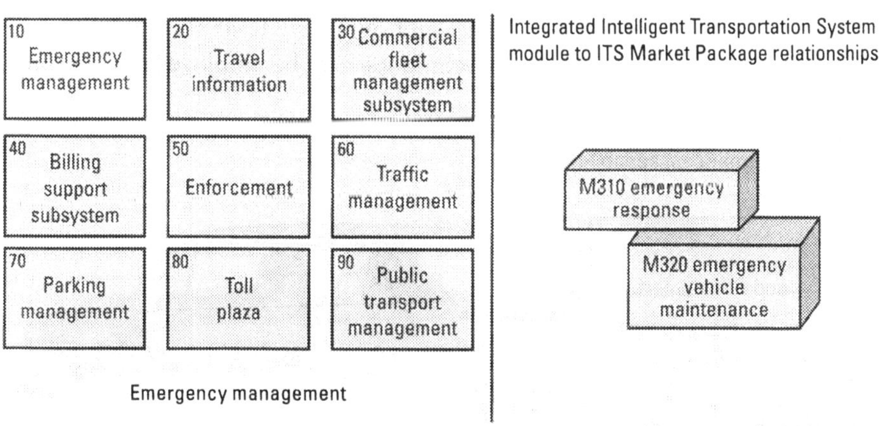


**Figure 20.13** Emergency management Market Packages.

A range of techniques has been envisioned, as this will ensure that the customer has a choice of service levels and cost of information access, enabling a

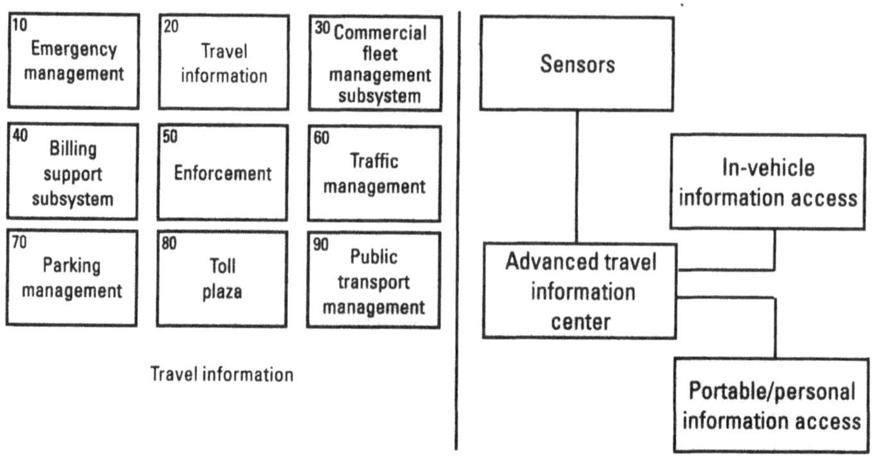


**Figure 20.14** Travel information subsystem composition.

large cross-section of the public to be addressed by the system. The envisioned techniques are described in Sections 20.8.1–20.8.6.

### 20.8.1 Basic ATIS

This would provide basic travel information to the driver within the vehicle. As shown in Figures 20.15 and 20.16, the device fitted to the vehicle would be toward the cheaper end of the market and only have basic functionality. For example, one-way radio broadcast techniques may be employed to distribute

Private in-vehicle subsystem
Basic traffic broadcast capability

**Figure 20.15** Basic advanced travel information systems.

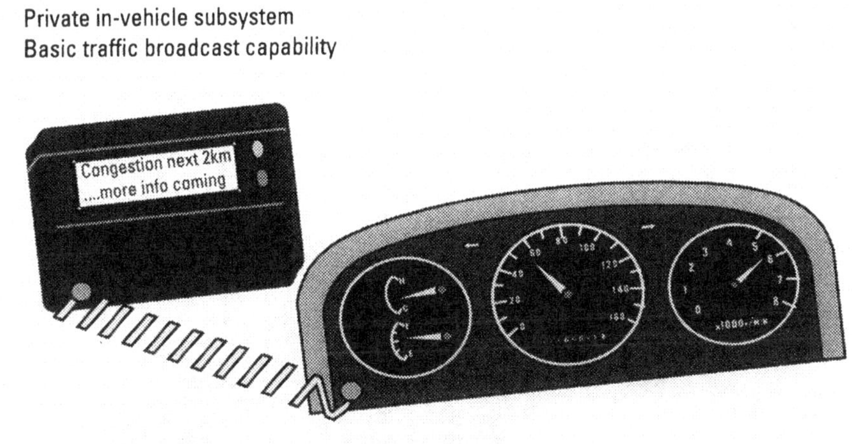


**Figure 20.16** Basic advanced travel information systems.

the travel information to vehicles. All vehicles would receive the same data, and there would be no possibility of responding to the travel information center.

### 20.8.2 Full Advanced Travel Information Systems

This variant of the "basic ATIS" option uses similar principles. It would utilize a sophisticated in-vehicle unit as illustrated in Figures 20.17 and 20.18. This would support a two-way communications link with the travel information

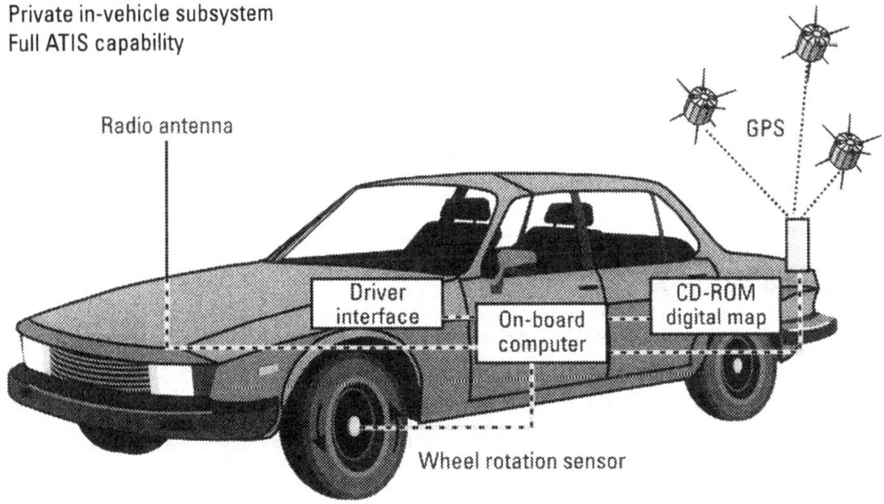


**Figure 20.17** Full advanced travel information systems.

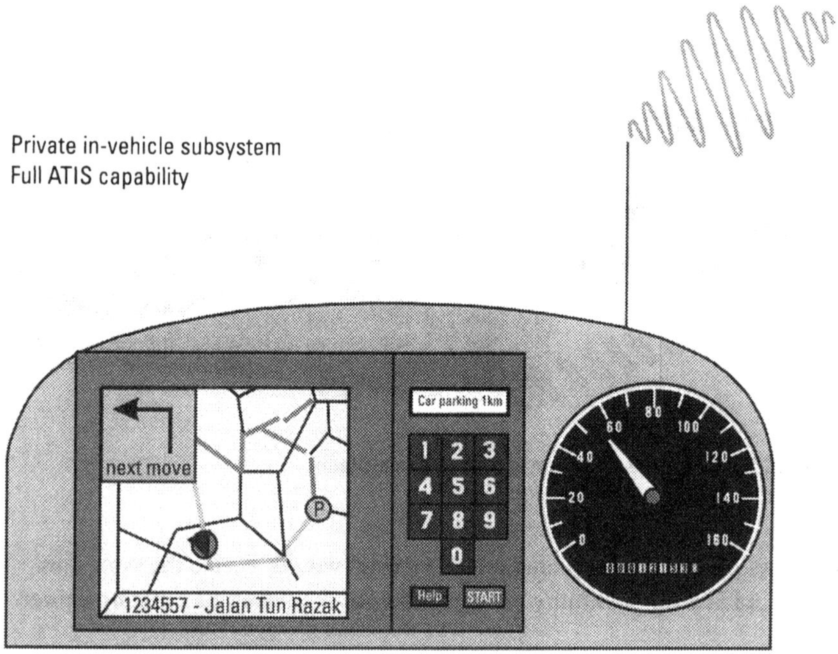


**Figure 20.18** Full advanced travel information systems.

center, providing up-to-date information to the driver. The uplink from the vehicle to the travel information center would be utilized to enable the driver to request updated information and to allow the vehicle to be used as a traffic probe. Using the traffic probe technique, the time taken for the vehicle to travel from one junction to another would be determined by the travel information center. This data, combined with data from other probe vehicles and other data sources, would be used to determine current network status and best routes for the current conditions.

### 20.8.3 Interactive Kiosk

This would provide information to travelers through a kiosk that may be located in such locations as metropolitan centers, office buildings, public transportation stations, and rest areas. Through an interactive process, the traveler may request information on a variety of topics such as transit schedules, route guidance, traffic information, and attractions and services. The information database would be frequently updated with current information to ensure accuracy of information.

### 20.8.4 Fax

A fax service is a low-cost, low-technology method of transmitting and receiving traveler information. The only equipment needed to transmit information is a standard fax machine. With this system, potential travelers may call in to receive pretrip travel information.

### 20.8.5 Telephone Hotline

This would provide travel information when it is needed, whether pretrip through a conventional telephone, or en route via cellular telephone. A telephone information system can target information to specific areas using voice response systems. This system would have no human operator and would be completely automated. Information could be updated frequently to ensure accuracy.

### 20.8.6 Home/Office Terminal

This would provide travel information via modem and a home/office terminal. Potential travelers would access pretrip information through the Internet while at home or work to make an educated decision on mode, time, and route of travel.

The ITS Market Packages incorporated into this subsystem are illustrated in Figure 20.19.

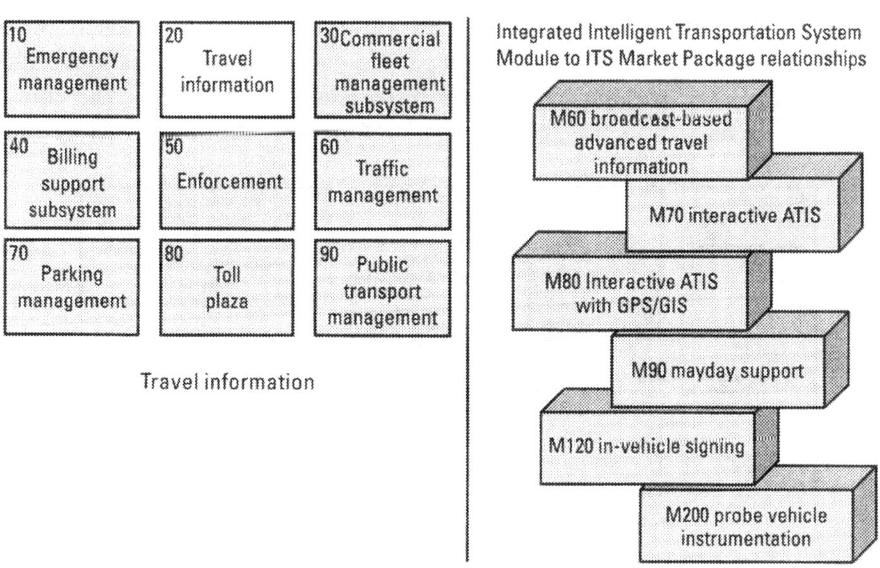


**Figure 20.19** Advanced traveler information Market Packages.

## 20.9 Subsystem 30—Commercial Fleet Management

This subsystem has a great deal in common with subsystem 10, emergency management system, in that the primary role served by both of these subsystems is to enable more effective and efficient management of vehicle fleets through the application of vehicle tracking, communications, and management information system technologies. In this case, it is likely that the in-vehicle system will utilize the same hardware as the emergency in-vehicle unit, but support different software with specific functionality for commercial operators. The emphasis will be placed on itinerary planning, cargo monitoring, and journey-time reliability, in contrast to the EMS context where response time reduction is the primary focus. Figure 20.20 shows the main elements of this subsystem. Figures 20.21 to 20.28 illustrate some potentially feasible design choices.

Figure 20.29 illustrates the ITS Market Packages that comprise this subsystem.

## 20.10 Subsystem 40—Billing Support

This is a multipurpose subsystem designed to support the automation of a range of financial transactions associated with both transportation related and nontransportation related applications. Figures 20.30–20.33 illustrate the main elements envisioned for this subsystem. In addition to the primary role of providing electronic financial transaction capability for transport applications such as cordon pricing, parking fee payment, expressway toll collection, and common stored value ticketing for public transportation, this also provides the

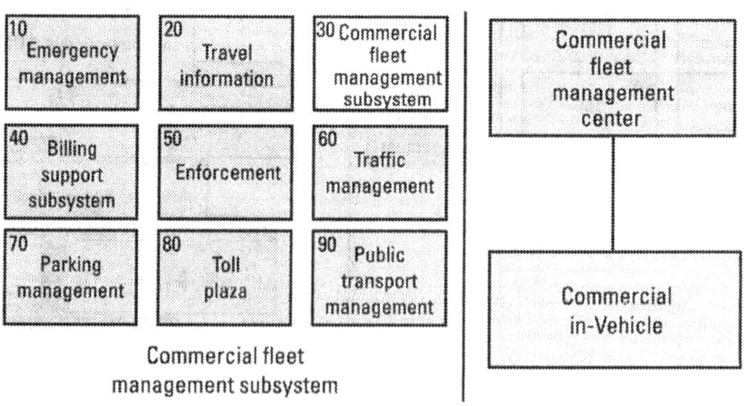


**Figure 20.20** Commercial fleet management system elements.

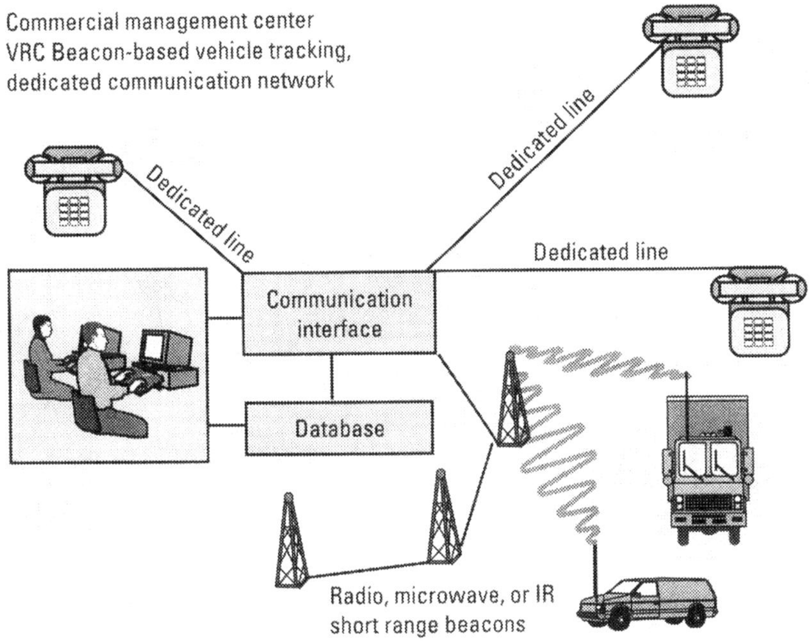


**Figure 20.21** Commercial management center—design choice 1.

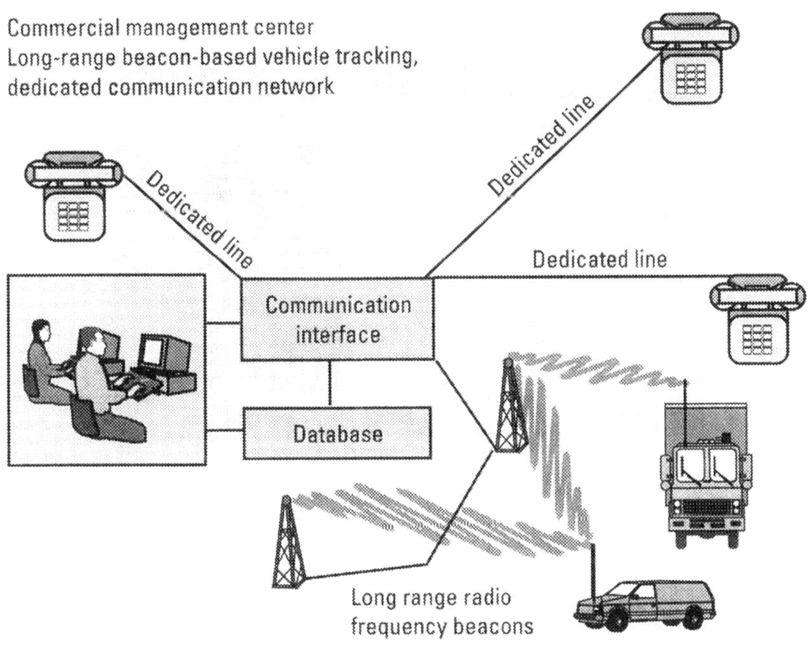


**Figure 20.22** Commercial management center—design choice 2.

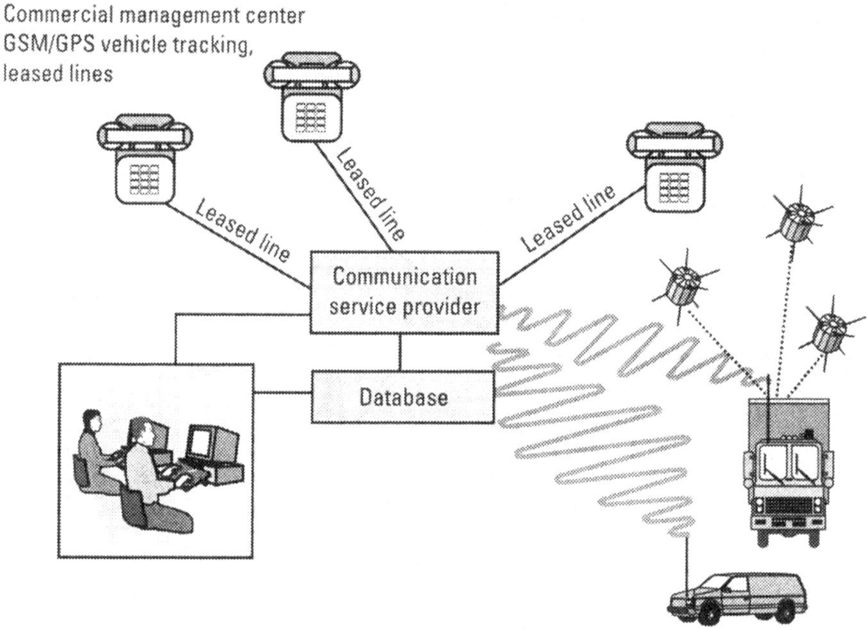


**Figure 20.23** Commercial management center—design choice 3.

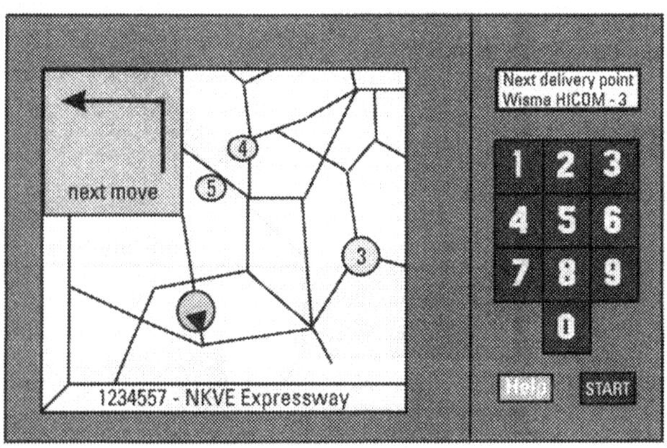


**Figure 20.24** Commercial in-vehicle unit display—design choice 1.

core for nontransport applications. In fact, the nontransport applications are very likely to make up the dominant part of the transaction flow in the longer term as banks and other financial institutions bring smart card applications online.

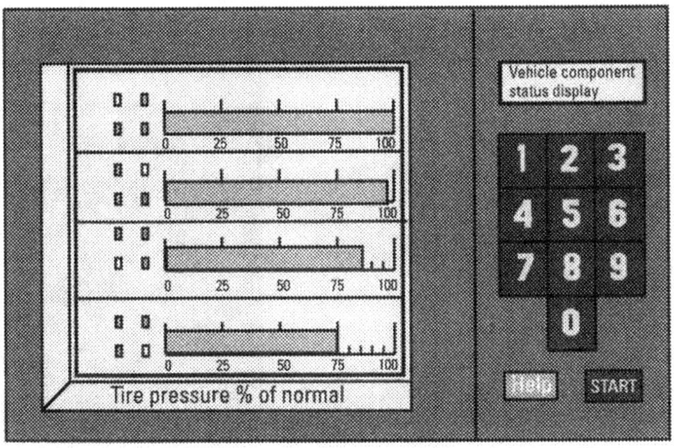


**Figure 20.25** Commercial in-vehicle unit display—design choice 2.

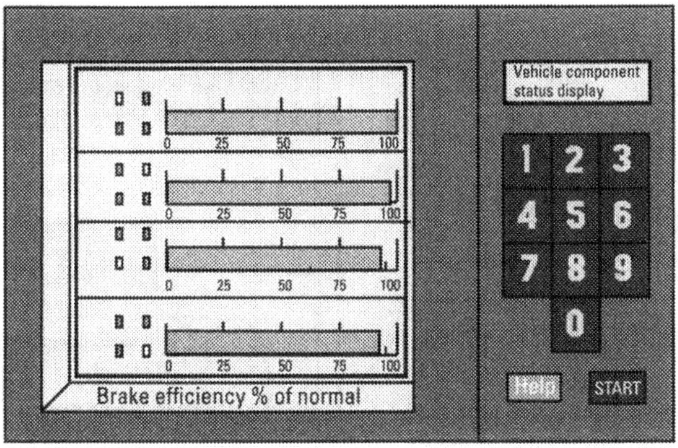


**Figure 20.26** Commercial in-vehicle unit display—design choice 3.

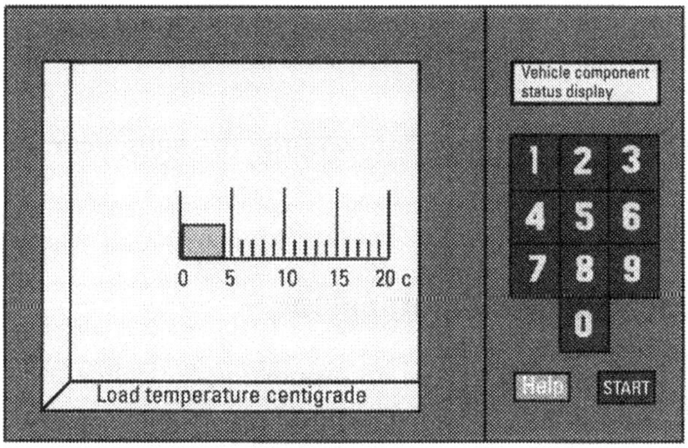


**Figure 20.27** Commercial in-vehicle unit display—design choice 4.

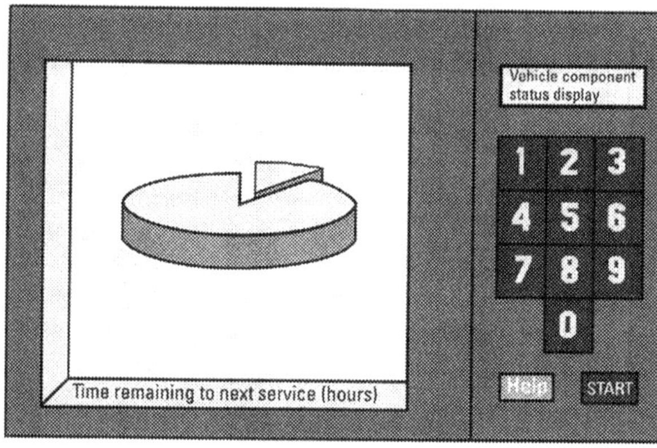


**Figure 20.28** Commercial in-vehicle unit display—design choice 5.

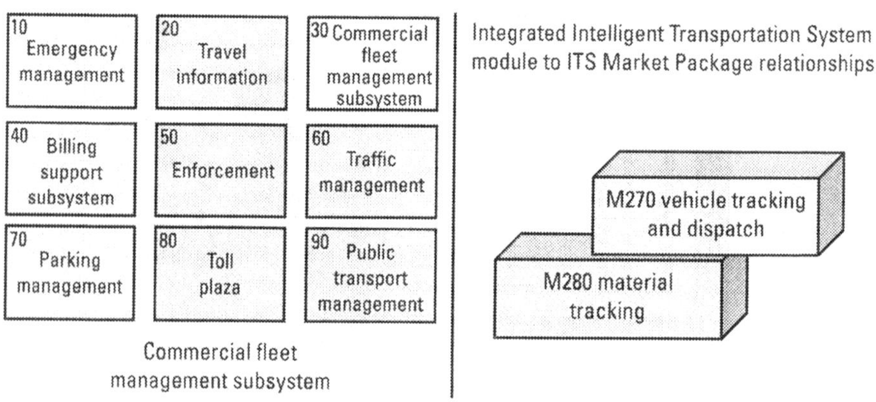


**Figure 20.29** Commercial fleet management ITS Market Packages.

The ITS Market Packages that comprise this subsystem are illustrated in Figure 20.33.

## 20.11 Subsystem 50—Enforcement

This subsystem has a specific focus on supporting the police and other enforcement agencies in the traffic- and transportation-related context. Figure 20.34 illustrates the two main elements of the subsystem.

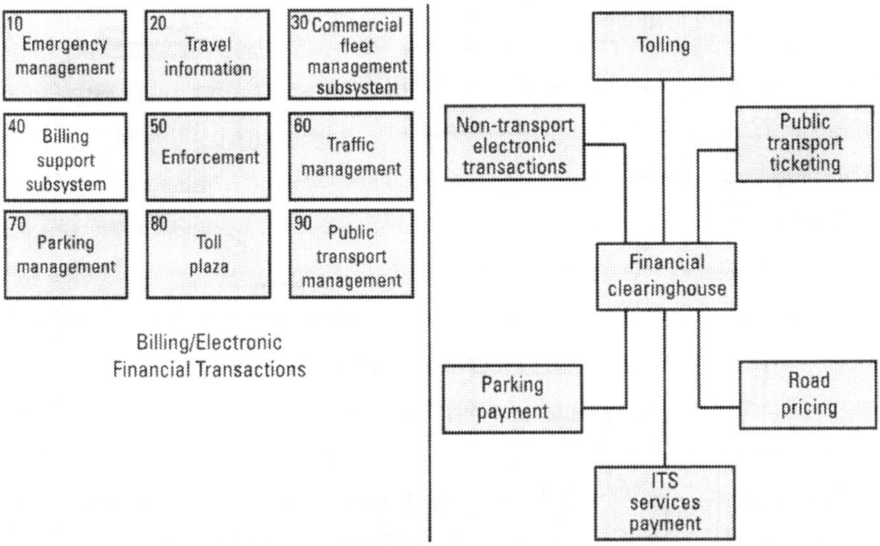


**Figure 20.30** Billing support subsystem elements.

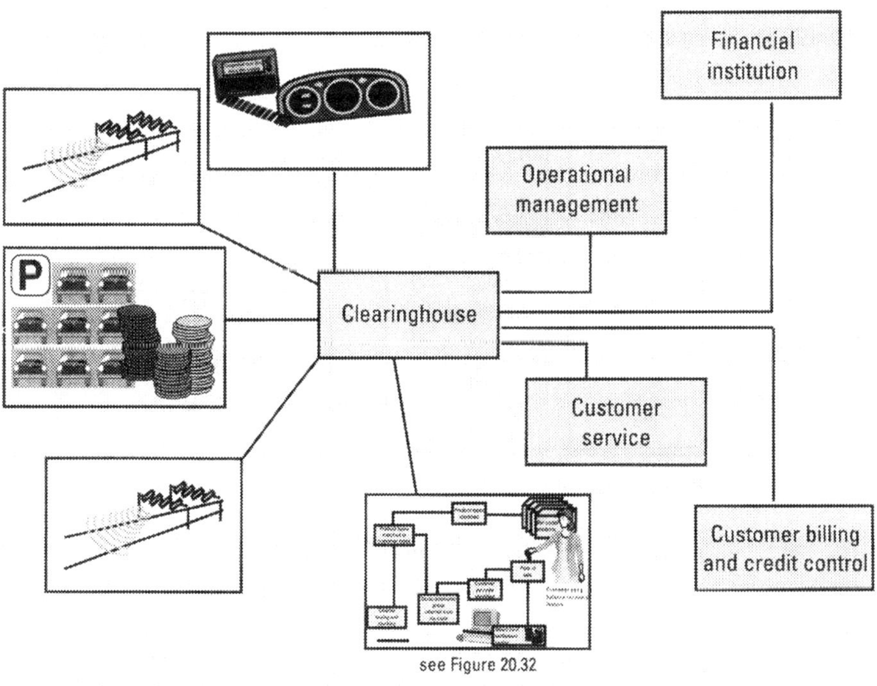


**Figure 20.31** Billing support subsystem elements.

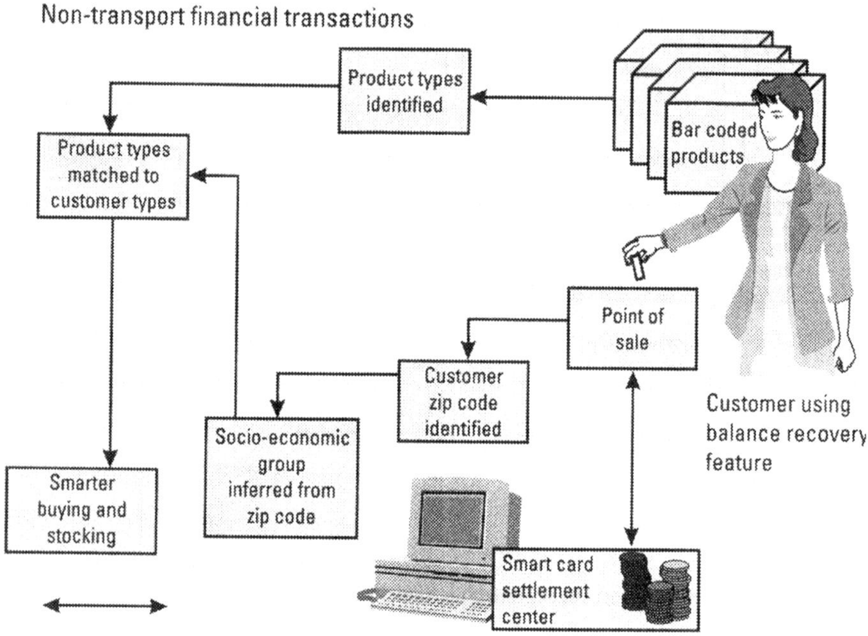


**Figure 20.32** Billing support subsystem elements.

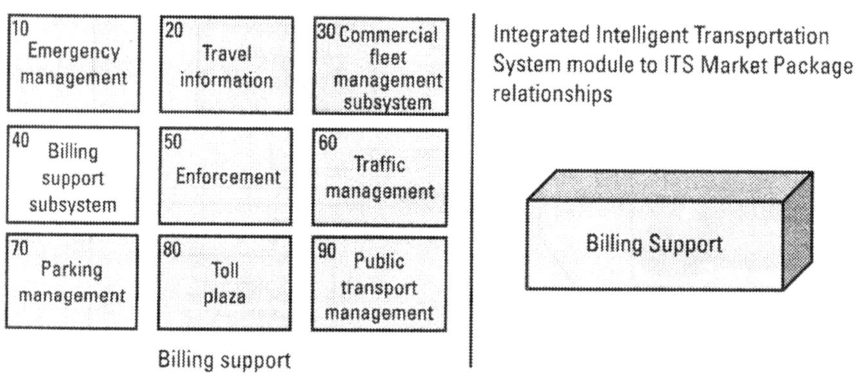


**Figure 20.33** Billing support subsystem ITS Market Packages.

- *Dedicated Traffic Sensors:* In addition to receiving and processing information from general-purpose traffic sensors, there will be a network of enforcement dedicated traffic sensors including the following:

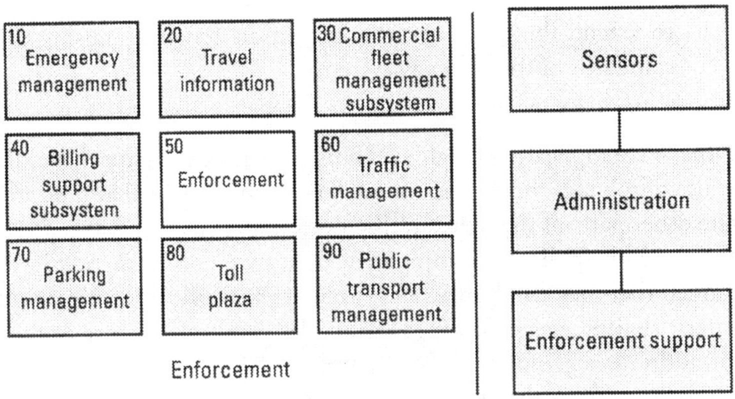


**Figure 20.34** Enforcement subsystem elements.

- *Video-based red light violation cameras:* This works on similar principles to the current still cameras but with direct connection to a control center and with sophisticated capabilities to recognize vehicle number plates automatically and match the plate to the owner through a vehicle licensing database.
- *Vehicle licensing database:* Although not envisioned as part of the core deployment activity for ITS, it is expected that this will be developed in parallel by the police and be utilized by the ITS. The emergency in-vehicle unit described under subsystem 10, emergency management system, would then access this database and provide police in the field with ownership information for use in enforcement processes.
- *Speed detection sensors:* These would automatically determine vehicle speed, compare it against the speed limit in force, and, if appropriate, send a request for a summons to be issued to the enforcement support system.
- *CCTV sensors:* These are capable of capturing full-motion video at selected points on urban surface streets and expressways. These would also be capable of automatic speed detection, vehicle counting, and classification. They are likely to be connected to the traffic information and management centers, but access would be available for police use in incident detection, verification, and classifications and public safety and security applications.
- *Automatic vehicle height, width, and weight sensors:* Installed at carefully selected sites on the highway network, these will enable police

to extend the range of their inspection activities, raising the safety standards of the highway system.

Other ITS applications such as the use of smart cards for driver licensing and the use of toll tags for storing vehicle inspection data will also be applied to automate other parts of the police traffic duties.

Although the police will implement their own enforcement subsystem, it is envisioned that they would play a key role in the traffic information and control centers, sharing access to other sensor data and taking part in the overall transportation management process. Figure 20.35 shows the ITS Market Packages that make up the subsystem.

## 20.12 Subsystem 60—Traffic Management

This subsystem encompasses a range of advanced traffic management functions aimed at increasing the efficiency and safety of both urban surface streets and expressway operations. Figure 20.36 illustrates the main elements of the subsystem.

It is envisioned that the following features would be incorporated into the detailed design:

- *Coordinated adaptive signal control:* Building on the existing and planned SCATS implementation, this would ensure that signal timings are appropriate to traffic flows and currently prevailing network conditions. It is anticipated that development work will include the

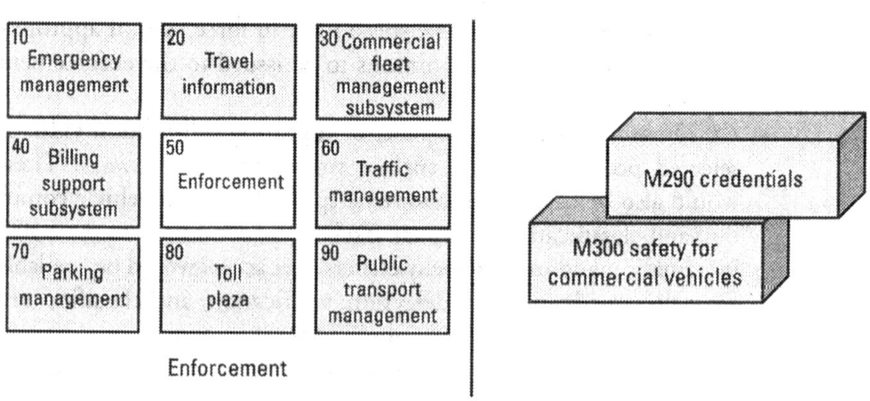


**Figure 20.35** Enforcement subsystem ITS Market Packages.

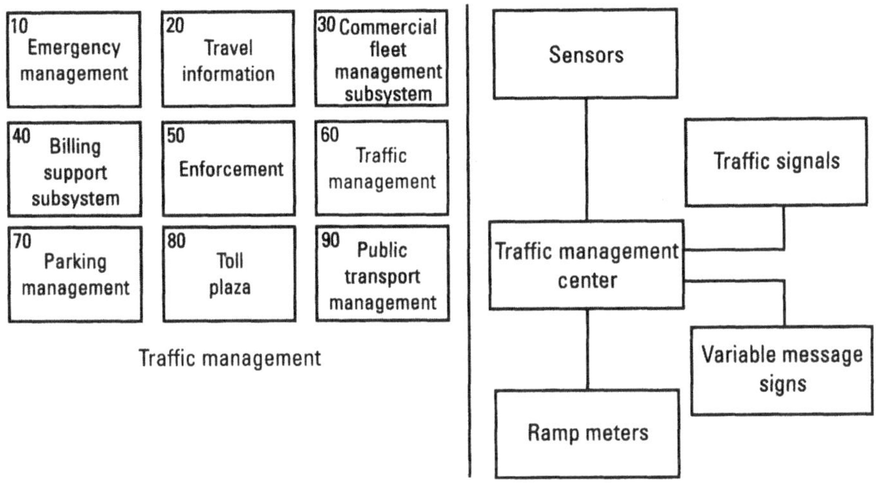


**Figure 20.36** Traffic management subsystem elements.

enhancement and expansion of the current traffic database, the provision of graphical user interfaces for traffic engineers to operate the system, and the establishment of a two-way communications link from the traffic management subsystem to the travel information subsystem. This latter feature will enable signal timing information to be used in route determination and probe data to be used in calculating traffic signal timings.

- *Advanced traffic sensors:* In addition to the existing inductive loop detection system, smart inductive loops with greater accuracy and additional capability will be installed. Other sensors based on radar, infrared, and other detection technologies will be deployed to carry out measurements on a range of parameters including the following:
    - Journey speed;
    - Time mean speed;
    - Space mean speed;
    - Emissions;
    - Queue lengths;
    - Delay;
    - Public transit vehicle detection;
    - Emergency vehicle detection.

These sensors will utilize a common communication backbone and share common communication protocol standards, enabling data to be transferred back to the traffic management center and devices to be polled regularly to check proper operation through telemetry techniques.

- *VMS:* These will be installed at important intersections and at significant decision points on both the urban surface street and freeway network. Connected to the same communication backbone as the sensor devices described above, they will enable the TMC to inform drivers on current traffic conditions, potential hazards ahead, advisory speed limits, and available diversion routes.
- *Ramp meters:* These have been used very successfully in the United States and the United Kingdom to optimize the capacity of heavily trafficked expressway-type roads. They work by "metering" the traffic wishing to enter the expressway to match the instantaneous capacity of the expressway.
- *Cordon pricing:* A major element in the implementation, this involves the installation of free-flow, multilane, electronic toll collection and enforcement equipment on all major roads approaching Jalan Tun Razak (Ring Road). An electronic cordon, or toll ring will be established that communicates with in-vehicle equipment and supports an electronic financial transaction involving the charging of a small fee to the driver for each cordon crossing. The application requires the installation of both roadside equipment and in-vehicle equipment.

  The roadside equipment will consist of two-way communications transceivers, a video-based enforcement subsystem, and a communications link to the TMC and the billing support clearinghouse. The in-vehicle unit may take a variety of forms ranging from a simple low-cost identification tag to a smart unit capable of supporting the use of a smart card for electronic payment. This latter arrangement has been assumed as other smart card applications are already planned for the area.

  It is worth noting at this point that the cordon pricing subsystem is considered to be part of the traffic management subsystem, since it has a capability for demand management through the use of variable pricing structures varying by time of day, by volume of traffic, or other criteria. While this capability will be available from the start of the implementation, it is envisioned that the cordon pricing subsystem will be initially operated as a revenue-generating system, rather than a demand management system. Use of demand management will be held in reserve for future use, should it be demonstrated as necessary to achieve a balance between private and public transportation modes.

Figure 20.37 illustrates the traffic management subsystem ITS Market Packages.

## 20.13 Subsystem 70—Parking Management

This subsystem incorporates the sensors, data processing, and information dissemination capability required to effectively manage on-street and off-street parking. This includes the use of VMS and in-vehicle and portable personal information devices to display locations of available parking spaces, parking charges and number of spaces available. It is envisioned that this information will be tailored to the driver's needs by taking account of ultimate destination and desired length of stay. Pretrip information devices will also carry this information to enable travelers to take account of likely parking availability when making trip choices. Figure 20.38 illustrates the main elements of the subsystem. Figure 20.39 illustrates the parking management subsystem ITS Market Packages.

This subsystem could also, in coordination with the billing support subsystem, manage the administration of parking fees and parking fines.

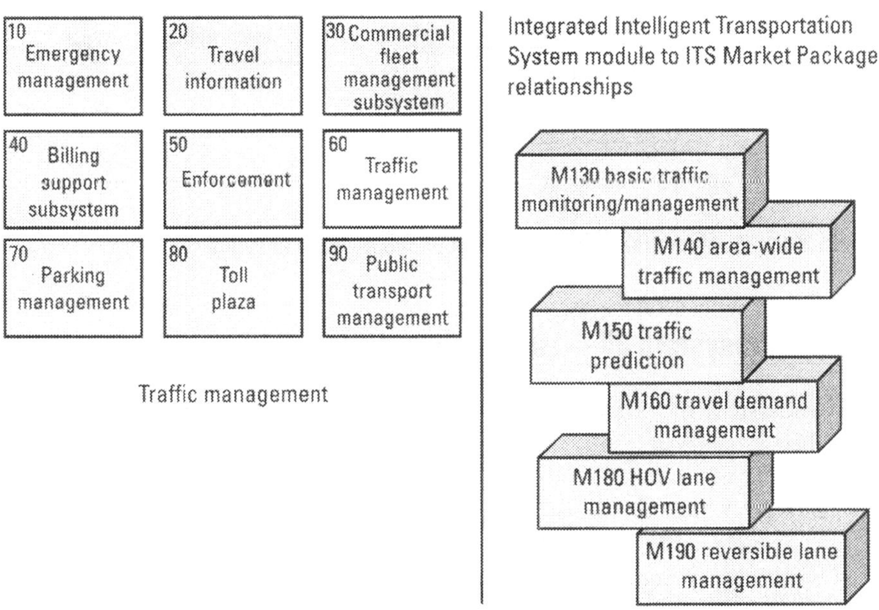


**Figure 20.37** Traffic management subsystem ITS Market Packages.

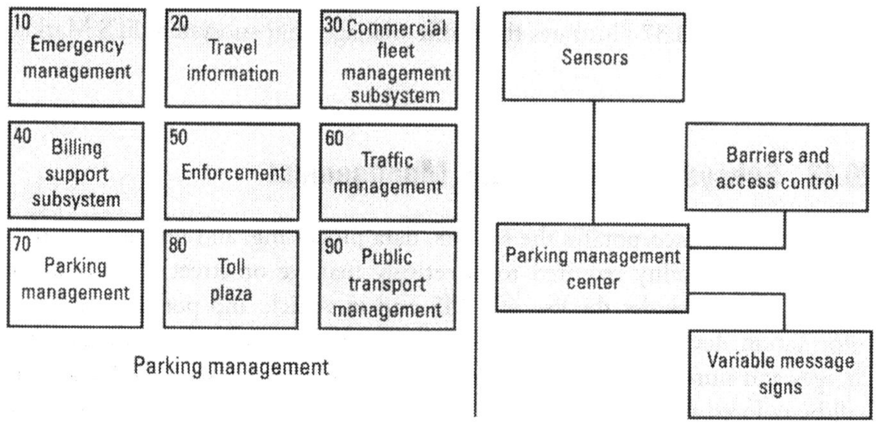


**Figure 20.38** Parking management subsystem—main elements.

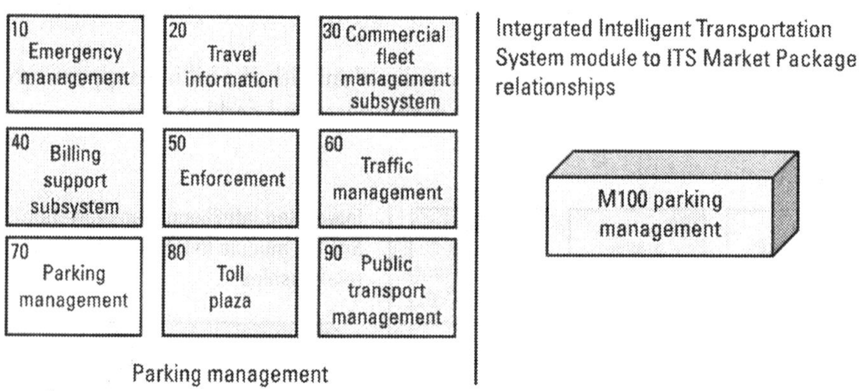


**Figure 20.39** Parking management subsystem—ITS Market Packages.

## 20.14 Subsystem 80—Toll Plaza

This subsystem has a similar functionality to the cordon pricing subsystem described in subsystem 60, traffic management. Electronic toll collection and enforcement equipment on all expressways will enable nonstop toll collection, reducing delay and emissions. Electronic "virtual toll plazas" will communicate with in-vehicle equipment and support electronic financial transactions associated with toll payment. The application requires the installation of both roadside equipment and in-vehicle equipment. Figure 20.40 illustrates the main elements of the subsystem. Figure 20.41 illustrates the toll plaza subsystem ITS Market Packages.

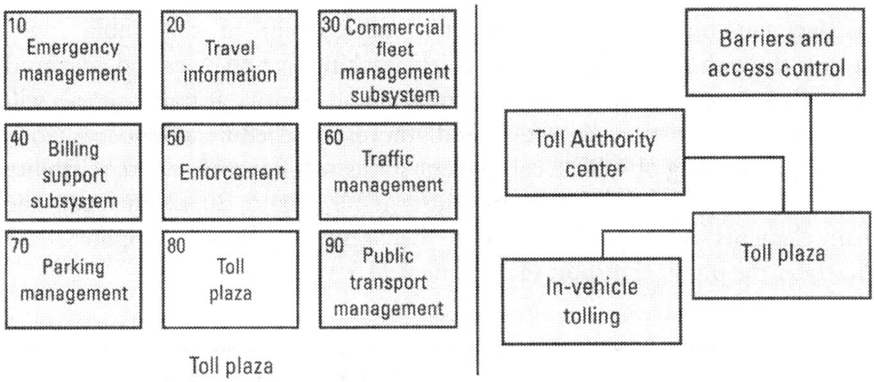


**Figure 20.40** Toll plaza subsystem elements.

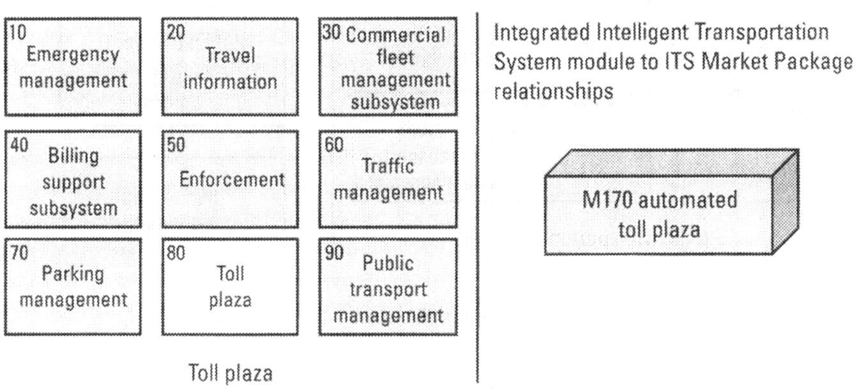


**Figure 20.41** Toll plaza subsystem ITS Market Packages.

The roadside equipment will consist of two-way communications transceivers, a video-based enforcement subsystem, and a communications link to the traffic management center and the billing support clearinghouse. The in-vehicle unit may take a variety of forms ranging from a simple low-cost identification tag to a smart unit capable of supporting the use of smart cards for electronic payment. This latter arrangement has been assumed as other smart card applications are already planned for the area.

## 20.15 Subsystem 90—Public Transport Management

This subsystem is illustrated in Figure 20.42. There are two primary functional groupings, public transport management and passenger information. The management functions enable effective deployment of the public transit fleet through the application of vehicle tracking technologies and advanced fleet management systems. Vehicle detection and priority at traffic signals will help to ensure on-time operation and improved schedule adherence, while remote monitoring of vehicle component status will improve service reliability and reduce required spare parts inventory levels. Figure 20.42 illustrates the main elements of the public transport management subsystem. Figure 20.43 illustrates the public transport management ITS Market Packages.

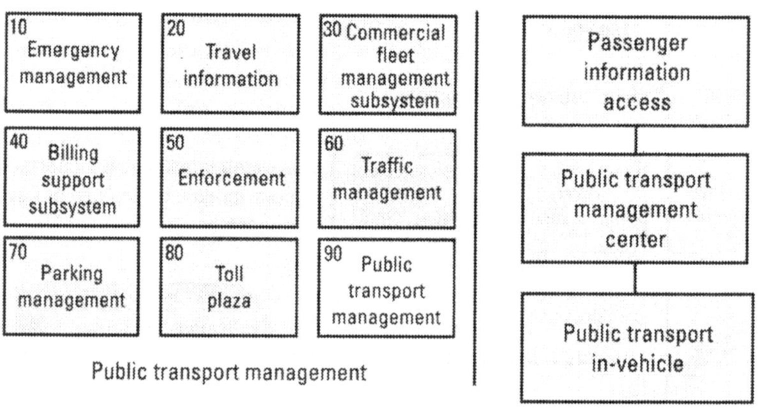


**Figure 20.42** Public transport management subsystem—main elements.

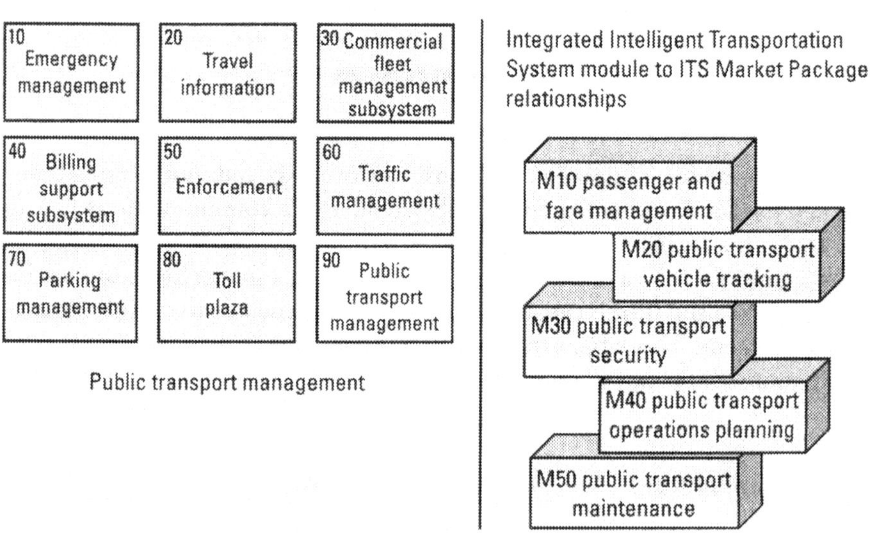


**Figure 20.43** Public transport management subsystem—ITS Market Packages.

Common stored-value ticketing techniques applied through the deployment of smart card technologies will provide operators with comprehensive demand data and trip pattern information, enabling services to be tailored to changing demand for travel.

On the passenger information side, vehicle tracking will enable smart bus stops to display an accurate estimate of the time to the next bus, along with route choice and fare information. A public transport management center will coordinate the data collection and information dissemination. On-board vehicle displays will inform passengers of network status and indicate disembarkation points.

## 20.16 Summary

In this chapter, we have described some of our real-life experiences in the application of a number of important concepts:

- Mission;
- Vision;
- ITS User Services;
- Measures of Effectiveness;
- Strawman ITS Future Big Picture;
- ITS Market Packages;
- ITS conceptual design.

These have been illustrated using three stories from previous projects with which we have been involved. We hope that we have helped to clarify our previous explanation of the various elements of the ITS Cooperative Development Methodology.

# 21

# Summary

We have always believed that the most effective spoken or written presentations consist of three important parts:

- "Telling them what you're going to tell them;"
- "Telling them;"
- "Telling them what you told them."

You can see this in action on most nights during the news programs. Well, we are now at the point where we should tell you what we have told you.

As we stated in Chapter 1, there are several things that we hoped this book would do for readers:

- Explain the cooperative approach to ITS development, planning, and implementation;
- Provide some assistance in dealing with the multidisciplinary nature of ITS; assist the transportation community to bridge the gap between transportation and system engineering;
- Promote the development and deployment of smarter ITS;
- Help stakeholders to consider, prepare, and deliver effective input to ITS development projects;
- Identify and address the "people" issues associated with ITS.

We started out by introducing ITS technologies in Chapter 2. This introduced you to the world of information and communications technologies that comprise ITS. This should have provided you with a basic understanding of the types of technologies, products, and services available to be applied to transportation problems. Our aim in that chapter was to equip you with the fundamental information required to get the best from later chapters and also to start you thinking about what technologies you would want to use in your region.

In Chapter 3 we introduced the concepts and philosophy that underlies the ITS Cooperative Development Methodology. We explained a number of essential concepts that represent the philosophy behind the methodology. We took the concepts and, based on our experiences in ITS planning, development, and implementation, configured them to the transportation context. Each of the concepts was introduced and explained in terms of the nature of the concept and why it has evolved to be important to ITS planning and development.

If you do not remember anything else in the book, we would like you to remember the essential concepts summarized in Sections 21.1–21.4.

## 21.1 The "What?/How?" Cycle

This describes the separation of "what?" from "how?" to enable a focus on user-driven, needs-based development and an avoidance of the leap-to-design solutions. This also involves the use of ITS User Services to encapsulate the "what" or the requirements and the use of ITS Market Packages to describe the products and services that could make up the "how" or the potential solutions.

## 21.2 ITS Future Big Picture

We discussed this in Chapter 11. The ITS Future Big Picture involves stepping back from the short-term project and implementations and making sure that we all understand the context for now and the future. Remember how we made use of layering to make it easier to see the features of the ITS Future Big Picture?

The technical layer describes the subsystems and the communications network between them. The institutional/organizational layer defines the "people" arrangements such as agreements, understandings, and procedures for working together. The commercial layer describes the money flow around the system and identifies who pays and who profits.

Collectively, all three layers provide the comprehensive ITS Future Big Picture required to frame short-term deployments.

## 21.3 ITS Implementation Strategy

We discussed this in Chapter 14. Assuming the ITS Future Big Picture as the destination and the Legacy Catalog as our picture of where we are today, we define a series of steps from today to tomorrow. This is known as the ITS Implementation Strategy.

## 21.4 ITS Cooperative Development Methodology

This is the central theme running throughout the book. It is initially discussed in Chapter 4, where we explain the ITS Cooperative Development Methodology process and provide an overview of each of the major steps in the process. Each step is then explained in more detail, in Chapters 5–19. These chapters contain practical advice and guidance based on our project experiences. In Chapter 20, we reinforce the practical aspects of the advice and guidance using stories from some of the projects with which we have been involved over the past three years.

We have tried to incorporate to the best of our knowledge and experience in the development and application of ITS into this book, using the ITS Cooperative Development Methodology as a central framework to hold it all together. We believe you will realize that successful utilization of the information and communications technologies which underpin ITS applications calls for a range of skills and experience drawn from multiple disciplines.

One of the essential ingredients for a successful ITS development project is the blend of skills and experience that is brought to bear on the various activities. Two of the major constituents of this blend are what we call "research and development types" and "47project management types." The former provide the technical expertise to recognize the technology capabilities and define solutions. The latter take on the role of keeping the project on track, delivering on time, on budget, and to the desired specification. However, they are completely different types. So what is the difference between researchers, developers, and project managers? In our experience, researchers—and we would definitely count most software developers we have met in this category—like the challenge of doing new things. They like to start a task without understanding the endpoint and want to do new things every day. The joy of the job is in doing things in new ways and being innovative. While this leads to incredible job satisfaction, it makes it very difficult to put accurate resource estimates against the work being done. Project managers want to have a predictable working life. They want to work with known quantities on processes that are mature and are completely understood. They are in the business of managing risk.

We have painted two extreme pictures of these character types to make a point. ITS attracts many research and development types because it is new, has many new challenges, and new ground to break. However, we will only get the best value for our money and achieve the synergies and benefits we seek if we can find a way to apply the rigor and discipline of a formal structured approach to managing the risk associated with ITS projects. This does not mean that we have to avoid anything new that might require some pioneering work. If we took that approach then very little ITS would ever be deployed. It simply means that we need to adopt methods, techniques, and tools that are designed to accommodate evolution and change along the way and are capable of providing the structure to effectively manage ITS deployments. This is why we have explained the ITS Cooperative Development Methodology to you over the course of this book. Our only objective in writing the book is to help the transportation professionals in our community as they try to harness the obvious power of information and communications technologies.

The ITS Cooperative Development Methodology is the best way we have found to manage the fact that the development of an ITS project is often a journey you start out on with little or no knowledge of the final destination. You usually have a broad idea about the desired outcome, but not much detail. The ITS Cooperative Development Methodology provides a flexible framework for managing the journey so that we do not go way off course but still have the flexibility to adapt in light of what we learn.

We have also found that ITS projects require the development of a very good understanding of cause and effect at the regional level. Due to the information and data flows that must be supported in order to attain the full range of benefits from ITS, new levels of integration and cooperation are required between agencies and between pubic and private sectors. It is a question of consequences—identifying the relationships between numerous actions and organizations, assessing how an action by one will affect another, and determining the most effective way for them to cooperate.

So what do you do now that you have read our book? Well, it really is up to you, but here's what we had in mind when we wrote it. We hoped to provide a structured collection of our thoughts and experiences gained through close involvement in the development of the National Architecture for ITS and several ITS planning and development projects around the world. We intend this to provide some value to a range of transportation professionals, including the following:

- State department of transportation and local government officials engaged in planning, development, and design of integrated ITS for their regions;

- Transportation consultants and system engineers and integrators engaged in the development and implementation of ITS;
- Members of interest groups or stakeholders participating in the development of a regional ITS Future Big Picture through involvement in consensus formation or consultation processes;
- ITS or transportation planning students taking courses in ITS.

The aim was to help apply the principles, methods, and tools used in the National Architecture for ITS development program, through the description of a practical process. We also hope that our experiences and ideas will act as a springboard for the reader to develop a customized or personal approach to each project. Given the variation in needs and the differing contexts within which ITS projects take place, we do not advocate a "cookie-cutter" approach. Our aim is not to provide a recipe or set of instructions to be followed but to provide practical guidance on developing your own specific approach. Therefore, we would expect that your next step would be to apply these principles and techniques in the development of your own approach. We hope that you will review and understand the process and steps illustrated and then adapt them to your own use. If you are a member of an interest group or a stakeholder participating in the development of a regional ITS Future Big Picture through involvement in consensus formation or consultation processes, then we hope that the information provided will enable you to take an active part in the ITS development process. Getting good results is like doing the tango—it takes two! The proactive participation of an informed, motivated stakeholder or client group is an essential element in the development of an ITS Future Big Picture. In particular, the effective use of the "what?/how?" cycle depends on the enthusiastic engagement of the stakeholder group in its role as the expert in "what?"

Of course, success also depends on the other half of the partnership—the consultant, developer, or system integrator. The quality of the consultant will have a great influence on the success or failure of the work. While a good consultant will carry out your commission in accordance with your stated requirements and specific instructions, a great consultant will go further, taking you through the following steps:

- Both parties will engage in an interactive dialog on the needs, issues, problems, and objectives to be addressed by the work.
- The consultant will provide relevant information to you about ITS and what has been learned from earlier work. This will be especially packaged for you and your stakeholders to assimilate and understand.

- You will teach the consultant about your context, environment, and objectives.
- There will be a period of "mutual learning."
- Your knowledge of the possibilities combined with the local knowledge will exceed that of the consultant.
- The consultant will switch to a support role, fetching desired information and conducting required studies on your behalf.

This will support the development of understanding required to produce an effective ITS Future Big Picture, ensuring that the outcome is positive, efficient, and in line with all our goals.

In the course of the book, we have introduced you to many new terms and maybe a whole new way to approach the planning and development of ITS. What we have explained has been based on our experience and to a large extent, on the work carried our during the development of the U.S. National Architecture for ITS. We feel both honored and privileged to have been closely involved in this program and would sincerely recommend the deliverables for further study.

Another next step that we would really encourage you to consider taking is to let us know what you think about the ITS Cooperative Development Methodology. Tell us about your perspectives and experiences in applying the principles and concepts we have described. While we believe that the ITS Cooperative Development Methodology is the best approach around, it can be improved through experience. One of the things we have learned in the course of our careers is "never stop learning." There is always another better way or improvement to be made.

We believe that this will be the case in the future development of institutional/organizational arrangements for ITS. Many of the projects with which we have been involved have ended up with a technical and commercial solution that has been adapted to fit the current institutional/organizational arrangements. This is due to the great concern about making changes to institutional/organizational structures. Perhaps as ITS matures and we understand more about the needs and consequences, there will be more consideration given to innovation in the institutional/organizational layer.

# About the Authors

**Bob McQueen** is internationally recognized as a leading expert in Intelligent Transportation Systems. He has provided high-level consultancy to public and private sector organizations in Europe, the United States, Southeast Asia, and the Middle East regarding strategic planning and system engineering for advanced transportation technology applications. In particular, he has a strong skill base and experience in the identification, selection, implementation, and evaluation of appropriate Intelligent Transportation Systems strategies as solutions, combined with a significant track record in the application of structured analysis techniques. His recent work has involved the development and application of systematic procedures for integrated Intelligent Transportation Systems at home and overseas. This has featured a user-driven approach for matching transportation policy objectives to technology capabilities.

Bob McQueen received a Higher National Certificate in Engineering from Glasgow College of Technology in 1977, his B.S. degree in Civil Engineering from University of Strathclyde, Glasgow in 1980, and went on to attain his M.S. degree in Highways and Transport from City University, London in 1992.

Bob was proactive in early work on the definition of both European and international standards for system architecture, terminology, and dedicated short range communication applications for Intelligent Transportation Systems. His track record incorporates extensive experience in conventional transportation planning and traffic engineering disciplines with a steady migration towards specialization in the application of information and communication technologies to transportation. This provides an excellent background for strategic planning, engineering and analysis work in development and implementation of advanced transport technologies, including public sector policy definition and private sector market analysis and opportunity evaluation.

**Judy McQueen** is president of Coherence Incorporated, providing ITS training and education. She is responsible for having developed an ITS awareness seminar and an ITS standards and interoperability workshop in conjunction with the Federal Highway Administration, ITS America, George Mason University, and the University of Michigan.

Judy McQueen received her B.A. degree in Education from the University of Arizona in 1967.

# Index

## Recent Titles in the Artech House ITS Library

John Walker and Chelsea White, *Series Editors*

*Advances in Mobile Information Systems,* John Walker, editor

*Incident Management in Intelligent Transportation Systems,* Kaan Ozbay, Pushkin Kachroo

*Intelligent Transportation Systems Architectures,* Bob McQueen and Judy McQueen

*Positioning Systems in Intelligent Transportation Systems,* Chris Drane and Chris Rizos

*Smart Highways, Smart Cars,* Richard Whelan

*Transport in Europe,* Christian Gerondeau

*Vehicle Location and Navigation Systems,* Yilin Zhao

*Wireless Communications for Intelligent Transportation Systems,* Scott D. Elliott, Daniel J. Dailey

www.ingramcontent.com/pod-product-compliance
Lightning Source LLC
LaVergne TN
LVHW011858200926
848484LV00002B/478